AN ASSESSMENT OF THE SBIR PROGRAM AT THE NATIONAL SCIENCE FOUNDATION

Committee for
Capitalizing on Science, Technology, and Innovation:
An Assessment of the Small Business Innovation Research Program

Policy and Global Affairs

Charles W. Wessner, Editor

NATIONAL RESEARCH COUNCIL
OF THE NATIONAL ACADEMIES

THE NATIONAL ACADEMIES PRESS
Washington, D.C.
www.nap.edu

THE NATIONAL ACADEMIES PRESS 500 Fifth Street NW Washington, DC 20001

NOTICE: The project that is the subject of this report was approved by the Governing Board of the National Research Council, whose members are drawn from the Councils of the National Academy of Sciences, the National Academy of Engineering, and the Institute of Medicine. The members of the committee responsible for the report were chosen for their special competences and with regard for appropriate balance.

This study was supported by Contract/Grant No. DASW01-02-C-0039 between the National Academy of Sciences and U.S. Department of Defense, NASW-03003 between the National Academy of Sciences and the National Aeronautics and Space Administration, DE-AC02-02ER12259 between the National Academy of Sciences and the U.S. Department of Energy, NSFDMI-0221736 between the National Academy of Sciences and the National Science Foundation, and N01-OD-4-2139 (Task Order #99) between the National Academy of Sciences and the U.S. Department of Health and Human Services. The content of this publication does not necessarily reflect the views or policies of the Department of Health and Human Services, nor does mention of trade names, commercial products, or organizations imply endorsement by the U.S. Government. Any opinions, findings, conclusions, or recommendations expressed in this publication are those of the author(s) and do not necessarily reflect the views of the organizations or agencies that provided support for the project.

International Standard Book Number-13:   978-0-309-10487-6
International Standard Book Number-10:   0-309-10487-4

Limited copies are available from the Policy and Global Affairs Division, National Research Council, 500 Fifth Street, NW, Washington, D.C. 20001; (202) 334-1529.

Additional copies of this report are available from the National Academies Press, 500 Fifth Street, NW, Lockbox 285, Washington, D.C. 20055; (800) 624-6242 or (202) 334-3313 (in the Washington metropolitan area); http://www.nap.edu.

THE NATIONAL ACADEMIES
Advisers to the Nation on Science, Engineering, and Medicine

The **National Academy of Sciences** is a private, nonprofit, self-perpetuating society of distinguished scholars engaged in scientific and engineering research, dedicated to the furtherance of science and technology and to their use for the general welfare. Upon the authority of the charter granted to it by the Congress in 1863, the Academy has a mandate that requires it to advise the federal government on scientific and technical matters. Dr. Ralph J. Cicerone is president of the National Academy of Sciences.

The **National Academy of Engineering** was established in 1964, under the charter of the National Academy of Sciences, as a parallel organization of outstanding engineers. It is autonomous in its administration and in the selection of its members, sharing with the National Academy of Sciences the responsibility for advising the federal government. The National Academy of Engineering also sponsors engineering programs aimed at meeting national needs, encourages education and research, and recognizes the superior achievements of engineers. Dr. Charles M. Vest is president of the National Academy of Engineering.

The **Institute of Medicine** was established in 1970 by the National Academy of Sciences to secure the services of eminent members of appropriate professions in the examination of policy matters pertaining to the health of the public. The Institute acts under the responsibility given to the National Academy of Sciences by its congressional charter to be an adviser to the federal government and, upon its own initiative, to identify issues of medical care, research, and education. Dr. Harvey V. Fineberg is president of the Institute of Medicine.

The **National Research Council** was organized by the National Academy of Sciences in 1916 to associate the broad community of science and technology with the Academy's purposes of furthering knowledge and advising the federal government. Functioning in accordance with general policies determined by the Academy, the Council has become the principal operating agency of both the National Academy of Sciences and the National Academy of Engineering in providing services to the government, the public, and the scientific and engineering communities. The Council is administered jointly by both Academies and the Institute of Medicine. Dr. Ralph J. Cicerone and Dr. Charles M. Vest are chair and vice chair, respectively, of the National Research Council.

www.national-academies.org

v

Linda F. Powers
Managing Director
Toucan Capital Corporation

Charles Trimble
CEO, *retired*
Trimble Navigation

Tyrone Taylor
President
Capitol Advisors
    on Technology, LLC

Patrick Windham
President
Windham Consulting

PROJECT STAFF

Charles W. Wessner
Study Director

Sujai J. Shivakumar
Senior Program Officer

McAlister T. Clabaugh
Program Associate

Jeffrey McCullough
Program Associate

David E. Dierksheide
Program Officer

RESEARCH TEAM

Zoltan Acs
University of Baltimore

Alan Anderson
Consultant

Philip A. Auerswald
George Mason University

Robert-Allen Baker
Vital Strategies, LLC

Robert Berger
Robert Berger Consulting, LLC

Grant Black
University of Indiana South Bend

Peter Cahill
BRTRC, Inc.

Dirk Czarnitzki
University of Leuven

Julie Ann Elston
Oregon State University

Irwin Feller
American Association for the
    Advancement of Science

David H. Finifter
The College of William and Mary

Michael Fogarty
University of Portland

Robin Gaster
North Atlantic Research

Albert N. Link
University of North Carolina

Benjamin Roberts
Harvard University

Rosalie Ruegg
TIA Consulting

Donald Siegel
University of California at Riverside

Paula E. Stephan
Georgia State University

Andrew Toole
Rutgers University

Nicholas Vonortas
George Washington University

Contents

APPENDIXES

Preface

Today's knowledge economy is driven in large part by the nation's capacity to innovate. One of the defining features of the U.S. economy is a high level of entrepreneurial activity. Entrepreneurs in the United States see opportunities and are willing and able to take on risk to bring new welfare-enhancing, wealth-generating technologies to the market. Yet, while innovation in areas such as genomics, bioinformatics, and nanotechnology present new opportunities, converting these ideas into innovations for the market involves substantial challenges.[1] The American capacity for innovation can be strengthened by addressing the challenges faced by entrepreneurs. Public-private partnerships are one means to help entrepreneurs bring new ideas to market.[2]

The Small Business Innovation Research (SBIR) program is one of the largest examples of U.S. public-private partnerships. Founded in 1982, the SBIR program was designed to encourage small business to develop new processes and products and to provide quality research in support of the many missions of the U.S. government. By including qualified small businesses in the nation's R&D (research and development) effort, SBIR grants are intended to stimulate innovative new technologies to help agencies meet the specific research and development needs of the nation in many areas, including health, the environment, and national defense.

---

[1]See Lewis M. Branscomb, Kenneth P. Morse, Michael J. Roberts, Darin Boville, *Managing Technical Risk: Understanding Private Sector Decision Making on Early Stage Technology Based Projects*, Gaithersburg, MD: National Institute of Standards and Technology, 2000.

[2]For a summary analysis of best practice among U.S. public-private partnerships, see National Research Council, *Government-Industry Partnerships for the Development of New Technologies: Summary Report*, Charles W. Wessner, ed., Washington, DC: The National Academies Press, 2002.

As the SBIR program approached its twentieth year of operation, the U.S. Congress asked the National Research Council to conduct a "comprehensive study of how the SBIR program has stimulated technological innovation and used small businesses to meet federal research and development needs" and to make recommendations on still further improvements to the program.[3] To guide this study, the National Research Council drew together an expert committee that included eminent economists, small businessmen and women, and venture capitalists, led by Dr. Jacques Gansler of the University of Maryland (formerly Undersecretary of Defense for Acquisition and Technology.) The membership of this committee is listed in the front matter of this volume. Given the extent of 'green-field research' required for this study, the Committee in turn drew on a distinguished team of researchers to, among other tasks, administer surveys and case studies, and develop statistical information about the program. The membership of this research team is also listed in the front matter of this volume.

This report is one of a series published by the National Academies in response to the congressional request. The series includes reports on the Small Business Innovation Research Program at the Department of Defense, the Department of Energy, the National Aeronautics and Space Administration, the National Institutes of Health, and the National Science Foundation—the five agencies responsible for 96 percent of the program's operations. It includes, as well, an Overview Report that provides assessment of the program's operations across the federal government. Other reports in the series include a summary of the 2002 conference that launched the study, and a summary of the 2005 conference on *SBIR and the Phase III Challenge of Commercialization* that focused on the Department of Defense and NASA.

PROJECT ANTECEDENTS

The current assessment of the SBIR program follows directly from an earlier analysis of public-private partnerships by the National Research Council's Board on Science, Technology, and Economic Policy (STEP). Under the direction of Gordon Moore, Chairman Emeritus of Intel, the NRC Committee on Government-Industry Partnerships prepared eleven volumes reviewing the drivers of cooperation among industry, universities, and government; operational assessments of current programs; emerging needs at the intersection of biotechnology and information technology; the current experience of foreign government partnerships and opportunities for international cooperation; and the changing roles of government laboratories, universities, and other research organizations in the national innovation system.[4]

---

[3]See the SBIR Reauthorization Act of 2000 (H.R. 5667-Section 108).

[4]For a summary of the topics covered and main lessons learned from this extensive study, see National Research Council, *Government-Industry Partnerships for the Development of New Technologies: Summary Report,* op. cit.

This analysis of public-private partnerships included two published studies of the SBIR program. Drawing from expert knowledge at a 1998 workshop held at the National Academy of Sciences, the first report, *The Small Business Innovation Research Program: Challenges and Opportunities,* examined the origins of the program and identified some operational challenges critical to the program's future effectiveness.[5] The report also highlighted the relative paucity of research on this program.

Following this initial report, the Department of Defense (DoD) asked the NRC to assess the Department's Fast Track Initiative in comparison with the operation of its regular SBIR program. The resulting report, *The Small Business Innovation Research Program: An Assessment of the Department of Defense Fast Track Initiative,* was the first comprehensive, external assessment of the Department of Defense's program. The study, which involved substantial case study and survey research, found that the SBIR program was achieving its legislated goals. It also found that DoD's Fast Track Initiative was achieving its objective of greater commercialization and recommended that the program be continued and expanded where appropriate.[6] The report also recommended that the SBIR program overall would benefit from further research and analysis, a perspective adopted by the U.S. Congress.

SBIR REAUTHORIZATION AND CONGRESSIONAL REQUEST FOR REVIEW

As a part of the 2000 reauthorization of the SBIR program, Congress called for a review of the SBIR programs of the agencies that account collectively for 96 percent of program funding. As noted, the five agencies meeting this criterion, by size of program, are the Department of Defense, the National Institutes of Health, the National Aeronautics and Space Administration, the Department of Energy, and the National Science Foundation.

Congress directed the NRC, via H.R. 5667, to evaluate the quality of SBIR research and evaluate the SBIR program's value to the agency mission. It called for an assessment of the extent to which SBIR projects achieve some measure of commercialization, as well as an evaluation of the program's overall economic and noneconomic benefits. It also called for additional analysis as required to support specific recommendations on areas such as measuring outcomes for

---

[5]See National Research Council, *The Small Business Innovation Research Program: Challenges and Opportunities,* Charles W. Wessner, ed., Washington, DC: National Academy Press, 1999.

[6]See National Research Council, *The Small Business Innovation Research Program: An Assessment of the Department of Defense Fast Track Initiative,* Charles W. Wessner, ed., Washington, DC: National Academy Press, 2000. Given that virtually no published analytical literature existed on SBIR, this Fast Track study pioneered research in this area, developing extensive case studies and newly developed surveys.

agency strategy and performance, increasing federal procurement of technologies produced by small business, and overall improvements to the SBIR program.[7]

ACKNOWLEDGMENTS

On behalf of the National Academies, we express our appreciation and recognition for the insights, experiences, and perspectives made available by the participants of the conferences and meetings, as well as by survey respondents and case study interviewees who participated over the course of this study. We are also very much in debt to officials from the leading departments and agencies. Among the many who provided assistance to this complex study, for this volume, we are especially in debt to Kesh Narayanan, Joseph Hennessey, and Ritchie Coryell of the National Science Foundation. Valuable, independent contributions and observations were provided by Roland Tibbetts, formerly of the National Science Foundation.

The Committee's research team deserves recognition for their instrumental role in the preparation and many revisions of this report. In that regard, special thanks are due to Rosalie Ruegg of TIA Consulting who served as the lead researcher for the NSF study. Her timely and insightful contributions played a key role in the committee's analysis. Without their collective efforts and close cooperation, amidst many other competing priorities, it would not have been possible to prepare this report. Among the many contributing Committee members, special thanks are due to Christina Gabriel, Kent Murphy, and Patrick Windham.

NATIONAL RESEARCH COUNCIL REVIEW

This report has been reviewed in draft form by individuals chosen for their diverse perspectives and technical expertise, in accordance with procedures approved by the National Academies' Report Review Committee. The purpose of this independent review is to provide candid and critical comments that will assist the institution in making its published report as sound as possible and to ensure that the report meets institutional standards for objectivity, evidence, and responsiveness to the study charge. The review comments and draft manuscript remain confidential to protect the integrity of the process.

We wish to thank the following individuals for their review of this report: Heidi Jacobus, Cybernet Systems Corporation; Brad Knox, Aflac Insurance; Jeanne Powell, National Institute of Standards and Technology; and Richard Wright, National Institute of Standards and Technology.

---

[7]Chapter 3 of the Committee's Methodology Report describes how this legislative guidance was drawn out in operational terms. National Research Council, *An Assessment of the Small Business Innovation Research Program: Project Methodology,* Washington, DC: The National Academies Press, 2004. Access this report at *<http://www7.nationalacademies.org/sbir/SBIR_Methodology_Report.pdf>*.

Although the reviewers listed above have provided many constructive comments and suggestions, they were not asked to endorse the conclusions or recommendations, nor did they see the final draft of the report before its release. The review of this report was overseen by Robert Frosch, Harvard University, and Robert White, Carnegie Mellon University. Appointed by the National Academies, they were responsible for making certain that an independent examination of this report was carried out in accordance with institutional procedures and that all review comments were carefully considered. Responsibility for the final content of this report rests entirely with the authoring committee and the institution.

Jacques S. Gansler   Charles W. Wessner

Summary

I. INTRODUCTION

The Small Business Innovation Research (SBIR) program was created in 1982 through the Small Business Innovation Development Act. As the SBIR program approached its twentieth year of operation, the U.S. Congress requested the National Research Council (NRC) of the National Academies to "conduct a comprehensive study of how the SBIR program has stimulated technological innovation and used small businesses to meet federal research and development needs" and to make recommendations with respect to the SBIR program. Mandated as a part of SBIR's reauthorization in late 2000, the NRC study has assessed the SBIR program as administered at the five federal agencies that together make up some 96 percent of SBIR program expenditures. The agencies, in order of program size, are the Department of Defense, the National Institutes of Health, the National Aeronautics and Space Administration, the Department of Energy, and the National Science Foundation.

Based on that legislation, and after extensive consultations with both Congress and agency officials, the NRC focused its study on two overarching questions.[1]

---

[1] Three primary documents condition and define the objectives for this study: These are the legislation—H.R. 5667, the NAS-Agencies *Memorandum of Understanding,* and the NAS contracts accepted by the five agencies. These are reflected in the Statement of Task addressed to the Committee by the Academies' leadership. Based on these three documents, the NRC Committee developed a comprehensive and agreed-upon set of practical objectives to be reviewed. These are outlined in the Committee's formal Methodology Report, particularly Chapter 3, "Clarifying Study Objectives." National Research Council, *An Assessment of the Small Business Innovation Research Program: Project Methodology,* Washington, DC: The National Academies Press, 2004, accessed at *<http://books.nap.edu/catalog.php?record_id=11097#toc>.*

1

First, how well do the agency SBIR programs meet four societal objectives of interest to Congress? That is: (1) to stimulate technological innovation; (2) to increase private-sector commercialization of innovations; (3) to use small business to meet federal research and development needs; and (4) to foster and encourage participation by minority and disadvantaged persons in technological innovation.[2] Second, can the management of agency SBIR programs be made more effective? Are there best practices in agency SBIR programs that may be extended to other agencies' SBIR programs?

To satisfy the congressional request for an external assessment of the program, the NRC analysis of the operations of the SBIR program involved multiple sources and methodologies. A large team of expert researchers carried out extensive NRC-commissioned surveys and case studies. In addition, agency-compiled program data, program documents, and the existing literature were reviewed. These were complemented by extensive interviews and discussions with program managers, program participants, agency "users" of the program, as well as program stakeholders.[3]

The study as a whole sought to understand operational challenges and to measure program effectiveness, including the quality of the research projects being conducted under the SBIR program, the challenges and achievements in commercialization of the research, and the program's contribution to accomplishing agency missions. To the extent possible, the evaluation included estimates of the benefits (both economic and noneconomic) achieved by the SBIR program, as well as broader policy issues associated with public-private collaborations for technology development and government support for high technology innovation.

Taken together, this study is the most comprehensive assessment of SBIR to date. Its empirical, multifaceted approach to evaluation sheds new light on the operation of the SBIR program in the challenging area of early-stage finance. As with any assessment, particularly one across five quite different agencies and departments, there are methodological challenges. These are identified and discussed at several points in the text. This important caveat notwithstanding, the scope and diversity of the study's research should contribute significantly to the understanding of the SBIR program's multiple objectives, measurement issues, operational challenges, and achievements.

---

[2]These congressional objectives are found in the Small Business Innovation Development Act (PL 97-219). In reauthorizing the program in 1992 (PL 102-564), Congress expanded the purposes to "emphasize the program's goal of increasing private-sector commercialization developed through Federal research and development and to improve the Federal government's dissemination of information concerning small business innovation, particularly with regard to women-owned business concerns and by socially and economically disadvantaged small business concerns."

[3]The Committee's methodological approach is described in National Research Council, *An Assessment of the Small Business Innovation Research Program: Project Methodology*, ibid. For a summary of potential biases in innovation survey responses, see Box A in Chapter 3 of this report.

II. OVERVIEW OF THE NATIONAL SCIENCE FOUNDATION'S SBIR PROGRAM

This report addresses the SBIR program operated by the National Science Foundation (NSF), which annually makes several hundred awards that total nearly $100 million. The NSF focuses on supporting research and education at the nation's universities. Nonetheless, it was the first federal organization to create a small business innovation research program. Roland Tibbetts is credited with creating the SBIR concept with a program for small business. Initiated, in 1977, it provided a model for the government-wide SBIR program launched in 1982. This precursor program was designated as the NSF's SBIR program in 1982.

The NSF's SBIR program has a number features that help distinguish it from the programs of other agencies. Unique features pertain to its history, size, mission and technology orientation, constituency, grant options, as well as its management and culture. These features are highlighted and elaborated below.

Box A Special Features of the NSF's SBIR Program

History: The NSF recognized early on that small businesses—like universities—can perform high quality, innovative research, and it developed the precursor of the SBIR program.

Program Size: With its nearly $100 million in annual grants, the NSF operates the smallest of the five large SBIR programs.

Mission: The NSF does not have a procurement mission; the program is oriented toward the private marketplace, but many of the technologies it funds also support other agency needs.

Technology Orientation: The program provides early-stage support for diverse technologies, including manufacturing and materials research.

Constituency: Rather than a single constituency, such as firms in aerospace or medicine or defense, the NSF has a broad constituency that cuts across industrial sectors.

Supplemental Grant Options: As an innovation, the program added Phase IIB grant supplements following a Phase II grant conditional on attraction of third-party financing.[a]

Program Management: The program is highly centralized, managed by staff with industry experience.

Culture: The NSF has a growing culture of program analysis, experimentation, and evaluation.

---

[a] NSF's Phase IIB supplements may be contrasted with the Defense Department's Fast Track initiative, which also encourages companies to obtain funds from third-party investors. However, the Fast Track initiative occurs at the front end of the grant for a faster transition from Phase I to Phase II, whereas NSF Phase IIB supplements occur at the end of the "regular" Phase II grant to further develop promising awards.

TABLE S-1 Number of NSF SBIR Grants, 1992-2005

Year	Total Awards	Phase I	Phase II	Phase IIB
1992	265	208	57	0
1993	308	256	52	0
1994	330	309	21	0
1995	349	301	48	0
1996	342	252	90	0
1997	383	261	122	0
1998	336	215	117	4
1999	346	236	89	21
2000	337	233	95	9
2001	324	219	91	14
2002	392	286	67	39
2003	538	437	77	24
2004	397	244	131	22
2005	309	149	132	28

SOURCE: NSF SBIR program.

The NSF's highly centralized SBIR program places substantial emphasis on the goal of commercialization. In recent congressional testimony, NSF SBIR officials said the program's primary focus is "the commercialization of research."[4] This study, however, considers the program's performance across three additional goals—stimulating technological innovation, using small business to meet federal research and development needs, fostering participation by minority and disadvantaged persons—as well as increasing private-sector commercialization of innovation.

Table S-1 shows the total number of NSF SBIR grants annually, as well as the number of each of the three types of SBIR grants: Phase I, Phase II, and Phase IIB. Between 1992 and 2005, the total number of grants fluctuated between 265 and 538 per year, and averaged 354. Over the entire period, Phase I grants accounted for the largest share of the total number at 73 percent. Phase II grants accounted for 24 percent of the total number of grants, and Phase IIB grants, which were started in 1998, accounted for only 3 percent of the total number per year. The three types of grants are defined and described below.

Figure S-1 shows the dollar amounts. In recent years, total grants have approached $100 million per year, with Phase II grants comprising the largest share in most years.

An NSF SBIR Phase I grant currently averages approximately $100,000 and lasts for six months. A Phase II grant ranges up to $500,000 and lasts for a period

---

[4]Testimony of Joseph Hennessey, "The Small Business Innovation Research Program: Opening Doors to New Technology," before the House Committee on Small Business's Subcommittee on Workforce, Empowerment and Government Programs, November 8, 2005.

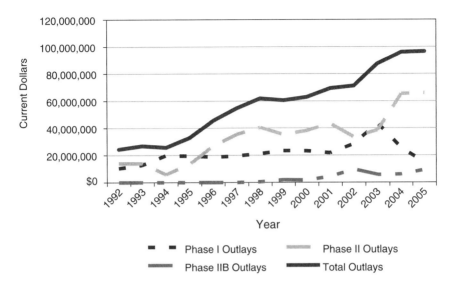

FIGURE S-1 Dollar Amounts of NSF SBIR Grants, 1992–2005 (Current Dollars). SOURCE: Based on data provided by the NSF SBIR program.

of up to two years. The NSF pioneered the use of the Phase IIB grant, which allows a firm to obtain a supplemental follow-on grant ranging from $50,000 to $500,000 provided the applicant is backed by $2 of third-party funding for every $1 of NSF funding provided. The Phase IIB grant is seen as a tool for promoting commercialization. It yields a different allocation of funding than would result from allocating all Phase II funding according to the initial Phase II selection process.

The NSF funded between 14 percent and 21 percent of the 1,000 to 2,000 Phase I proposals received each year over the period 1994 to 2005. It funded between 17 percent and 61 percent of the several hundred Phase II proposals received each year over the same period, and between 41 percent and 61 percent in the last few years of this period.

Many of the small companies that have received the NSF's SBIR program assistance have developed novel and promising technologies, new products for market, new processes, and new capabilities. The NSF recently identified a few of these companies and technologies as illustrative of companies with "big ideas" under development. The NSF SBIR-funded technologies of these companies range from educational and medical tools, to nanoengineered powders for environmental cleanup, to new algorithms for improving information searches, to new devices for converting low levels of radiation into electricity.

III. SUMMARY OF KEY FINDINGS AND RECOMMENDATIONS

Findings

The core finding of the study is that the SBIR program is sound in concept and effective in practice. It can also be improved. Currently, the program is delivering results that meet most of the congressional objectives. Specifically, the program is:

- **Stimulating Technological Innovation**
  o **Generating Knowledge.** SBIR is contributing to the nation's store-house of public scientific and technological knowledge. This knowledge is embodied in data, scientific and engineering publications, patents and licenses, presentations, analytical models, algorithms, new research equipment, reference samples, prototype products and processes, spin-off companies, "human capital" (greater know-how, expertise, etc.), and new capability for further innovative activity. Publications and patenting activity occur with considerable frequency.[5]
  o **Creating and Disseminating Intellectual Capital.** Extensive licensing activities of Phase II awardees attested to the fact that useful intellectual capital has been created and disseminated. For example, the NRC Phase II Survey showed one-fifth of Phase II projects reported they had reached licensing agreements with U.S. companies and investors, and another fifth reported they had ongoing negotiations with U.S. companies and investors on licensing agreements.
  o **Building Networks with Universities**. Both the NRC surveys and the case studies showed extensive networking between NSF SBIR-funded projects and universities. University faculty and students used the NSF SBIR program to establish businesses, start projects, and work on projects.
  o **Moving Technology from Universities Toward the Market**. The NSF SBIR program has facilitated technology transfer out of universities. Fourteen percent of the NRC Phase II Survey projects were based on technology originally developed at a university by a project participant. Five percent of NSF Phase II Survey projects were based on technology licensed from a university.

---

[5]The NRC Phase II survey reported averages of 1.66 scientific publications and 0.67 patents per surveyed project. See Table 7.2-1 in Chapter 7, Section 2, for additional survey information on the NSF's SBIR program. The underlying distribution of patents and publications reported is skewed, with some companies reporting none and some reporting relatively high numbers.

- **Increasing Private-Sector Commercialization of Innovations**
  - o **Achieving Commercialization.** Despite the fact that the agency itself normally does not acquire the results of SBIR-funded projects, a significant portion of NSF's SBIR projects commercialize successfully or are making progress toward commercialization. For example, one-fifth of survey respondents indicated that their project had resulted in products, processes, or services that were in use and still active. A little more than a quarter of respondents indicated that the project was continuing technology development in the post-Phase II period. Of course, few individual SBIR projects lead directly to "home runs" in the commercial sense. Nonetheless, the NRC Phase II Survey shows that small firms believe that the NSF's SBIR program helped them to enter commercial markets.
  - o **Project Initiation.** The SBIR awards play a significant role in initiating the development of technologies and products that are subsequently commercialized. When asked if their companies would have undertaken the projects had there been no SBIR grant, approximately two-thirds of respondents answered either probably not (43 percent) or definitely not (24 percent).
  - o **A Small Percentage of Projects Account for Most Successes.** As is typical for other private and public technology programs, a relatively few projects account for the majority of sales and licensing revenue from NSF SBIR recipients.[6] This highly skewed performance distribution among projects is an inherent characteristic of early-stage investment—one that is impossible to avoid if innovation is to be promoted.

- **Using Small Businesses to Meet Federal Research and Development Needs**
  - o **Mission Alignment.** The NSF's SBIR program funding is closely aligned with the agency's broader mission and is contributing broadly to federal research and development procurement needs.
  - o **Meeting Agency Procurement Needs.** The NSF SBIR program helps

---

[6]Among the 162 projects surveyed, just 8 projects—each of which had $2.3 million or more in sales—accounted for over half the total reported sales dollars for the surveyed projects. The project with the highest reported amount had $4.8 million in sales. Similarly, the results for sales by licensees of those survey projects' technologies were highly skewed by a single licensee that accounted for over half the total licensee sales dollars, amounting to $200 million or more in licensee sales. Similar results were revealed in previous National Research Council assessments of early-stage innovation awards. See National Research Council, *The Advanced Technology Program: Assessing Outcomes*, Charles W. Wessner, ed., Washington, DC: National Academy Press, 2001. See also National Research Council, *The Small Business Innovation Research Program: An Assessment of the Department of Defense Fast Track Initiative*, Charles W. Wessner, ed., Washington, DC: National Academy Press, 2000.

to meet the procurement needs of federal agencies. The NRC Phase II Survey found that sales of NSF Phase II–funded technologies go to multiple markets with broad and diversified customer bases.

- **Fostering Participation by Minority and Disadvantaged Persons in Technological Innovation.**
  - o **Open to New Entrants.** SBIR has a high proportion of new entrants each year, rising to nearly two-thirds in 2003. Overall, between 1996 and 2003, 54 percent of the Phase I grants went to new entrants. Only 9 percent of selected firms had previously received more than 20 Phase I grants.[7]
  - o **Participation Rates of Women and Minorities.** Women and minorities also participated in projects as principal investigators, with 21 percent of Phase II projects surveyed reporting either a woman, a minority, or a minority woman as the principal investigator. [8]
  - o **Lower Success Rates.** Success rates for woman- and minority-owned firms applying for Phase I awards are significantly lower than for other firms.[9] Levels of participation by woman- and minority-owned businesses also continue to lag other groups. The cause of these lower participation rates is unclear. They may reflect the low representation of women and minorities in high technology firms.[10] These lower suc-

---

[7]See Table 4.2-8 and Figure 4.2-20.

[8]Analysis of data provided by NSF shows the number of Phase I proposals from and grants received by woman-owned businesses fell from 1994 through 2005. With the exception of a bump up in 2002 and 2003, there is no upward trend. The number of Phase II proposals from and grants received by woman-owned businesses from 1995 through 2005 exhibits no clear trend. Finally, the number of Phase IIB grants received by woman-owned businesses annually from 1998 through 2005 shows no obvious trend. See Chapter 4, Section 4.2.5, in this report for graphs of the data.

[9]From 1995 through 2005, woman-owned businesses submitted 12.2 percent of Phase I proposals and received 9.5 percent of Phase I grants. They submitted 8.8 percent of Phase II proposals and received 7.5 percent of Phase II grants. Minority-owned businesses (including minority women) submitted 16 percent of Phase I proposals and received 13.5 percent of Phase I grants. They submitted 12.9 percent of Phase II proposals and received 13.7 percent of Phase II grants.

[10]White males, who comprise 40 percent of the nation's overall workforce, hold 68 percent of all science, engineering, and technology jobs. In contrast, white women, who comprise 35 percent of the national workforce, hold only 15 percent of these positions, and only 10 percent of the 2 million scientists and engineers in the United States in a recent year were women. African Americans and Hispanics, who comprise almost 21 percent of the American workforce, represent just 6 percent of the science, engineering, and technology workforce. (Congressional Commission on the Advancement of Women and Minorities in Science, Engineering and Technology Development. Findings of the Commission as reported in SSTI Weekly Digest, April 6, 2001; and U.S. Bureau of Labor Statistics.) However, it should be noted that not all minority groups have low representation in science, engineering, and technology fields relative to their representation in the U.S. population Indicative of shifting representation among minority groups, there was a strong increase during the 1990s in the percentage of doctorate degrees and jobs in science and engineering going to foreign-born individuals,

cess rates may also reflect obstacles that women and minorities face in pursuing careers in science and engineering.[11]

o **An Innovative Program**. The NSF's SBIR program operates with a limited administrative budget and legislated limits on funds for commercialization assistance. Nonetheless, with a number of valuable supporting functions and innovative approaches (such as Phase IIB to commercialization), the NSF's SBIR program office has made an impressive effort in developing a well-run program with a number of supporting functions and growing evaluation effort.

o **Professional Staff.** The NSF's SBIR program is generally effective in achieving its goals and has benefited from its strong, centralized management and talented program managers.

o **More Assessment Is Needed.** It is important to recognize the inherent challenges of early-stage funding of high technology companies with new but unproven ideas. All projects will not succeed. Nevertheless, greater efforts to rigorously document and evaluate current achievements and the impact of program innovations could contribute to improved program output.

Recommendations

- **Improving Program Operations.**
  o **Retain Program Flexibility**. First and foremost, it is essential to retain and encourage the flexibility that has enabled NSF SBIR program management to develop an innovative and effective multiphase program.[12]
  o **Conduct Regular Evaluations.** Regular, rigorous program evaluation is essential for quality program management and accountability. Accordingly, NSF program management should give greater attention and resources to the systematic evaluation of the program, supported

---

particularly those from India, China, and the Philippines. (NSF Science and Engineering Indicators 2006, "The U.S. S&E Labor Force.") By like token, the representation in the NSF's SBIR program is uneven among different minority groups.

[11]Academics represent an important future pool of applicants, firm founders, principal investigators, and consultants. Recent research shows that owing to the low number of women in senior research positions in many leading academic science departments, few women have the chance to lead a spinout. "Underrepresentation of female academic staff in science research is the dominant (but not the only) factor to explain low entrepreneurial rates amongst female scientists." See Peter Rosa and Alison Dawson, "Gender and the commercialization of university science: academic founders of spinout companies," *Entrepreneurship & Regional Development*, Volume 18, Issue 4 July 2006, pages 341–366.

[12]See Recommendation I, Chapter 2.

by reliable data, and should seek to make the program as responsive as possible to the needs of small company applicants.[13]

o **Improve Processes.** The NSF should ensure that solicitation topics are broadly defined and that the topic definition process is bottom-up, and take steps to ensure the necessary flexibility to permit firms to receive relatively prompt access to Phase I solicitations. Finally, the NSF should increase its use of technically competent reviewers with strong technical expertise and strong business understanding for both Phase I and Phase II selection.[14]

o **Increase Management Funding for SBIR.** To enhance program utilization, management, and evaluation, consideration should be given to the provision of additional program funds for management and evaluation. Additional funds might be allocated internally within the existing NSF budget, drawn from the existing set-aside for the program, or by increasing the set-aside for the program, currently at 2.5 percent of external research budgets. The NSF spends some $100 million a year on SBIR, and the return on this investment could be enhanced with a modest addition to funds for management and evaluation.[15]

- **Continue to Increase Private-Sector Commercialization.**[16]
  o **Support Commercialization Assistance**. The NSF should increase support for commercialization assistance as resources permit.
  o **Encourage Continued Experimentation.** The NSF should continue to promote its positive initiative with Phase IIB awards, refining the tool as experience suggests and raising the number and amount of these awards as third-party funding permits.

- **Improve Participation and Success by Women and Minorities.**[17]
  o **Encourage Participation.** The NSF should develop targeted outreach to improve the participation rates of woman- and minority-owned firms, and strategies to improve their success rates based on causal factors determined by analysis of past proposals and feedback from the affected groups.
  o **Improve Data Collection and Analysis.** The NSF should arrange for an independent analysis of a sample of past proposals from woman- and minority-owned firms and from other firms (to serve as a control group). This will help identify specific factors accounting for the lower

---

[13]See Recommendation IV, Chapter 2.
[14]See Recommendations V and II-b in Chapter 2.
[15]See Recommendation VIII in Chapter 2.
[16]See Recommendation II in Chapter 2.
[17]See Recommendation III in Chapter 2.

success rates of woman- and minority-owned firms, as compared with other firms, in having their Phase I proposals granted.

o **Extend Outreach to Younger Women and Minority Students**. The NSF should immediately encourage and solicit women and under-represented minorities working at small firms to apply as principal investigators (PIs) and senior co-investigators (Co Is) for SBIR awards and track their success rates.

1

Introduction

1.1 SMALL BUSINESS INNOVATION RESEARCH PROGRAM CREATION AND ASSESSMENT

Created in 1982 by the Small Business Innovation Development Act, the Small Business Innovation Research (SBIR) program was designed to stimulate technological innovation among small private-sector businesses while providing the government with cost-effective new technical and scientific solutions to challenging mission problems. The SBIR program was also designed to help stimulate the U.S. economy by encouraging small businesses to market innovative technologies in the private sector.[1]

As the SBIR program approached its twentieth year of existence, the U.S. Congress requested that the National Research Council (NRC) of the National Academies conduct a "comprehensive study of how the SBIR program has stimulated technological innovation and used small businesses to meet Federal research and development needs," and make recommendations on improvements to the program.[2] Mandated as a part of the SBIR program's renewal in 2000, the NRC study has assessed the SBIR program as administered at the five federal agencies that together make up 96 percent of SBIR program expenditures. The agencies

---

[1] The SBIR legislation drew from a growing body of evidence, starting in the late 1970s and accelerating in the 1980s, which indicated that small businesses were assuming an increasingly important role in both innovation and job creation. This evidence gained new credibility with the Phase I empirical analysis by Zoltan Acs and David Audretsch of the US Small Business Innovation Data Base, which confirmed the increased importance of small firms in generating technological innovations and their growing contribution to the U.S. economy. See Zoltan Acs and David Audretsch, *Innovation and Small Firms*, Cambridge MA: MIT Press, 1990.

[2] See Public Law 106-554, Appendix I—H.R. 5667, Section 108.

are, in decreasing order of program size, the Department of Defense (DoD), the National Institutes of Health (NIH), the National Aeronautics and Space Administration (NASA), the Department of Energy (DoE), and the National Science Foundation (NSF).

The NRC Committee assessing the SBIR program was not asked to consider if the SBIR program should exist or not—Congress has affirmatively decided this question on three occasions.[3] Rather, the Committee was charged with providing assessment-based findings to improve public understanding of the program as well as recommendations to improve the program's effectiveness.

1.2 SBIR PROGRAM STRUCTURE

Eleven federal agencies are currently required to set aside 2.5 percent of their extramural research and development (R&D) budget exclusively for SBIR contracts. Each year these agencies identify various R&D topics, representing scientific and technical problems requiring innovative solutions, for pursuit by small businesses under the SBIR program. These topics are bundled together into individual agency "solicitations"—publicly announced requests for SBIR proposals from interested small businesses. A small business can identify an appropriate topic it wants to pursue from these solicitations and, in response, propose a project for an SBIR grant. The required format for submitting a proposal is different for each agency. Proposal selection also varies, though peer review of proposals on a competitive basis by experts in the field is typical. Each agency then selects the proposals that are found best to meet program selection criteria and awards contracts or grants to the proposing small businesses.

As conceived in the 1982 Small Business Development Act, the SBIR program's grant-making process is structured in three phases:

- *Phase I* grants essentially fund feasibility studies in which award winners undertake a limited amount of research aimed at establishing an idea's scientific and commercial promise. Today, the legislation anticipates Phase I grants as high as $100,000.[4]
- *Phase II* grants are larger—typically about $750,000—and fund more extensive R&D to further develop the scientific and commercial promise of research ideas.
- *Phase III*. During this phase, companies do not receive further SBIR awards. Instead, grant recipients should be obtaining additional funds from a procurement program at the agency that made the award, from

---

[3]These are the 1982 Small Business Development Act and the subsequent multiyear reauthorizations of the SBIR program in 1992 and 2000.

[4]With the agreement of the Small Business Administration, which plays an oversight role for the program, this amount can be higher in certain circumstances (e.g., drug development at NIH) and is often lower with smaller SBIR programs (e.g., the EPA or the Department of Agriculture).

private investors, or from the capital markets. The objective of this phase is to move the technology from the prototype stage to the marketplace.

Obtaining Phase III support is often the most difficult challenge for new firms to overcome. In practice, agencies have developed different approaches to facilitate SBIR grantees' transition to commercial viability; not least among them are additional SBIR grants.

Previous NRC research has shown that firms have different objectives in applying to the program. Some want to demonstrate the potential of promising research, but they may not seek to commercialize it themselves. Others think they can fulfill agency research requirements more cost-effectively through the SBIR program than through the traditional procurement process. Still others seek a certification of quality (and the investments that can come from such recognition) as they push science-based products toward commercialization.[5]

1.3 SBIR REAUTHORIZATIONS

The SBIR program approached reauthorization in 1992 amidst continued concerns about the U.S. economy's capacity to commercialize inventions. Finding that "U.S. technological performance is challenged less in the creation of new technologies than in their commercialization and adoption," the National Academy of Sciences at the time recommended an increase in SBIR funding as a means to improve the economy's ability to adopt and commercialize new technologies.[6]

Following this report, the Small Business Research and Development Enhancement Act (P.L. 102-564), which reauthorized the SBIR program until September 30, 2000, doubled the set-aside rate to 2.5 percent.[7] This increase in the percentage of R&D funds allocated to the program was accompanied by a stronger emphasis on encouraging the commercialization of SBIR-funded technologies.[8] Legislative language explicitly highlighted commercial potential

---

[5]See Reid Cramer, "Patterns of Firm Participation in the Small Business Innovation Research Program in Southwestern and Mountain States," in National Research Council, *The Small Business Innovation Research Program: An Assessment of the Department of Defense Fast Track Initiative*, Charles W. Wessner, ed., Washington, DC: National Academy Press, 2000.

[6]See National Research Council, *The Government Role in Civilian Technology: Building a New Alliance*, Washington, DC: National Academy Press, 1992, pp. 29.

[7]For fiscal year 2003, this has resulted in a program budget of approximately $1.6 billion across all federal agencies, with the Department of Defense (DoD) having the largest SBIR program, at $834 million, followed by the National Institutes of Health (NIH), at $525 million. The DoD's SBIR program is made up of 10 participating components: Army, Navy, Air Force, Missile Defense Agency (MDA), Defense Advanced Research Projects Agency (DARPA), Chemical Biological Defense (CBD), Special Operations Command (SOCOM), Defense Threat Reduction Agency (DTRA), National Imagery and Mapping Agency (NIMA), and the Office of the Secretary of Defense (OSD). NIH counts 23 separate institutes and agencies making SBIR awards, many with multiple programs.

[8]See Robert Archibald and David Finifter, "Evaluation of the Department of Defense Small Busi-

as a criterion for awarding SBIR grants. For Phase I awards, Congress directed program administrators to assess whether projects have "commercial potential," in addition to scientific and technical merit, when evaluating SBIR applications.

The 1992 legislation mandated that program administrators consider the existence of second-phase funding commitments from the private sector or other non-SBIR sources when judging Phase II applications. Evidence of third-phase follow-on commitments, along with other indicators of commercial potential, was also to be sought. Moreover, the 1992 reauthorization directed that a small business's record of commercialization be taken into account when evaluating its Phase II application.[9]

The Small Business Reauthorization Act of 2000 (P.L. 106-554) extended the SBIR program until September 30, 2008. It called for this assessment by the National Research Council of the broader impacts of the program, including those on employment, health, national security, and national competitiveness.[10]

1.4 STRUCTURE OF THE NRC STUDY

This NRC assessment of the SBIR program has been conducted in two phases. In the first phase, at the request of the agencies, a research methodology was developed by the NRC. This methodology was then reviewed and approved by an independent National Academies panel of experts.[11] Information about the program was also gathered through interviews with SBIR program administrators and during two major conferences where SBIR officials were invited to describe program operations, challenges, and accomplishments.[12] These conferences highlighted the important differences in each agency's SBIR program goals, practices,

---

ness Innovation Research Program and the Fast Track Initiative: A Balanced Approach," in National Research Council, *The Small Business Innovation Reseearch Program: An Assessment of the Department of Defense Fast Track Initiative*, op. cit. pp. 211–250.

[9]GAO report had found that agencies had not adopted a uniform method for weighing commercial potential in SBIR applications. See U.S. General Accounting Office, 1999, *Federal Research: Evaluations of Small Business Innovation Research Can Be Strengthened*, AO/RCED-99-114, Washington, DC: U.S. General Accounting Office.

[10]The current assessment is congruent with the Government Performance and Results Act (GPRA) of 1993: *<http://govinfo.library.unt.edu/npr/library/misc/s20.html>*. As characterized by the GAO, GPRA seeks to shift the focus of government decision making and accountability away from a preoccupation with the activities that are undertaken—such as grants dispensed or inspections made—to a focus on the results of those activities. See *<http://www.gao.gov/new.items/gpra/gpra.htm>*.

[11]The SBIR methodology report is available on the Web. Access at *<http://www7.nationalacademies. org/sbir/SBIR_Methodology_Report.pdf>*.

[12]The opening conference on October 24, 2002, examined the program's diversity and assessment challenges. For a published report of this conference, see National Research Council, *SBIR: Program Diversity and Assessment Challenges*, Charles W. Wessner, ed., Washington, DC: The National Academies Press, 2004. The second conference, held on March 28, 2003, was titled "Identifying Best Practice." The conference provided a forum for the SBIR program managers from each of the five agencies in the study's purview to describe their administrative innovations and best practices.

and evaluations. The conferences also explored the challenges of assessing such a diverse range of program objectives and practices using common metrics.

The second phase of the NRC study implemented the approved research methodology. The Committee deployed multiple survey instruments and its researchers conducted case studies of a wide profile of SBIR firms. The Committee then evaluated the results and developed both agency-specific and overall findings and recommendations for improving the effectiveness of the SBIR program. The final report includes complete assessments for each of the five agencies and an overview of the program as a whole.

1.5 SBIR ASSESSMENT CHALLENGES

At its outset, the NRC's SBIR study identified a series of assessment challenges that must be addressed. As discussed at the October 2002 conference that launched the study, the administrative flexibility found in the SBIR program makes it difficult to make cross-agency assessments. Although each agency's SBIR program shares the common three-phase structure, the SBIR concept is interpreted uniquely at each agency. This flexibility is a positive attribute in that it permits each agency to adapt its SBIR program to the agency's particular mission, scale, and working culture. For example, the NSF operates its SBIR program differently than DoD because "research" is often coupled with procurement of goods and services at DoD but rarely at NSF. Programmatic diversity means that each agency's SBIR activities must be understood in terms of their separate missions and operating procedures. This commendable diversity makes an assessment of the program as a whole more challenging.

A second challenge concerns the linear process of commercialization implied by the design of SBIR's three-phase structure.[13] In the linear model illustrated in Figure 1-1, innovation begins with basic research supplying a steady stream of fresh and new ideas. Among these ideas, those that show technical feasibility become innovations. Such innovations, when further developed by firms, become marketable products driving economic growth.

As the NSF's Joseph Bordogna observed at the launch conference, innovation almost never takes place through a protracted linear progression from research to development to market. Research and development drives technological innovation, which, in turn, opens up new frontiers in R&D. True innovation, Bordogna noted, can spur the search for new knowledge and create the context in which the next generation of research identifies new frontiers. This nonlinearity, illustrated in Figure 1-2, makes it difficult to rate the efficiency of SBIR program. Inputs do not match up with outputs according to a simple function. Figure 1-2, while

---

[13]This view was echoed by Duncan Moore: "Innovation does not follow a linear model. It stops and starts." National Research Council, *SBIR: Program Diversity and Assessment Challenges*, Charles W. Wessner, ed., Washington, DC: The National Academies Press, 2004 p. 24.

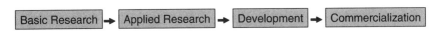

FIGURE 1-1 The Linear Model of Innovation.

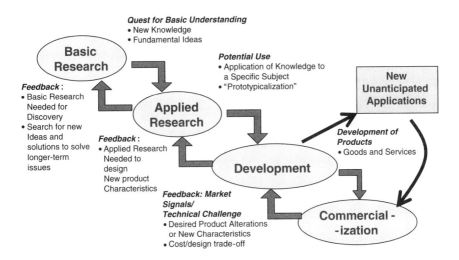

FIGURE 1-2 A Feedback Model of Innovation.

more complex than Figure 1-1, is itself a highly simplified model. For example, feedback loops can stretch backward or forward by more than one level.

A third assessment challenge relates to the measurement of outputs and outcomes. Program realities can and often do complicate the task of data gathering. In some cases, for example, SBIR recipients receive a Phase I award from one agency and a Phase II award from another. In other cases, multiple SBIR awards may have been used to help a particular technology become sufficiently mature to reach the market. Also complicating matters is the possibility that for any particular grantee, an SBIR award may be only one among other federal and nonfederal sources of funding. Causality can thus be difficult, if not impossible, to establish. The task of measuring outcomes is made harder because companies that have garnered SBIR awards can also merge, fail, or change their names before a product reaches the market. In addition, principal investigators or other key individuals can change firms, carrying their knowledge of an SBIR project with them. A technology developed using SBIR funds may eventually achieve commercial success at an entirely different company than that which received the initial SBIR award.

Complications plague even the apparently straightforward task of assessing commercial success. For example, research enabled by a particular SBIR award may take on commercial relevance in new, unanticipated contexts. At the launch conference, Duncan Moore, former Associate Director of Technology at the White House Office of Science and Technology Policy (OSTP), cited the case of SBIR-funded research in gradient index optics that was initially considered a commercial failure when an anticipated market for its application did not emerge. Years later, however, products derived from the research turned out to be a major commercial success.[14] Today's apparent dead end can be a lead to a major achievement tomorrow. Lacking clairvoyance, analysts cannot anticipate or measure such potential SBIR benefits.

Gauging commercialization is also difficult when the product in question is destined for public procurement. The challenge is to develop a satisfactory measure of how useful an SBIR-funded innovation has been to an agency mission. A related challenge is determining how central (or even useful) SBIR awards have proved in developing a particular technology or product. In some cases, the Phase I award can meet the agency's need—completing the research with no further action required. In other cases, surrogate measures are often required. For example, one way of measuring commercialization success is to count the products developed using SBIR funds that are procured by an agency such as the DoD. In practice, however, large procurements from major suppliers are typically easier to track than products from small suppliers such as SBIR firms. Moreover, successful development of a technology or product does not always translate into successful "uptake" by the procuring agency. Often, the absence of procurement may have little to do with the product's quality or the potential contribution of the SBIR program.

Understanding failure is equally challenging. By its very nature, an early-stage program such as SBIR should anticipate a high failure rate. The causes of failure are many. The most straightforward, of course, is *technical failure*, where the research objectives of the award are not achieved. In some cases, the project can be a technically successful but a commercial failure. This can occur when a procuring agency changes its mission objectives and hence its procurement priorities. NASA's new Mars Mission is one example of a *mission shift* that may result in the cancellation of programs involving SBIR awards to make room for new agency priorities. Cancelled weapons system programs at the Department of Defense can have similar effects. Technologies procured through SBIR may also fail in the *transition to acquisition*. Some technology developments by small businesses do not survive the long lead times created by complex testing and certification procedures required by the Department of Defense. Indeed, small firms encounter considerable difficulty in penetrating the "procurement thicket"

---

[14]Duncan Moore, "Turning Failure into Success,"in National Research Council, *SBIR: Program Diversity and Assessment Challenges,* op. cit., p. 94.

that characterizes defense acquisition.[15] In addition to complex federal acquisition procedures, there are strong disincentives for high-profile projects to adopt untried technologies. Technology transfer in commercial markets can be equally difficult. A *failure to transfer to commercial markets* can occur even when a technology is technically successful if the market is smaller than anticipated, competing technologies emerge or are more competitive than expected, if the technology is not cost competitive, or if the product is not adequately marketed. Understanding and accepting the varied sources of project failure in the high-risk, high-reward environment of cutting-edge R&D is a challenge for analysts and policy makers alike.

This raises the issue concerning the standard on which SBIR programs should be evaluated. An assessment of SBIR must take into account the expected distribution of successes and failures in early-stage finance. As a point of comparison, Gail Cassell, Vice President for Scientific Affairs at Eli Lilly, has noted that only one in ten innovative products in the biotechnology industry will turn out to be a commercial success.[16] Similarly, venture capital funds often achieve considerable commercial success on only two or three out of twenty or more investments.[17]

In setting metrics for SBIR projects, therefore, it is important to have a realistic expectation of the success rate for competitive awards to small firms investing in promising but unproven technologies. Similarly, it is important to have some understanding of what can be reasonably expected—that is, what constitutes "success" for an SBIR award—and some understanding of the constraints and opportunities successful SBIR awardees face in bringing new products to market. From the management perspective, the rate of success also raises the

---

[15]For a description of the challenges small businesses face in defense procurement, the subject of a June 14, 2005, NRC conference and one element of the congressionally requested assessment of SBIR, see National Research Council, *SBIR and the Phase III Challenge of Commercialization*, Charles W. Wessner, ed., Washington, DC: The National Academies Press, 2007. Relatedly, see remarks by Kenneth Flamm on procurement barriers, including contracting, overhead, and small firm disadvantages in lobbying in National Research Council, *SBIR: Program Diversity and Assessment Challenges*, op. cit., pp. 63–67.

[16]Gail Cassell, "Setting Realistic Expectations for Success," in National Research Council, *SBIR: Program Diversity and Assessment Challenges,* op. cit., p. 86.

[17]See John H. Cochrane,"The Risk and Return of Venture Capital,"*Journal of Financial Economics* 751:3–52, 2005. Drawing on the VentureOne database, Cochrane plots a histogram of net venture capital returns on investments that "shows an extraordinary skewness of returns. Most returns are modest, but there is a long right tail of extraordinary good returns. Fifteen percent of the firms that go public or are acquired give a return greater than 1,000 percent! It is also interesting how many modest returns there are. About 15 percent of returns are less than 0, and 35 percent are less than 100 percent. An IPO or acquisition is not a guarantee of a huge return. In fact, the modal or 'most probable' outcome is about a 25 percent return." See also Paul A. Gompers and Josh Lerner, "Risk and Reward in Private Equity Investments: The Challenge of Performance Assessment," *Journal of Private Equity* 1 (Winter 1977): 5–12. Steven D. Carden and Olive Darragh, "A Halo for Angel Investors," *The McKinsey Quarterly* 1 (2004), also show a similar skew in the distribution of returns for venture capital portfolios.

question of appropriate expectations and desired levels of risk taking. A portfolio that always succeeds would not be pushing the technology envelope. A very high rate of "success" would thus, paradoxically, suggest an inappropriate use of the program. Understanding the nature of success and the appropriate benchmarks for a program with this focus is therefore important to understanding the SBIR program and the approach of this study.

2

Findings and Recommendations

NATIONAL RESEARCH COUNCIL (NRC) STUDY FINDINGS

I. **The National Science Foundation's (NSF) Small Business Innovation Research (SBIR) program is adding to the storehouse of public scientific and technological knowledge. The program contributes in many important ways, for example, by:**

a. **Generating Knowledge**. The study showed that projects were yielding a variety of knowledge outputs. Contributions to knowledge are embodied in data, scientific and engineering publications, patents and licenses of patents, presentations, analytical models, algorithms, new research equipment, reference samples, prototype products and processes, spin-off companies, "human capital" (greater know-how, expertise, etc.), and new capability for further innovative activity. Publications and patenting activity occurred with considerable frequency.[1]

Research quality is difficult to measure, and the value of knowledge created takes time to manifest and has been shown to be highly variable among projects. It can be concluded, however, that the NSF's SBIR program in its statements of purpose, goals, and criteria—hence, intentions—consistently emphasizes research quality and knowledge creation. Furthermore, the program's peer-review selection process appears to have integrity in applying the research-merit criterion. Nothing was found by

---

[1]The NRC Phase II Survey reported averages of 1.66 scientific publications and 0.67 patents per surveyed project. The underlying distribution of patents and publications reported is skewed, with some companies reporting none and some reporting relatively high numbers.

the study suggestive of shortcomings in research quality or knowledge creation. (This point is elaborated upon in Chapter 7, Section 7.1.1.)

b. **Creating and Disseminating Intellectual Capital**. A short-run result of the program's technological innovation is knowledge outputs. These are important in meeting the NSF's goal of funding research that leads to broader impact, because they provide paths by which others may use the program's knowledge gains to achieve additional benefits. There is evidence that the program is producing the kind of knowledge outputs that are typically associated with innovation, such as patents, copyrights, and publications. Extensive licensing activities of Phase II awardees attested to the fact that useful intellectual capital has been created and disseminated. For example, the NRC Phase II Survey showed 20 percent of Phase II projects reporting they had reached licensing agreements with U.S. companies and investors, and 21 percent reporting they had ongoing negotiations with U.S. companies and investors on licensing agreements. (For more detail, see Chapter 7, including Table 7.2-2.)

c. **Building Networks with Universities**. Both the NRC surveys and the case studies showed extensive networking between NSF SBIR-funded projects and universities. University faculty and students used the NSF SBIR program to establish businesses, start projects, and work on projects. University staff and faculty often assisted with proposal preparation, provided facilities and equipment, and made ongoing contributions to the intellectual capital underpinning company innovations. Faculty also often served as proposal reviewers. (For more details, see Chapter 7, including Table 7.2-5, and also the case studies in Appendix D.)

d. **Moving Technology from Universities Toward the Market**. The NSF SBIR program has facilitated transfer of technology from universities. Fourteen percent of the NRC Phase II Survey projects were based on technology originally developed at a university by a project participant. Five percent of Phase II Survey projects were based on technology licensed from a university. (This effect is discussed further in Chapter 7, including Table 7.2-5.)

e. **Broadening the Scope and Speed of Research**. The NSF's SBIR program has enabled firms to broaden their research scope and to accelerate research. Two-thirds of participating firms reported they "definitely" or "probably" would not have undertaken their projects without an NSF SBIR grant. For those who "definitely" or "probably" would have undertaken the project, 84 percent reported that the project would have been narrower in scope without the NSF SBIR grant. Most also said their projects would have been delayed without the NSF SBIR grant. (For more details, see Table 5.2-8.)

f. **Testing Ideas and Building Capacity**. The case studies provided concrete examples of how the NSF's SBIR program has enabled small businesses to: test creative ideas; develop new capabilities and build capacity; grow intellectual property; gain additional market credibility; and use that enhanced credibility to attract further investment funding while maintaining control of their firms until technical progress enabled them to form more effective partnerships. (These points are discussed further in Section 5.2.4 and Appendix D.)

g. **Conducting High-Risk Research**. There is evidence that the projects funded were desirably characterized by high technical risk. Technical failure or difficulties were frequently given as a reason by NSF SBIR grantees for not moving to the Phase II stage and for discontinuing Phase II projects once begun. The risk profile of projects is significant to knowledge creation because overcoming challenging scientific and technical hurdles associated with high technical risk is synonymous with increasing the knowledge base. (For more on the technical riskiness of projects, see Sections 5.2.2 and 5.2.3, and Table 5.2-9.)

h. **Increasing the Program's Knowledge Outputs May Be Possible.** It appears that the NSF's SBIR program tends to represent "broader impact" as synonymous with "direct commercial sales" and may not place sufficient emphasis on the encouragement of publishing and other modes of knowledge dissemination by the funded projects. It is recognized that encouragement of knowledge dissemination must occur within the constraints of companies needing to protect their ability to commercialize. Nonetheless, it appears that the NSF's SBIR program may be able to increase knowledge outputs from the program by signaling to grantees that it places a high value on knowledge outputs, such as publications, papers, presentations, and patents, in addition to commercial outputs. At the same time, it appears that the program is producing knowledge outputs and therefore has the opportunity to compile data on knowledge outputs comprehensively and systematically as part of an improved data collection plan. (See Sections 7.1.2 and 7.2.)

II. **NSF SBIR program funding is closely aligned with the NSF's broader mission and is contributing broadly to federal research and development procurement needs. The program serves as a means of:**

a. **Promoting Scientific Progress**. Congress created the NSF in 1950 "to promote the progress of science; to advance the national health, prosperity, and welfare; [and] to secure the national defense." While grants to university researchers continue to be the most common tool the NSF uses to carry out its mission, the SBIR program is another useful tool for NSF to advance science and the nation's health, prosperity, welfare, and defense. (For more discussion of this point, see Section 4.1.)

b. **Helping the Transition of Technology from Universities Toward the Market**. NSF SBIR grants further the agency's academic goals and activities by aiding the transfer of technology out of universities and by helping academic researchers build the small firms that convert their ideas and inventions into commercial products. (These effects are discussed further in Sections 6.3 and 7.2.6 and are illustrated by example in Appendix D.)

c. **Meeting Agency Procurement Needs**. The NSF SBIR program helps meet the procurement needs of federal agencies. The NRC Phase II Survey found that sales of NSF Phase II–funded technologies go to multiple markets with broad and diversified customer bases. For example, while 57 percent of sales by NSF Phase II survey participants consisted of new and improved products for domestic civilian markets, 20 percent of sales were meeting federal agency procurement needs. (See Section 6.3.2 and Table 6.3-1 for more details.)

III. **The NSF's SBIR program supports a diverse array of small businesses. At the same time, there is a need to improve the analysis of the success rates of woman- and minority-owned businesses, with the goal of increasing their participation in the program.[2]**

a. **Diversity**. Small businesses participating in the NSF's SBIR program are diverse across a number of dimensions. Businesses funded by the NSF are diverse in terms of their age and size, in the technologies they are pursuing, in their form of ownership, in their origin and past history, in their business strategies, in their geographical locations, in their market-sector orientation, in their progress toward commercial success, and in a host of other ways. NSF's program provides a big tent under which small, innovative businesses across the country can and do develop. (The case studies in Appendix D illustrate the considerable diversity among SBIR-funded firms.)

---

[2]There are inconsistencies in the use of designations. The terminology used here follows that of the Small Business Administration's description of the SBIR program, which uses the terms "women-owned" and "minority-owned" small businesses. (See *SBA.gov/sbir/indexwhatwedo.html*.) However, for purposes of this study, "minority-owned" and "disadvantaged" are used interchangeably, although it is recognized that the terms are not interchangeable in federal contracting.

b. **Continuous Influx of Firms New to the Program.** The program has demonstrated a continuous influx of firms new to the program. From 1996 through 2003, the period examined, more than half of each year's SBIR award recipients had not previously received an NSF SBIR grant, with the exception of one year. In the most recent year examined, 2003, more than 60 percent of grantees were new to the SBIR program. As the program continuously approves applications from these newly proposing firms, the population of small firms helped by the SBIR program grows. (Table 4.2-8 provides additional details on newly proposing firms.)

c. **A Commercial Enabler for Small Firms.** Although few individual SBIR projects lead directly to "home runs" in the commercial sense, many SBIR projects funded by federal agencies contribute to agency missions and to the broader economy. Both the NRC Phase II Survey and the set of case studies showed that small firms believe that the NSF's SBIR program helped them enter commercial markets.

d. **Role in Project Initiation.** The SBIR awards play a key role in initiating commercially oriented research. When asked if their companies would have undertaken the projects had there been no SBIR grant, approximately two-thirds of respondents answered either probably not (43 percent) or definitely not (24 percent). (See Table 5.2-8.)

e. **Awards as Enablers.** Case study firms called the NSF's SBIR program an "enabler" and a "lifeline." They said the SBIR awards enabled them to:

  - start a company;
  - survive while they were trying to refine their ideas to the point where they could attract private money;
  - build capacity;
  - develop new technologies and improve, renew, and lower the cost of existing technologies;
  - build on and add other sources of funding;
  - grow an intellectual property portfolio essential to commercial success;
  - retain control of their firms; and
  - pursue multiple paths to commercialization, including contract research.

  (For details, see Sections 5.2.2, 5.2.4, and 5.2.5, Table 5.2-11, and Appendix D.)

f. **Flexibility for Small Companies**. Because the NSF generally cannot provide formal financial support into Phase III by contracting with grantees, its SBIR recipients do not have the Phase III funding options often available to recipients of SBIR funding from the procurement agencies.[3] On the other hand, the NSF funds a variety of technology areas, which allows firms more leeway in aligning their research with market demand. (For more details, see Section 6.2.)

IV. **While women and minorities have benefited from the NSF's SBIR program, both as company owners and as principal investigators, there is room for improvement.**

a. **Participation Rates of Women and Minorities.** Levels of participation by woman- and minority-owned businesses have continued to lag other groups.

- From 1995 through 2005, woman-owned businesses submitted 12.2 percent of Phase I proposals and received 9.5 percent of Phase I grants. They submitted 8.8 percent of Phase II proposals and received 7.5 percent of Phase II grants.
- Minority-owned businesses (including minority women) submitted 16 percent of Phase I proposals and received 13.5 percent of Phase I grants. They submitted 12.9 percent of Phase II proposals and received 13.7 percent of Phase II grants.

Woman and minorities also participated in projects as principal investigators, with 21 percent of Phase II projects surveyed reporting either a woman, a minority, or a minority woman as the principal investigator. (For elaboration on participation by women and minorities as business owners and as principal investigators, see Section 4.2.5.)

b. **Room for Improvement.** The cause of the lower participation rates is unclear. It may reflect the low representation of women and minorities in high technology firms.[4] These lower success rates may also reflect

---

[3]As noted, the NSF's SBIR program does make available additional support to selected firms through its Phase IIB activities.

[4]White males, who comprise 40 percent of the nation's overall workforce, hold 68 percent of all science, engineering, and technology jobs. In contrast, white women, who comprise 35 percent of the national workforce, hold only 15 percent of these positions, and only 10 percent of the 2 million scientists and engineers in the United States in a recent year were women. African Americans and Hispanics, who comprise almost 21 percent of the American workforce, represent just 6 percent of the science, engineering, and technology workforce. (Congressional Commission on the Advancement of Women and Minorities in Science, Engineering and Technology Development. Findings of the Commission as reported in SSTI Weekly Digest, April 6, 2001; and U.S. Bureau of Labor

obstacles that women and minorities face in pursuing careers in science and engineering.[5] Moreover, it seems noteworthy that their participation does not seem to be trending upward despite the recent gains these groups have made in other areas.

It is a source of concern that the success rates for woman- and minority-owned firms at Phase I are below those of other firms.[6] It is also troubling that the number of NSF SBIR proposals and grants to woman- and minority-owned businesses do not show a positive trend over the last decade when gains by these groups were being made in other areas. There may be room for improved participation in the program for woman- and minority-owned firms. Further analysis is required to determine what steps, if any, are required.

c. **More Examination Needed.** To understand the causal factors behind these statistics for woman- and minority-owned firms requires further investigation beyond the scope of this study, particularly at the Phase I proposal review stage. Are there systematic shortcomings in the application process that could be addressed through customized outreach and targeted training? Are there problems in proposal review and selection that could be addressed? Finally, are there actions that could proactively encourage greater participation by woman- and minority-owned firms? (See Section 4.2.5.)

Statistics.) However, it should be noted that not all minority groups have low representation in science, engineering, and technology fields relative to their representation in the U.S. population Indicative of shifting representation among minority groups, there was a strong increase during the 1990s in the percentage of doctorate degrees and jobs in science and engineering going to foreign-born individuals, particularly those from India, China, and the Philippines. (National Science Foundation, *Science and Engineering Indicators 2006*, "The U.S. S&E Labor Force," Arlington, VA: National Science Foundation, 2006.) By like token, the representation in the NSF's SBIR program is uneven among different minority groups.

[5]Academics represent an important future pool of applicants, firm founders, principal investigators, and consultants. Recent research shows that owing to the low number of women in senior research positions in many leading academic science departments, few women have the chance to lead a spinout. "Underrepresentation of female academic staff in science research is the dominant (but not the only) factor to explain low entrepreneurial rates amongst female scientists." See Peter Rosa and Alison Dawson, "Gender and the commercialization of university science: academic founders of spinout companies," *Entrepreneurship & Regional Development* 18(4):341–366.

[6]The lower success rates of woman- and minority-owned firms in obtaining Phase I grants were found by tests for statistical significance to be highly significant. The probabilities of getting the t-test results if there were no differences in success rates of woman- and minority-owned firms versus all firms at the Phase I stage was 0.000 at a level of significance of 0.05. The results of the tests of differences at Phase II were weaker; at the Phase II stage, the probabilities of getting the t-test results if there were no differences in success rates were less than 0.23 for both woman- and minority-owned firms. (For more details, see Section 4.2.5, particularly Figures 4.2-12, 4.2-13 and 4.2-18.)

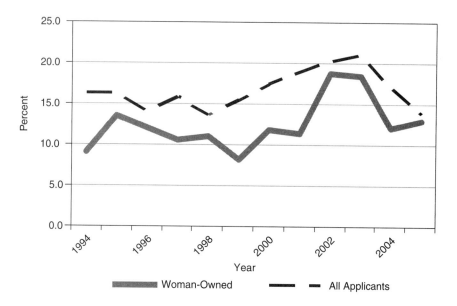

FIGURE 2-1 Comparative Success Rates for Woman-owned versus
All Applicants in Having Their Phase I Proposals Approved, 1994–2005.
SOURCE: Developed from data provided by the NSF SBIR program.
NOTE: Differences are statistically significant.

V. **Although the agency itself normally does not acquire the results of SBIR-
   funded projects, NSF projects do achieve commercialization.[7,8]**

   a. **NSF SBIR Commercialization Outcomes:** The NRC Phase I and Phase
      II Surveys sought to identify whether some form of commercialization
      activity had occurred for projects and firms contacted. In interpreting
      these data, it should be noted that they include firms that may have had
      very limited sales (e.g., several units for test purposes); only a rela-
      tively small proportion of firms have high commercial success rates, as
      described below.

---

[7]The lack of a procurement function tends to limit commercialization at the NSF in comparison
with procurement agencies, such as the Defense Department.

[8]The NSF's SBIR program has recently emphasized its goal to "commercialize research." Interest
in promoting commercialization appeared to intensify with the centralization of the program within a
single office. This emphasis on commercialization contrasts with the concentration of most of NSF's
resources on academic research. The SBIR program holds that the small share of resources that goes
to business should emphasize research for business purposes. The committee finds this position entirely reasonable and applauds the program's efforts to foster commercialization.

- Of the Phase I projects that did not result in a direct follow-on Phase II grant, nearly half reported sales, expected sales, or some other form of progress.
- Of the Phase II projects included in the survey:
  - o Twenty-two percent reported that the referenced projects had resulted in commercial products, processes, or services in use and were still active. (Recall that this figure includes projects for which sales were quite limited.)
  - o Nineteen percent reported that commercialization was under way.
  - o Ten percent reported that they achieved sales, licensing, or additional funding before discontinuing the projects. (Note again that this figure includes projects for which sales were quite small.)
  - o Twenty-eight percent indicated that they were continuing technology development after completing the Phase II research.
- Altogether, 69 percent of respondents for Phase II projects indicated some form of progress toward either commercial or continuing technological development. (For more details, see Sections 5.2.1, 5.2.2, and 5.2.3 and Figures 5.2-3 and 5.2-4.)

b. **A Small Percentage of Projects Account for Most Successes**.

- As is typical for other private and public technology innovation programs, a relatively few projects have accounted for the majority of sales and licensing revenue from NSF SBIR recipients.[9] (For details, see Section 5.2.)
- **Outcomes Are Skewed.** A smaller percentage of projects will likely achieve large growth and huge sales revenues, i.e., be commercial "home runs."[10] These patterns are similar to those found in other private and public investments in high-risk advanced technology investments, where many research projects must be seeded to yield a few commercial "home runs."[11]

---

[9]Similar results were revealed in previous National Research Council assessments of early-stage innovation awards. See National Research Council, *The Advanced Technology Program: Assessing Outcomes*, Charles W. Wessner, ed., Washington, DC: National Academy Press, 2001. See also National Research Council, *The Small Business Innovation Research Program: An Assessment of the Department of Defense Fast Track Initiative*, Charles W. Wessner, ed., Washington, DC: National Academy Press, 2000.

[10]Among the 162 projects surveyed, just 8 projects—each of which had $2.3 million or more in sales—accounted for over half the total reported sales dollars for the surveyed projects. The project with the highest reported amount had $4.8 million in sales. Similarly, the results for sales by licensees of those survey projects' technologies were highly skewed by a single licensee that accounted for over half the total licensee sales dollars, amounting to $200 million or more in licensee sales.

[11]As with investments by angel investors or venture capitalists, SBIR awards result in highly con-

- Meanwhile, many small successes together comprise a potentially important component of the nation's innovative capability.

c. **Balancing Technological Innovation and Success in Commercialization.**

- In reviewing findings on commercialization, it is important to keep in mind the inherent tension between the program's goal to stimulate advanced technological innovation and its goal to increase private-sector commercialization.
- Some would argue that a research program achieving a high percentage of commercial successes would suggest a failure to attack sufficiently challenging technical problems. It is also important to keep in mind that much can be learned from technical failure; unsuccessful projects can suggest what paths not to follow and identify hurdles that remain to be overcome.

d. **Commercialization as a Core Goal.** In recent Senate testimony, a senior program official affirmed that the program's focus has become "the commercialization of research."[12] To increase commercialization, the program has taken several steps that represent a change from earlier practice. The program now:

- requires a business plan as part of Phase II proposals;
- provides commercialization assistance training and networking assistance to award-recipient firms;

---

centrated sales, with a few awards accounting for a very large share of the overall sales generated by the program. These are appropriate referent groups, though not an appropriate group for direct comparison, not least because SBIR awards often occur earlier in the technology development cycle than where venture funds normally invest. Nonetheless, returns on venture funding tend to show the same high skew that characterizes commercial returns on the SBIR awards. See John H. Cochrane, "The Risk and Return of Venture Capital," *Journal of Financial Economics* 75(1) 2005:3–52. Drawing on the VentureOne database Cochrane plots a histogram of net venture capital returns on investments that "shows an extraordinary skewness of returns. Most returns are modest, but there is a long right tail of extraordinary good returns. Fifteen percent of the firms that go public or are acquired give a return greater than 1,000 percent! It is also interesting how many modest returns there are. About 15 percent of returns are less than 0, and 35 percent are less than 100 percent. An IPO or acquisition is not a guarantee of a huge return. In fact, the modal or 'most probable' outcome is about a 25 percent return." See also Paul A. Gompers and Josh Lerner, "Risk and Reward in Private Equity Investments: The Challenge of Performance Assessment," *Journal of Private Equity* 1 (Winter 1977):5–12. Steven D. Carden and Olive Darragh, "A Halo for Angel Investors," *The McKinsey Quarterly* 1 (2004), also show a similar skew in the distribution of returns for venture capital portfolios.

[12]Testimony of Joseph Hennessey, "The Small Business Innovation Research Program: Opening Doors to New Technology," before the House Committee on Small Business's Subcommittee on Workforce, Empowerment and Government Programs, November 8, 2005.

- uses Phase IIB awards to encourage firms to seek third-party financing and to provide a market test of commercial potential;
- designs its solicitation topics to be more in line with industry interests; and
- uses more reviewers with business experience in the proposal selection process. (For elaboration on NSF's commercialization efforts, see Chapter 5, and for more on the Phase IIB program, see Section 8.5.)[13]

e. **Commercialization May Be Further Increased**. For example, the NRC Phase II Survey found that Phase II projects were often discontinued for market-related reasons; in particular, companies cancelled projects because market demand was too small or the product/process/service was not competitive. It may be possible for the program to reduce the percentage of projects going forward into Phase II with insufficient market potential by continuing and intensifying improvements to outreach, commercialization assistance, and project review/selection. This would require additional resources for commercialization assistance, which are constrained at 1992 levels. (See Chapter 5, particularly Table 5.2-9, and Chapter 8, Section 8.4.4.)

VI. Firms that have won multiple SBIR awards were a point of study.

a. **The Single Award Assumption.** Some companies that have received numerous SBIR awards have been criticized as SBIR "mills." Implicit in this criticism is the assumption that one or a few awards should be sufficient for a company to launch a product and become "independent" from the SBIR program. In some cases, initial awards may, in fact, be sufficient to attract private investors and enable a company to develop exclusively through private funding.

b. **Developing Technologies.** For companies pursuing technologies with limited markets (e.g., NASA or the DoD) or, in the case of the NSF, technologies that may have broad applications, a single set of SBIR awards may not be sufficient to develop the technology to the point that it is attractive for procurement (by another agency) or for funding by private capital markets.[14]

---

[13]As noted above, the program recently has taken steps to increase the rate of commercialization. At the time of this study's surveys, however, there had been insufficient time to determine if these steps have increased rates of commercialization. The results of such efforts may begin to show up in post-survey data.

[14]For companies pursuing multiple technology areas, multiple grants may entail no overlap. For example, Luna Innovations, Inc., has won SBIRs from multiple agencies for multiple technology

c. **Role of Multiple Awards.** The case study results support the hypotheses that a single grant is seldom adequate to bring an idea to market readiness and that multiple funding sources, multiple projects, and multiple technologies are frequently needed for success in commercialization. At the same time, single grants can and do have a significant impact, opening a line of inquiry or revealing an opportunity.

d. **Diversity of Firm Objectives.** As noted above, the NSF program attracts quite a large percentage of firms new to the program each year—some 60 percent in 2003. Nonetheless, not all firms are new participants. Some firms that complete work successfully win again. Indeed, some companies utilize the program as a means of exploring technologies that they believe will have commercial promise, while others seek to resolve research questions of relevance to government missions. Some of these companies are essentially contract research organizations, where fulfilling the terms of SBIR solicitations is an important part of their business model, one that also meets government needs.

e. **Addressing Agency Missions.** With regard to frequent award winners, a key issue is the research quality and cost. Small companies often have lower overheads than some other alternative sources of research, e.g., national laboratories and large companies that often receive many federal contracts. Companies that are providing quality research that meets agency solicitation requirements should not be subject to an arbitrary limit.

f. **Discontinue Nonperforming Frequent Award Winners.** However, in cases where a company has received a number of past SBIR grants for which the research has been of poor quality or the company has otherwise failed to carry through on its proposal, agency management should (and does) reduce or eliminate awards to the company unless and until there is compelling evidence favoring additional grants. In the case where a company is good at writing proposals but poor at execution, this may require an intervention after the peer review process. (According to NSF's program descriptions, it applies such intervention as necessary.)

---

areas, including multiple Phase I, Phase II, and Phase IIB SBIR grants from the NSF. For every $1 of SBIR funding received, Luna has generated at least $2 in non-SBIR funding. It has spun off six companies—each specializing in a different technical field ranging from optical devices to life sciences to advanced materials. It has used the SBIR program as a funding source to get a number of ideas off the ground and move them along a market-focused pipeline. Founded in the early 1990s, Luna was listed on the NASDAQ in June 2006.

VII. **The NSF's SBIR program is well run, largely effective at achieving program goals, and has benefited from its strong, centralized management and talented program managers.[15]**

a. **Effective and Respected Management.** The NSF's SBIR program benefits from a centralized, well-coordinated management team that is widely seen as effective in managing the program. (See Sections 4.3 and 8.9.1.)

b. **Skilled Program Managers.** Program managers have firsthand industry experience. Case study firms spoke positively of the program's management team, describing the program managers as "highly motivated," "knowledgeable," "flexible," and "effective." The firms described the turnover among program staff as lower than at other SBIR programs and characterized the lower turnover as an advantage to program participants.[16] (See Section 8.9.1 and Appendix D.)

c. **Funding Across a Broad Spectrum of Technology Solicitations.** Case study firms spoke of the program's willingness to fund a variety of technologies, including manufacturing projects and those with long development cycles, such as advanced materials. The firms noted that the program's relatively broad topic definitions allow them leeway to align their projects with in-house strengths and market demands. (For more details, see Sections 5.2.4 and Appendix D.)

d. **Responsive to External Critiques.** The NSF's SBIR program regularly receives and responds to external advice. It receives external oversight from its Advisory Committee and from its Committee of Visitors (COV), which meets triannually. Continuous program improvements, reflecting management's focus on program improvement, were observed in the course of the study. (For more details, see Section 8.13.2.)

e. **Active Outreach and Training.** The NSF's SBIR program runs active outreach, training, mentoring, and networking programs, and it has taken a leadership role in recent national SBIR conferences.

- The program has sponsored SBIR spring and fall national conferences in recent years. It participates in state SBIR workshops and has scheduled one of its national conferences each year in an EPSCoR state (a state participating in the "Experimental

---

[15]The effectiveness of this management approach at the NSF should not necessarily be extrapolated to other agencies or departments (e.g., Defense) that have programs ten times larger with widely distributed budgets.

[16]More recently the program has experienced staff turnover.

Program to Stimulate Competitive Research and Institutional Development Awards").[17] (For more details, see Section 8.2.)

- In 1999, the NSF's SBIR program began a commercialization assistance workshop for Phase I awardees, and since 2001 the program has contracted for a Commercialization Planning Assistance program for Phase I awardees using Dawnbreaker, Inc., and (later) Foresight Science & Technology, Inc. (For more details, see Section 8.4.)

- After implementing its commercialization assistance activities, program officials reported improvements in Phase II proposals. In 2005 the program began sponsoring participation by Phase II award recipients in an annual "Opportunity Forum." (For more details, see Sections 8.2 and 8.4.)

VIII. Notwithstanding the program's strong program management, the Committee identified a number of areas that may benefit from change. There are several issues that should be addressed to increase performance.

a. **Inflexible and Inadequate Management Budget.** Resources for program administration and commercialization assistance are inadequate. Program administrative budgets reportedly do not vary with changes in SBIR funding. The administrative budget has fallen both in constant dollars and as a percentage of the total program size. Increasingly, the program has had to do more with less, which over time may negatively impact performance and appears to be a constraint in initiating new initiatives, including some mentioned below, e.g., assessment. (See Sections 8.12 and 8.4.4.)

b. **Limited Assessment Efforts.** The program's past efforts at assessment, while commendable, were limited because the evaluations have been ad hoc rather than systematic; internal and/or limited in availability; constrained by firms not meeting post-project reporting goals; and hampered by ad hoc definitions of "success." The program's inadequate budget for evaluation is one reason for the limited amount of assessment.

c. **Inadequate Attention to, or Discouragement of, Knowledge Outputs.** There appear opportunities for the program to provide a more consistent message to program participants about the importance of knowledge outputs, including scientific publications, to encourage knowledge outputs,

---

[17]Currently, EPSCoR is aimed at 25 states, Puerto Rico, and the U.S. Virgin Islands—jurisdictions that have historically received lesser amounts of federal R&D funding.

and to compile a comprehensive set of indicators of knowledge creation and dissemination for evaluation purposes.

d. **Limited Electronic Data Collection.** Systematic and reliable compilation of data in electronic form useful for program management and evaluation has, until recently, been limited. (See Sections 4.2.2 and 8.8.)

e. **"Topic Gap" and Need for Transparent, Bottom-up Topic Selection Process.** Despite the NSF's broad solicitation topics, there may be a "topic gap" that arises when firms must wait until their topic is featured in a solicitation. According to program staff, firms may have to wait up to 18 months for a particular topic area to be part of an NSF solicitation. Long intervals between topic solicitations may prevent some firms from taking advantage of time-limited opportunities. Furthermore, the selection of topics may lack transparency and represent more of a top-down than a bottom-up process. (See Section 8.1, particularly 8.1.3.)

f. **Controversial Application of "Additional Factors."** The program's use of "additional factors" for proposal selection lacks transparency. These factors are mentioned briefly in program guidance to applicants, but there is little or no explanation as to how they are used. While some additional factors, such as "past commercialization efforts by the firm when previous grants exist," appear valid, others seem more problematic. Some interviewed firms were particularly troubled by the additional factor, "excessive concentration of grants in one firm." While the NSF's desire to avoid excessive concentration of grants may have merit, firms claim that they do not know how "excessive concentration" is defined. Moreover, this factor is applied after firms have incurred the cost of submitting a proposal, and several surveyed firms believe that it is applied unevenly among firms. It should be noted that the firms interviewed did not necessarily object to a limit on how many grants they could receive from the NSF, but they did want to know in advance if there is a limit, what the limit is if there is one, and how the rule will be evenly applied across firms. (For additional discussion, see Section 8.3.3.)

g. **Limits on the Number of Proposals per Company.** The NSF limits firms to four proposals per solicitation. Larger eligible companies that have more areas of research and/or more scientists feel constrained by this limit. Given that a goal of the SBIR program is to promote innovative activity among small firms, and given that the act of developing proposals has itself been found to stimulate innovation, the program's emphasis should be to encourage firms to develop SBIR proposals. On the other hand, the program's administrative budget is limited, there appears to be no shortage of grant-worthy proposals with the limit in effect, and the limit works against higher concentrations of grants to particular firms. (For elaboration, see Section 8.1.3 and Appendix D.)

NRC STUDY RECOMMENDATIONS

Recognizing the accomplishments already achieved, the Committee recommends that the NSF reinforce its efforts to improve commercialization. Greater efforts, including outreach and better data collection, should be made to raise participation by women and minorities. Better documentation and evaluation are also recommended in order to improve program output and to facilitate program management. Consideration should be given to the provision of additional program funds to realize these objectives. Above all, every effort should be made to preserve program flexibility.

I. **Preserve program flexibility.** SBIR agencies and departments, the Small Business Administration, and the Congress should seek to ensure that program adjustments and refinements do not reduce the flexibility with which the program is administered.

   a. The SBIR program is effective across the agencies partly because a "one-size-fits-all" approach has not been imposed.
   b. This flexible approach should be continued, subject to appropriate monitoring, across the departments and agencies in order to adapt the program to their evolving needs and to improve its operation and output.

II. **Continue to increase private-sector commercialization of innovations derived from federal research and development.**

   a. Continue to promote Phase IIB awards, refining the tool as experience suggests and raising the number and amount of these awards as third-party funding permits.
   b. Increase use of technically competent reviewers with strong technical expertise and strong business understanding for both Phase I and Phase II selection.
   c. Increase support for commercialization assistance as resources permit.
   d. While recognizing the difficulty of assessing markets for new-to-the-market and disruptive innovations, reduce as possible the substantial share of projects for which it is determined during Phase I that perceived market demand is inadequate for the grantee to continue pursuing the innovation even if Phase I is technically successful.

III. **Improve participation and success of women and minorities.**

a. **Encourage Participation.** Develop targeted outreach to improve the participation rates of woman- and minority-owned firms and strategies to improve their success rates based on causal factors determined by analysis of past proposals (see item IIa.) and feedback from the affected groups.

b. **Improve Data Collection and Analysis.** Arrange for an independent analysis of a sample of past proposals from woman- and minority-owned firms and from other firms (to serve as a control group) for the purpose of identifying specific factors accounting for the lower success rates of woman- and minority-owned firms as compared with other firms in having their Phase I proposals granted.

c. **Extend Outreach to Younger Women and Minorities.** The NSF should immediately encourage and solicit women and underrepresented minorities working at small firms to apply as principal investigators (PIs) and senior co-investigators (Co-Is) for SBIR awards and track their success rates.

1. **Encourage Emerging Talent.** The number of women and, to a lesser extent, minorities graduating with advanced scientific and engineering degrees has been increasing significantly over the past decade, especially in the biomedical sciences. This means that many of the women and minority scientists and engineers with the advanced degrees usually necessary to compete effectively in the SBIR program are relatively young and may not yet have arrived at the point in their careers where they own their own companies. However, they may well be ready to serve as principal investigators and/or senior co-investigators on SBIR projects.

2. **Track Success Rates.** The Committee also strongly encourages the NSF to gather the data that would track women and minority principal investigators and to ensure that SBIR is an effective road to opportunity for these PIs as well as for women- and minority-owned firms. The success rates of women and minority PIs and Co-Is is the traditional measure of their participation in the non-SBIR research grants funded by nonmission research agencies like NIH and the NSF. It is also a very appropriate measure of women and minority participation in the SBIR program. After all, experience as a PI or Co-I on a successful SBIR program may well give a woman or minority scientist or engineer the personal confidence and standing with agency program officers that encourage them to found their own SBIR firms.

IV. Regular, rigorous program evaluation is essential for quality program management and accountability. Accordingly, the NSF program management should give greater attention and resources to the systematic evaluation of the program supported by reliable data and should seek to make the program as responsive as possible to the needs of small company applicants.

    a. To this end, the program should undertake a systematic compilation of performance measures for program activities, outputs, and outcomes measured against specific goals. In order for the SBIR program to be agile and responsive to applicants, performance measures should include how quickly proposal reviews take place and funds are moved to grantee companies.

    b. The evaluation should incorporate both internal efforts and arms-length external efforts, and evaluation studies should be made publicly available in an annual report.

V. Ensure that solicitation topics are broadly defined, that the definition process is bottom-up, and take steps to ensure the necessary flexibility to permit firms to receive relatively prompt access to Phase I solicitations.

    a. Consider adding an "other category" within a solicitation, if needed, to keep it open to all promising topics.

    b. Alternatively, explore the feasibility of a pilot program designed specifically to decrease the delay between solicitations for given topics.

VI. Recognizing that transparency in project selection procedures is important to the perceived fairness of a public funding program, the NSF's SBIR program should adopt the following steps:

    a. Revisit the "additional factors" affecting award decisions and decide which ones are worth keeping, and then make these factors explicit in program guidance, in terms of their definitions and how, when, and by whom they are to be judged.

    b. In particular, clarify additional factor #3—"Excessive concentration of grants in one firm"—in terms of how it is defined and applied, and ensure that firms can determine prior to proposing if they will be affected.

    c. NSF management should ensure that all project selection factors are clear, transparent, and evenly applied.

VII. The use of quotas to reduce multiple winners should be reconsidered.

    a. In the case of multiple award winners who qualify in terms of the selection criteria, the acceptance/rejection decision should be based on their performance on past grants rather than on the number of grants received. Firms able to provide quality solutions to solicitations should not be excluded, a priori, from the program except on clear and transparent criteria (e.g., quality of research and/or commercialization performance).

    b. Avoid imposing quotas on applications unless there are compelling reasons to do so. Specifically, remove the limit of four on the number of proposals per solicitation per firm. Such limitations run the risk of limiting innovative ideas and of unnecessarily restricting the opportunity among prospective principal investigators in the larger eligible small companies.

    c. In order to continue to attract new entrants and to help avoid concentrations in existing companies, the NSF should continue its focus on outreach to ensure a high rate of proposals from firms new to the program.

VIII. Increase management funding for SBIR. To enhance program utilization, management, and evaluation (as described above), consideration should be given to the provision of additional program funds. There are three ways by which this might be achieved:

    a. Additional funds might be allocated internally, within the existing NSF budget, with reference to the management funding provided for comparable NSF programs, keeping in mind the special requirements of SBIR applications, solicitations, evaluations, selection, monitoring, reporting, outreach, commercialization services, site visits, and other functions related to the normal and effective operation of the program.

    b. Funds might be drawn from the existing set-aside for the program to carry out these activities.[18]

    c. The set-aside for the program, currently at 2.5 percent of external research budgets, might be increased, with the goal of providing additional resources to maximize the program's return to the nation.[19]

---

[18]Under current legislation, funds drawn from the SBIR "set-aside" cannot be used for these purposes. They are almost exclusively allocated for awards.

[19]Each of these options has its advantages and disadvantages. For the most part, the departments, institutes, and agencies responsible for the SBIR program have not proved willing or able to make additional management funds available. Without direction from the Congress, they are unlikely to do so. With regard to drawing funds from the program for evaluation and management, current legislation does not permit this and would have to be modified; therefore the Congress has clearly intended program funds to be for awards only. The third option, involving a modest increase to the program, would also require legislative action and would perhaps be more easily achievable in the event of an

overall increase in the program. In any case, the Committee envisages an increase of the "set-aside" of perhaps 0.03 to 0.05 percent on the order of $35 million to $40 million per year or, roughly, double what the Navy currently makes available to manage and augment its program. In the latter case (0.05 percent), this would bring the program "set-aside" to 2.55 percent, providing modest resources to assess and manage a program that is approaching an annual spending of some $2 billion. Whatever modality adopted by the Congress, without additional resources the Committee's call for improved management, data collection, experimentation, and evaluation may prove moot.

3

Scope and Methodology

3.1 STUDY SCOPE

This National Research Council (NRC) study focuses on the National Science Foundation's (NSF's) Small Business Innovation Research (SBIR) program, examining the program's performance in meeting each of the following four SBIR legislated objectives:

1. Stimulate technological innovation;
2. Increase commercialization of innovation;
3. Use small business to meet federal R&D needs;
4. Foster participation by minority and disadvantaged persons in technological innovations.

The study also examines the effectiveness of the NSF's management of its SBIR program, including resources, topic definition, solicitations, proposal selection, commercialization assistance, and general oversight of the program.

3.2 METHODOLOGY DESIGN

This study of the NSF's SBIR program and the study's findings and recommendations are based on a research methodology designed by the Committee, approved by an independent panel of experts, and described in more detail in the study's "Methodology Report."[1] The study uses multiple methods to allow

---

[1]National Research Council, *An Assessment of the Small Business Innovation Research Program: Project Methodology*, Washington, DC: The National Academies Press, 2004.

cross-checking and confirmation of findings that emerge from any single method. Multiple methods are also necessary to examine the several objectives of the SBIR program, which vary in their relative tractability to quantitative and qualitative approaches. The core methodologies used in generating the data underlying the findings of this report are: surveys of firms that participated in the SBIR program; interviews with NSF SBIR officials and program managers; review of program documents and other relevant literature; analysis of program data; and case studies of firms.

3.3 METHODS USED

3.3.1 Surveys

Three surveys were sent to SBIR firms: (1) Phase II Survey; (2) Firm Survey; and (3) Phase I Survey.[2]

The first listed survey was focused on a sample of Phase II NSF SBIR grants. This Phase II Survey had 162 respondents. The focus of the Phase II Survey was on specific grants and their commercial outcomes.

The second survey explored the SBIR program's influence on small businesses more generally. Responding to this survey were 137 companies identified primarily as NSF Phase II grant recipients. The survey asked questions about the background experience of each firm's founders, the receipt of Phase I and Phase II awards, and the influence of the SBIR program on firm founding, commercial progress, and growth.

The third survey was focused on a sample of Phase I NSF SBIR grants, primarily for the purpose of identifying and learning more about those Phase I grants that did not receive a direct follow-on Phase II grant. The Phase I Survey had 248 respondents, of which 135 received Phase I grants that were not followed by a direct follow-on Phase II grant, and 113 received Phase I grants that were followed by a direct follow-on Phase II grant. Those that did receive a follow-on Phase II grant were asked if they received assistance and from whom.

The methodologies of these three surveys are described in more detail in Appendixes B and C. These descriptions reveal the initial sample sizes, adjustments to the samples, response rates, and statistics on response rates. These appendixes also provide the survey instruments as well as survey results for individual questions. The number of respondents to each question and the base numbers used to calculate percentage responses are also given. In considering these firm survey results, it is worth keeping in mind the possibility of response biases that could significantly affect the survey results. For example, it may be possible that some of the firms that could not be found have been unsuccessful and folded.

---

[2]For the NRC Phase II Survey, 35 percent of the overall sample of 457 responded. This represents 48 percent of the awards contacted. For the NRC Phase I Survey, the response rate was 10 percent of a sample size of 2,458. See Appendixes B and C.

It may also be possible that unsuccessful firms were less likely to respond to the survey. (See Appendixes B and C for details about the survey methodologies, survey instruments, and responses to individual questions.)

NSF SBIR Program Manager Survey. A separate survey was sent to the relatively small group of NSF SBIR program managers. NSF SBIR officials declined having their program managers complete the survey and provided a single response to this survey prepared by the NSF SBIR program's senior advisor, Dr. Joseph Hennessey.

3.3.2 Interviews with NSF SBIR Managers

Multiple interviews were conducted with NSF SBIR program officials and program managers. The interviews explored various aspects of program operations and how they have changed over time, past and current management practices, and possible program modifications.

3.3.3 Review of Program Documents and Data

Documents of the NSF SBIR program that were reviewed for the study included internal copies of evaluation studies, related data tables and success stories, survey and interview guides, success stories known as "nuggets," annual reports, reports by the Committee of Visitors, staff presentations, program guidelines, solicitation materials, examples of funded projects, and other materials.

The NSF SBIR Office provided data on proposals and grants by individual records. Analysis of these data informed the program description and often figured in the analysis of program outputs.

3.3.4 Case Studies of Firms

Case studies were prepared for ten selected companies located in seven states: California, Ohio, Tennessee, Minnesota, Arizona, Vermont, and Michigan. The companies are developing ten different technologies from a variety of disciplines, including software, electrochemical processes, information technology for pest monitoring and control, electronics, manufacturing processes, and nanomaterials.

The oldest company is more than 20 years old and the youngest was just 2 years old at the time of the interview. The smallest company has about a dozen employees, the largest nearly 150.

Among the case-study companies are a university spin-off, a large company spin-off, a small company spin-off, a company started by a graduate student, a company started by university faculty, a company started by a retired large company executive, several companies started by entrepreneurs leaving other com-

Box A
Multiple Sources of Bias in Survey Response

Large innovation surveys involve multiple sources of bias that can skew the results in both directions. Some common survey biases are noted below.

- **Successful and more recently funded firms are more likely to respond.** Research by Link and Scott (2005) demonstrates that the probability of obtaining research project information by survey decreases for less recently funded projects and that it increases the greater the award amount.[a] Nearly 40 percent of respondents in the NRC Phase II Survey began Phase I efforts after 1998, partly because the number of Phase I awards increased, starting in the mid-1990s, and partly because winners from more distant years are harder to reach. They are harder to reach as time goes on because small businesses regularly cease operations, are acquired, merge, or lose staff with knowledge of SBIR awards.

- **Success is self-reported.** Self-reporting can be a source of bias, although the dimensions and direction of that bias are not necessarily clear. In any case, policy analysis has a long history of relying on self-reported performance measures to represent market-based performance measures. Participants in such retrospective analyses are believed to be able to consider a broader set of allocation options, thus making the evaluation more realistic than data based on third-party observation.[b] In short, company founders and/or principal investigators are, in many cases, simply the best source of information available.

- **Survey sampled projects at firms with multiple awards.** Projects from firms with multiple awards were underrepresented in the sample because they could not be expected to complete a questionnaire for each of dozens or even hundreds of awards.

- **Failed firms are difficult to contact.** Survey experts point to an "asymmetry" in their ability to include failed firms for follow-up surveys in cases where the firms no longer exist.[c] It is worth noting that one cannot necessarily infer that the SBIR project failed; what is known is only that the firm no longer exists.

- **Not all successful projects are captured.** For similar reasons, the NRC Phase II Survey could not include ongoing results from successful projects in firms that merged or were acquired before and/or after commercialization of the project's technology. The survey also did not capture projects of firms that did not respond to the NRC invitation to participate in the assessment.

- **Some firms may not want to fully acknowledge SBIR contribution to project success.** Some firms may be unwilling to acknowledge that they received important benefits from participating in public programs for a variety of reasons. For example, some may understandably attribute success exclusively to their own efforts.

- **Commercialization lag.** While the NRC Phase II Survey broke new ground in data collection, the amount of sales made—and indeed, the number of projects that generate sales—are inevitably undercounted in a snapshot survey taken at a single point in time. Based on successive data sets collected from the National

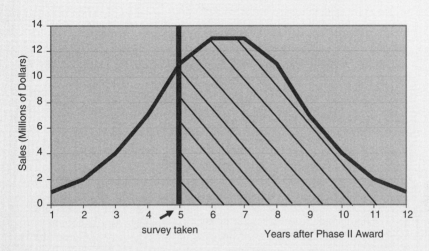

FIGURE A-1 Survey Bias due to Commercialization Lag.

Institutes of Health SBIR award recipients, it is estimated that total sales from all responding projects will likely be on the order of 50 percent greater than can be captured in a single survey.[d] This underscores the importance of follow-on research based on the now-established survey methodology.

---

[a] Albert N. Link and John T. Scott, *Evaluating Public Research Institutions: The U.S. Advanced Technology Program's Intramural Research Initiative*, London: Routledge, 2005.

[b] While economic theory is formulated on what is called 'revealed preferences,'—meaning individuals and firms reveal how they value scarce resources by how they allocate those resources within a market framework—quite often, expressed preferences are a better source of information, especially from an evaluation perspective. Strict adherence to a revealed preference paradigm could lead to misguided policy conclusions because the paradigm assumes that all policy choices are known and understood at the time that an individual or firm reveals its preferences and that all relevant markets for such preferences are operational. See Gregory G. Dess and Donald W. Beard, "Dimensions of Organizational Task Environments," *Administrative Science Quarterly* 29 (1984):52–73, and Albert N. Link and John T. Scott, *Public Accountability: Evaluating Technology-Based Institutions*, Norwell, MA: Kluwer Academic Publishers, 1998.

[c] Link and Scott, *Evaluating Public Research Institutions: The U.S. Advanced Technology Program's Intramural Research Initiative* op. cit.

[d] Data from the National Institutes of Health indicates that a survey taken two years later would reveal very substantial increases in both the percentage of firms reaching the market and the amount of sales per project. See National Research Council, *An Assessment* of *the Small Business Innovation Research Program at the National Institutes of Health*, Charles W. Wessner, ed., Washington, DC: The National Academies Press, forthcoming.

panies, and one firm started by a professor-husband and entrepreneur-wife team. The case studies include two woman-owned companies, one of which is actively operated by a woman who is also a member of a minority group.

The sample includes companies that were able to commercialize a product very quickly and companies whose technologies are taking considerable time to commercialize. Commercialization strategies of these firms include licensing agreements, contract research, sales of product produced in-house, commercialization partnerships, as well as the sale of technology or the company itself to a larger company.

The companies' annual revenues range from $2 million to about $24 million. This includes companies whose share of annual revenue contributed by the SBIR program and other government grants comprise as little as 4 percent and as much as 70 percent.

The ten companies selected for the case studies do not represent a random sample. The companies were selected to provide companies of different ages and size, pursuing different technologies, located in different parts of the country, with differing forms of ownership, and with some, although varying degrees of, commercial success. Some of the companies are university spin-offs; some are company spin-offs; some are neither. Some received many SBIR grants; some relatively few. Some continue to obtain a high percentage of their funding from government sources; others have reduced the percentage to low numbers. To illustrate different types of successful outcomes, six of the cases selected were identified by NSF as "stars." In addition, all of these cases selected are revenue-earning SBIR-funded companies. The process for selecting the case studies is explained in detail in Appendix D.

4

Overview of the NSF SBIR Program

4.1 A BRIEF HISTORY OF THE NSF'S SBIR PROGRAM

4.1.1 The Nation's First SBIR Program

Between 1977 and 1980, the NSF operated the only federal research grant program solely for small businesses. It was called "SBIR."

The Department of Defense (DoD) started a parallel small business initiative in 1980. Hence, there were two small business initiatives operating on July 22, 1982, when Congress passed the Small Business Innovation Development Act authorizing the SBIR program.

Upon passage of that act, the NSF designated its 1977 precursor program the agency's SBIR program, and it became the nation's first. The program appears to have kept essentially the same philosophy, focus, and features of the original NSF initiative. Then and now, the NSF SBIR program's emphasis on promoting small business and entrepreneurial research differs from that of the rest of the agency, which focuses on the support of academic research. The NSF's SBIR program also continues to differ in important ways from the SBIR programs operated by the other major R&D agencies.

4.1.2 Factors That Led to Establishment of the NSF's SBIR Program

The NSF established the precursor SBIR program in response to efforts of the small business community to secure research funding. Small companies were at a disadvantage in seeking research funding from the NSF because of the strong orientation of the agency toward the university community. Many entrepreneurs thought that the NSF believed that quality research was synonymous with academic research. Small companies reportedly were also at a disadvantage when competing for funding at DoD and NASA because those agencies' procurement process was oriented toward large, complex systems and the companies that could supply them. The NSF's precursor SBIR program was established in part to respond to complaints by the small business community that it was being shut out of government funding for innovative research and for procurement and in part due to the conviction of key individuals within the NSF that small business represented an untapped resource.

4.1.3 The NSF's Early Emphasis on Commercialization

As is often the case, the right people in the right place at the right time played a critical role in shaping the NSF's response to the complaints by the small business community. These key individuals were Roland Tibbetts, who was instrumental in establishing the precursor SBIR program and rightly receives credit for developing the SBIR concept along with Ritchie Coryell. Interestingly both individuals had business training and shared the view that small companies could produce innovative, high quality research that could contribute to the nation's science base. Furthermore, they believed that the small companies with the drive and know-how to commercialize were likely to be the same companies that could perform high quality research—a view that led the NSF program to have from the start a strong emphasis on commercialization That has continued and increased to the present time.[1]

According to Mr. Coryell, the view driving NSF's formulation of its small business initiative was that the agency already had a substantial emphasis on academic research in its grant programs. To provide a real counterpoint, the new program would target small companies and stress commercialization while also promoting innovation and high quality research.[2] These dual goals were considered completely compatible by the developers of the original SBIR program.[3] The NSF SBIR program's emphasis on commercialization, then and now, contrasts strongly with NSF's otherwise strong orientation towards funding academic research as a means of contributing to the nation's science and technical knowledge base.

---

[1]Interview with Ritchie Coryell, NSF SBIR Program Staff, on October 23, 2003. (Note that Mr. Coryell has since retired.)

[2]Ibid.

[3]More than a decade later, the experience of the Advanced Technology Program lends support to the idea that many small businesses, including the extremely small, often combine capacities for high quality, innovative research with exceptional business acumen, such that it is unnecessary for the program to choose between these strengths in order to engage with small businesses. For a description of ATP processes and achievements, see "The Advanced Technology Program: Assessing Outcomes," National Academy Press, 2001.

4.2 NSF SBIR DEMOGRAPHICS

This section uses data tables and graphs to provide an overview of the NSF's SBIR program in terms of the number of proposals received and the number and dollar amounts of grants made. It looks at the data by grant phase, by type of applicant, and by state. [4] The period covered is 1992 to 2005, in most cases. There are certain subcategories for which data were not available back to 1992 or were not extended through 2005. Appendix A contains data tables from which tables and graphs used in this section are drawn.

4.2.1 Description of the NSF's SBIR Grants

To recap, the entry level Phase I grant lasts six months and is currently capped at $100,000. Those who receive a Phase I grant from the NSF are eligible to apply for a follow-on NSF Phase II grant in an amount up to $500,000 for a period of up to two years. The NSF's grant arrangement is essentially the same as that of other agency programs up to this point. Past this point, however, the NSF's SBIR program has developed special features.

In 1998, the NSF introduced an innovation in its Phase II grant to promote increased commercialization. It split the Phase II grant into two parts: a "regular" Phase II grant of up to $500,000 and extending over two years, and a "supplemental" Phase II grant, called a "Phase IIB," for which the grantee could apply as a follow-on to the initial Phase II grant. The Phase IIB grant provided a minimum of $50,000 and a maximum of $250,000 (until November 2003—see below). The Phase II grant also extends the research time by a year. The NSF's SBIR program essentially holds back what could otherwise be part of Phase II funding and grants it as a supplement to those Phase II grantees who show evidence of commercialization potential as indicated by their ability to attract third-party funding.

To receive supplementary funding under the Phase IIB grant, the grantee must have completed one year of work on the initial Phase II grant (or receive special permission from the NSF SBIR program officer) and must show hard evidence that it has secured $2 of funding from other sources for each dollar of supplementary funding provided by the NSF. As can be inferred from the minimum grant size of $50,000, this means the grantee must have a minimum of $100,000 in funding from other sources to be eligible for a Phase IIB grant.

After November 1, 2003, the NSF's SBIR program increased the maximum Phase IIB supplement to $500,000 by adding a "Phase IIB+" supplemental grant of up to $250,000 for an additional year to the existing $250,000 maximum

---

[4]The NSF did not provide data showing a breakout of firms by size of applicants and grantees. A single designation was given of "small business." Hence, it has not been possible as yet to develop profiles of proposers and grantees by their size.

Phase IIB grant. The Phase IIB+ grant carries the same two-to-one requirement for third-party funding, but it adds a stipulation that the Phase IIB+ third-party funding cannot come from a federal source.

Together, Phase II, Phase IIB, and Phase IIB+ cannot last longer than four years. Thus, the combined maximum grant for Phase II and related Phase IIB/IIB+ supplements has increased from $750,000 to $1 million. (For additional discussion of the Phase IIB/IIB+ supplement, see Section 8.5.)

4.2.2 Annual Dollar Outlays and Number of Grants

As explained above, grants and associated dollar outlays are made by the NSF SBIR program office through three funding tools: Phase I, Phase II, and Phase IIB/IIB+ grants. The number of grants and dollar outlays by type change over time in response to multiple factors, including the following: (1) changes in the NSF R&D budget; (2) changes in the SBIR program percentage that is applied to the NSF R&D budget to compute the SBIR program's annual budget; (3) the timing and types of solicitations held and company responses to them; and (4) the prior distribution of funding among Phase I, Phase II, and Phase IIB/IIB+. According to NSF program officials, there is no formulaic approach to the allocation of funding among the types of grants. The distribution of outlays in a given year is to some extent a function of prior decisions. For example, a spike in Phase I outlays in one year drives a lagged spike in Phase II outlays.

The study selected the time period for coverage as 1992–2005. The NSF has not converted all of its historical data to electronic form. When certain sub-categories of data were not available for the entire period, the period covered is stated. In several other cases, such as state data, charts and tables were prepared prior to release of the 2005 data, and they were not updated if the existing data adequately addressed the subject.

Phase I Grants. Figure 4.2-1 shows total NSF funding for Phase I SBIR grants from 1992 through 2005. Two jumps in funding for Phase I grants are apparent: from 1992 to 1994, when Phase I outlays almost doubled, rising from $10.3 to $19.8 million between 1992 and 1997, and from 2001 to 2003, when the funding amount again almost doubled, rising from $21.8 million to $43.4 million. Between 1994 and 2001, outlays for Phase I fluctuated but remained close to $20 million.

The spike in Phase I funding in 2003 was, according to NSF managers, a strategic decision to respond to current events. With the collapse of the venture capital bubble and the events of September 11, 2001, small businesses found that traditional sources of private funding had dried up. Responding to this situation, the NSF increased its acceptance rate, maintaining that rate even as the number of proposals increased dramatically in 2003. Not being able to sustain this level of activity with available funds, the NSF scaled back by decreasing the success

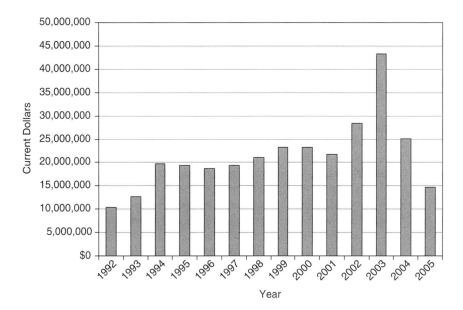

FIGURE 4.2-1 Annual Outlays for Phase I NSF SBIR Grants, 1992–2005 (Current Dollars).
SOURCE: Developed from data provided by NSF SBIR program.

rate and by going to an eighteen-month cycle to reduce the number of proposals. Reflecting these changes since 2003, outlays for Phase I grants dropped substantially—to $25.1 million in 2004, and to $14.7 million in 2005.

Figure 4.2-2 combines a bar chart showing the number of Phase I grants over the period 1992 to 2005 with a line chart showing the average size of Phase I grants. The line chart of Figure 4.2-2 shows the increase in the average grant size from 1992 to 1998, from approximately $50,000 to approximately $100,000 (reflecting changes in maximum grant size). Since 1998, the average Phase I grant size has remained essentially unchanged in current dollars, which means that it has declined steadily in constant dollars.

Phase II Grants. Figure 4.2-3 shows total NSF outlays on Phase II grants (excluding Phase IIB grants) over the same fourteen years, 1992–2005. From 1992 to 1995, annual Phase II outlays were consistently below $15 million. A steep rise occurred between 1995 and 1998, when annual outlays for Phase II grants rose to $40 million. Between 1998 and 2003, total annual outlays fluctuated between $35 million and $43 million. A surge in dollar outlays for Phase II grants occurred in 2004 and continued into 2005, following the surge in dollar outlays for Phase I grants in 2003.

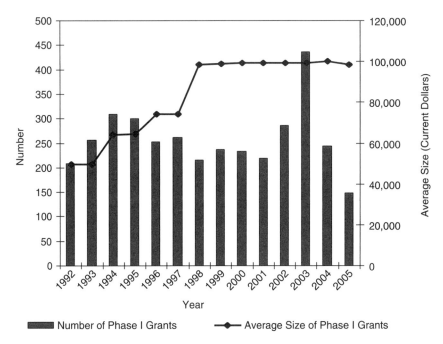

FIGURE 4.2-2 Number and Average Size of NSF SBIR Phase I Grants (Current Dollars), 1992–2005.
SOURCE: Developed from data provided by NSF SBIR program.

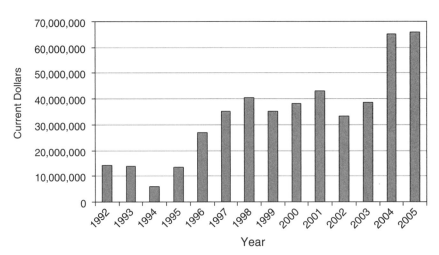

FIGURE 4.2-3 Annual Outlays for Phase II NSF SBIR Grants, 1992–2005 (Current Dollars)
SOURCE: Developed from data provided by NSF SBIR program.
NOTE: Excludes Phase IIB grants.

Figure 4.2-4 combines a bar chart showing the yearly number of Phase II grants and a line chart showing the average size of Phase II grants. The number of grants increased sharply from 1994 to 1997, fell through 2003, and then surged again in 2004. The average Phase II grant size ranged between $245,000 in 1992 and $289,000 in 1997, then rose to about $400,000 in 2000, and since 2002 has been close to the maximum allowed size of $500,000. Again, the dollar amounts are given in current dollars.

Phase IIB Grants. Figure 4.2-5 shows outlays for Phase IIB grants from 1998, when they first began, through 2005. The amount spent on these grants reached $9.6 million in 2002, dropped in 2003 and 2004, and rose again to $9.3 million in 2005.

Table 4.2-1 summarizes the number, total dollar amount, average size, and range of Phase IIB grants from 1998 to 2005. In 1998, these supplemental grants averaged about $100,000 the first year, and only four Phase IIB grants were approved. During the second year, 1999, the average grant size was about the same, but more grants (twenty one) were made. In 2000, the average grant size nearly doubled, but only nine grants were approved. Between 2001 and 2005, the average grant size fluctuated, ranging from $235,000 to $330,000, and the number of grants fluctuated between fourteen and thirty-nine.

In total, the NSF's SBIR program approved 161 Phase IIB grants from 1998 through 2005. The smallest Phase IIB grant was $51,000 and the largest was

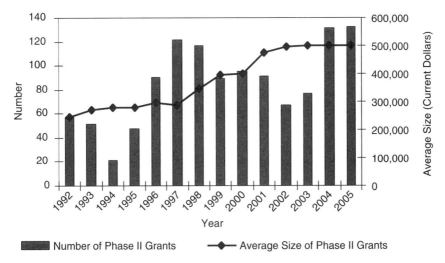

FIGURE 4.2-4 Number and Average Size (Current Dollars) of NSF SBIR Phase II grants, 1992–2005.
SOURCE: Developed from data provided by NSF SBIR program.
NOTE Excludes Phase IIB grants.

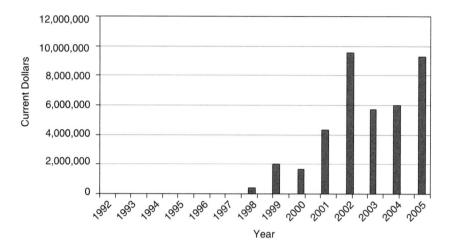

FIGURE 4.2-5 Annual Outlays for NSF SBIR Phase IIB Grants, 1992–2005 (Current Dollars).
SOURCE: Developed from data provided by NSF SBIR program.
NOTE: Includes both Phase IIB and IIB+ grants.

TABLE 4.2-1 Phase IIB Awards from Start in 1998 to 2005

Year	Number of Awards	Total Funding ($)	Average Award Size ($)	Range ($)
1998	4	399,944	99,986	99,944 to 100,000
1999	21	1,998,574	95,170	59,785 to 100,000
2000	9	1,660,191	184,466	115,000 to 200,000
2001	14	4,302,682	307,334	127,532 to 350,800
2002	39	9,588,580	245,861	65,239 to 386,360
2003	24	5,690,294	237,096	79,923 to 350,000
2004	22	6,025,436	273,883	56,250 to 500,000
2005	28	9,260,464	330,731	51,125 to 500,000

SOURCE: Developed from data provided by NSF SBIR program.

NOTE: Dollar amounts are in current dollars.

$500,000. Over this period, the program provided nearly $39 million in these supplemental grants to those Phase II projects for which third-party financing had been raised. The amount of funding issued under this supplemental program was constrained by the ability of Phase II grantees to obtain firm commitments from third parties for funding.

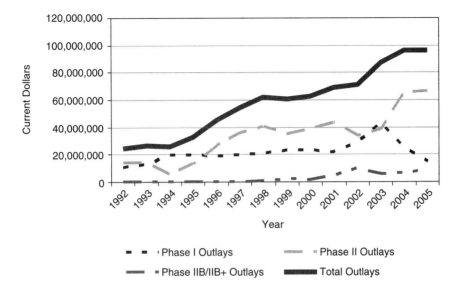

FIGURE 4.2-6 Outlays by Grant Phase, NSF SBIR, 1992–2005 (Current Dollars).
SOURCE: Developed from data provided by NSF SBIR program.

All Grants. Figure 4.2-6 shows the total annual dollar outlays for the NSF's SBIR program from 1992 through 2005. While outlays for each type of grant increased over this period, large increases in Phase II spending drove the significant jump in total outlays. Most recently, in 2004 and 2005, the trend in total NSF SBIR outlays has been flat.

4.2.3 Applications and Success Rates

The NSF funded between 14 percent and 21 percent of the approximately 1,000 to 2,000 Phase I proposals received each year over the period 1994–2005. The highest funding rates occurred in 2002 and 2003.

According to the NSF's response to the NRC Program Manager Survey, 80 percent of Phase I recipients in a recent year applied for a Phase II grant.[5] However, 68 percent of the firms in the Phase I Survey sample, taken over a range of years, said they applied for a direct follow-on Phase II grant.[6] Whatever the true figure, an average of close to 200 Phase II proposals have been submitted yearly from 1994 to 2005, and the NSF annual acceptance rate has ranged from 17 per-

---

[5]Note that Dr. Joseph Hennessey, NSF SBIR program senior advisor, alone responded to the NRC Program Manager Survey.

[6]NRC Phase I Survey.

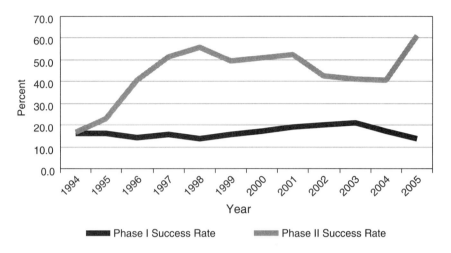

FIGURE 4.2-7 Success Rates of Applicants for Phase I and II Grants, 1994–2005.
SOURCE: Developed from data provided by NSF SBIR program.

cent to 61 percent over the same period. The lower annual acceptance rates, however, all came during the early years of the program. Since 1997, acceptance rates have for the most part ranged between 40 percent and 60 percent. Figure 4.2-7 compares the success rates for Phase I and II grants from 1994 through 2005.

According to one program official, 30 percent of Phase II recipients in a recent year applied for a Phase IIB grant and 25 percent of Phase II recipients received Phase IIB grants.[7] All companies that found outside funding had their Phase IIB applications approved. Companies who are unable to meet the third-party financing requirement simply withdraw their proposal.[8] Allowing for a one-year lag, the number of Phase IIB grants as a percentage of Phase II awards ranged from 3 percent in 1998 to a maximum of 42 percent in 2002. In 2005, the number of Phase IIB grants was 21 percent of 2004 Phase II grants.

4.2.4 Geographical Location of Grants and Applicants

Table 4.2-2 shows the geographical distribution of Phase I grants from 1992 through 2004. The top two recipient states, California and Massachusetts, accounted for more than 1,000 grants, approximately 30 percent of the total. These two states, plus the next eight states—Colorado, New York, Texas, New

---

[7]NRC Program Manager Survey.
[8]Comments provided by NSF Program Officials at a meeting with NRC research team members on August 17, 2005.

TABLE 4.2-2 Distribution of Phase I NSF SBIR Grants by State in Descending Order of Grants Received, 1992–2003

State	Grants	Percent of All	State	Grants	Percent of All
CA	656.0	19.0	MO	28	0.8
MA	476.0	13.8	IN	25	0.7
CO	192.0	5.6	WI	26	0.8
NY	183.0	5.3	AR	24	0.7
TX	140.0	4.0	DE	22	0.6
NJ	129.0	3.7	HI	21	0.6
VA	123.0	3.6	OK	20	0.6
CT	119.0	3.4	ME	20	0.6
MD	115.0	3.3	VT	18	0.5
OH	105.0	3.0	KS	16	0.5
PA	88.0	2.5	SC	15	0.4
AZ	81.0	2.3	WY	13	0.4
NM	78.0	2.3	SD	12	0.3
MI	69.0	2.0	AK	12	0.3
MN	63.0	1.8	ND	12	0.3
WA	63.0	1.8	WV	11	0.3
FL	60.0	1.7	LA	11	0.3
IL	53.0	1.5	ID	9	0.3
NC	53.0	1.5	IA	8	0.2
OR	46.0	1.3	KY	7	0.2
GA	44.0	1.3	MS	7	0.2
UT	40.0	1.2	NE	6	0.2
AL	34.0	1.0	NV	6	0.2
TN	30.0	0.9	PR	3	0.1
MT	32.0	0.9	DC	2	0.1
NH	29.0	0.8	RI	2	0.1

SOURCE: NSF SBIR program.

NOTE: Included are the District of Columbia and Puerto Rico.

Jersey, Connecticut, Virginia, Maryland, and Ohio—received more than 2,000 grants, 65 percent of the total. That these states received the most grants is not a surprise; they all have established high-tech clusters. The bottom 20 states accounted for fewer than 200 grants and less than 6 percent of total grants.

Table 4.2-3 shows the distribution of Phase II grants by state. Again, California and Massachusetts received a large share of the grants—approximately 40 percent. And, in this case, the same technologically advanced states made up the top ten (though not in the same order) and took an even bigger share of the total number of grants—74 percent. The bottom 20 states, each of which received less than one-tenth of 1 percent of grants, accounted for less than 9 percent of 821 Phase II grants.

TABLE 4.2-3 Distribution of Phase II Grants by State in Descending Order of Grants Received, 1992–2004

State	Grants	Percent of All	State	Grants	Percent of All
CA	228	27.8	NH	10	<0.05
MA	139	16.9	MT	12	<0.05
CO	57	6.9	VT	8	<0.05
NY	56	6.8	DE	8	<0.05
VA	39	4.8	WI	8	<0.05
CT	38	4.6	HI	6	<0.05
MD	36	4.4	ND	7	<0.05
TX	35	4.3	OK	6	<0.05
OH	32	3.9	SD	5	<0.05
NJ	29	3.5	WY	6	<0.05
MI	24	2.9	AK	4	<0.05
NM	24	2.9	IN	4	<0.05
MN	21	2.6	KS	5	<0.05
PA	21	2.6	ME	4	<0.05
IL	20	2.4	SC	6	<0.05
WA	20	2.4	AR	6	<0.05
AZ	17	2.1	KY	3	<0.05
FL	17	2.1	MO	5	<0.05
TN	15	1.8	ID	2	<0.05
GA	15	1.8	LA	3	<0.05
OR	14	1.7	MS	1	<0.05
UT	13	1.6	NE	1	<0.05
NC	12	1.5	NV	1	<0.05
AL	11	1.3	RI	1	<0.05
			DC	1	<0.05
			IA	1	<0.05

SOURCE: NSF SBIR program.

The states that had the most Phase I grants did not necessarily have the highest success rates in terms of percentage of proposals funded. Similarly, the states with fewer Phase I grants did not necessarily have the lowest success rates. For example, Nebraska and Wyoming had high success rates, but they were near the bottom in terms of total grants. Wyoming had a higher success rate than California, which had the most proposals funded. The sheer number of proposals submitted from companies in the states with the most Phase I grant winners accounted for these states' large share of grants, not a higher-than-average success rate.

Because Phase I is prerequisite to Phase II, the states with many Phase I grant winners are positioned to be large winners in Phase II and Phase IIB. However,

TABLE 4.2-4 Distribution of NSF SBIR Phase IIB Grants by State in Descending Order of Grants Received, 1998–2005

State	Grants	Percent of All	State	Grants	Percent of All
CA	36	22.4	DE	2	1.2
MA	21	13.0	KS	2	1.2
NY	11	6.8	NH	2	1.2
NC	7	4.3	NM	2	1.2
OH	7	4.3	TN	2	1.2
MN	6	3.7	UT	2	1.2
FL	5	3.1	VT	2	1.2
MD	5	3.1	WI	2	1.2
VA	5	3.1	AK	1	0.6
AL	4	2.5	AZ	1	0.6
IL	4	2.5	HI	1	0.6
MI	4	2.5	MT	1	0.6
PA	4	2.5	ND	1	0.6
GA	3	1.9	OK	1	0.6
TX	3	1.9	OR	1	0.6
WA	3	1.9	RI	1	0.6
WY	3	1.9	SC	1	0.6
CO	2	1.2	SD	1	0.6
CT	2	1.2			

SOURCE: NSF SBIR program.

as may be seen in Table 4.2-4, the largest winners in Phase II are not always the largest winners in Phase IIB. Thus, splitting off part of the funding that could have been awarded in lump sums to Phase II grantees and instead awarding it in Phase IIB grants based on the ability of the grantees to obtain third-party support changes the geographical distribution of SBIR funding.[9]

4.2.5 Women and Minorities as Applicants, Grantees, and Principal Investigators

A statutory goal of the SBIR program is to foster and encourage participation by women and minorities in technological innovation.

Woman-Owned Businesses. Figure 4.2-8 shows the number of Phase I proposals from and grants received by woman-owned businesses from 1994 through 2005. With the exception of the bump-up in 2002 and 2003, there is clearly no upward trend. In other words, while woman-owned businesses do participate in the NSF's SBIR program, that participation does not seem to be increasing over time.

---

[9]At this point, the geographical effort is based on small numbers of Phase IIB grants.

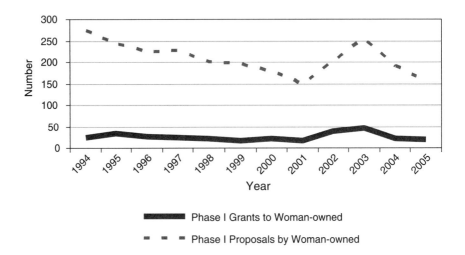

FIGURE 4.2-8 Numbers of Phase I Grants to and Proposals from Woman-owned Firms, 1994–2005.
SOURCE: Developed from data provided by the NSF SBIR program.

Figure 4.2-9 shows the number of Phase II proposals from and grants received by woman-owned businesses from 1995 through 2005. Note the erratic patterns, with no clear trend in either data series.

Figure 4.2-10 shows the number of Phase IIB grants received by woman-owned businesses annually from 1998 through 2005. In this case, there is no separate proposal data series because proposals that do not meet the third-party funding requirement are withdrawn. The numbers are small and show no obvious trend. Woman-owned businesses receive a few Phase IIBs each year.

The comparative success rates of woman-owned businesses relative to all participating businesses in obtaining grants are shown in Figure 4.2-11. Woman-owned businesses received 10 percent of NSF Phase I grants from 1994 through 2005. Their annual share ranged from 7 percent to 13 percent. The average size of the Phase I grants to woman-owned businesses was approximately the same as that received by businesses overall.

Woman-owned businesses received 7 percent of Phase II grants over the period 1995 to 2005, with their annual share ranging from 2 to 14 percent. Again, there was no significant difference between the size of Phase II grants to woman-owned and to businesses overall.

For Phase IIB grants, the numbers were small and the share of Phase IIB grants given to woman-owned businesses fluctuated, ranging from 0 percent to

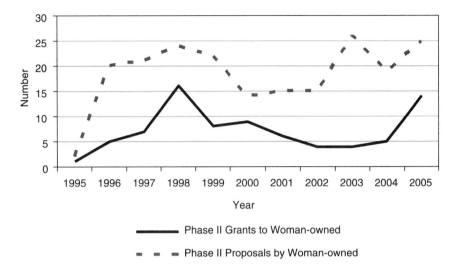

FIGURE 4.2-9 Number of Phase II Grants to and Proposals from Woman-owned Firms, 1995–2005.
SOURCE: Developed from data provided by the NSF SBIR program.

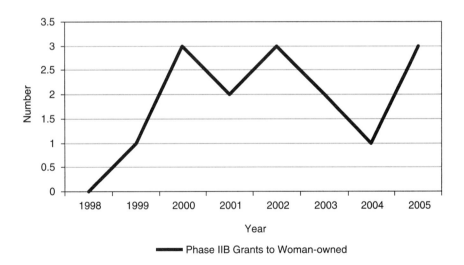

FIGURE 4.2-10 Number of Phase IIB Grants to Woman-owned Firms, 1998–2005.
SOURCE: Developed from data provided by the NSF SBIR program.

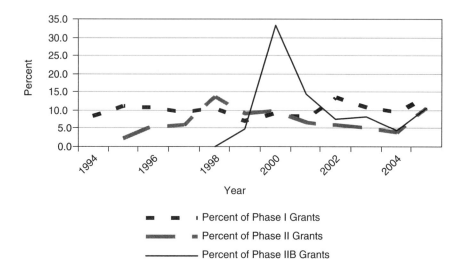

FIGURE 4.2-11 Shares of Phase I, Phase II, and Phase IIB NSF SBIR Grants to Woman-owned Businesses, 1994–2005.
SOURCE: Developed from data provided by the NSF SBIR program.

33 percent. Overall, woman-owned businesses received 9 percent of Phase IIB grants from 1998 through 2005.

Figure 4.2-12 shows that in every year since 1994, woman-owned businesses have been less successful than applicants in general in having their Phase I applications approved. Over the period 1995–2005, woman-owned businesses contributed 12 percent of all Phase I applications (2,069/16,897), and received 9.5 percent of all Phase I grants. In half the years, the success rate of woman-owned businesses was 70 percent or less the rate for all applicants.[10] Statistical tests of significance showed the differences in success rates between woman-owned and all firms at Phase I to be highly significant, with the probability of obtaining the t-test results to be 0.000 if there were no real differences in success rates for woman-owned firms.[11]

Figure 4.2-13 shows that woman-owned businesses also had a lower success rate than businesses overall in getting Phase II proposals approved. In all except three of the ten years from 1995 through 2005, the success rate of woman-owned businesses fell below that of all businesses in having their Phase II applications

---

[10]The comparison would be even more unfavorable for woman-owned companies if the total of all applicants were adjusted to remove woman-owned firms.

[11]The results of a paired t-test was t = −5.93 with 11 degrees of freedom, which was significantly less than t = −1.795 at a level of significance of 0.05.

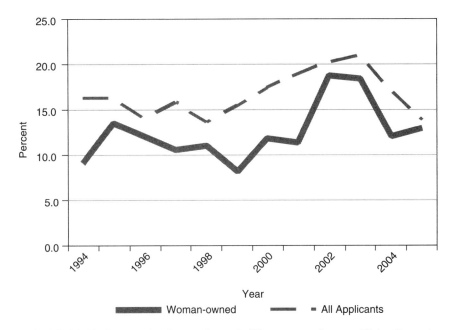

FIGURE 4.2-12 Comparative Success Rates for Woman-owned versus All Applicants in Having Their Phase I Proposals Approved, 1994–2005.
SOURCE: Developed from data provided by the NSF SBIR program.

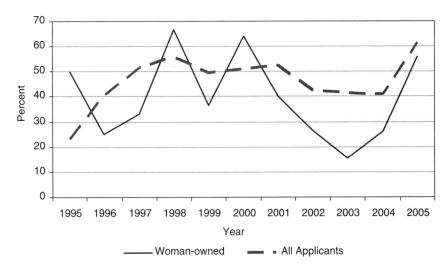

FIGURE 4.2-13 Comparative Success Rates for Woman-owned versus All Applicants in Having Their Phase II Proposals Approved, 1995–2005.
SOURCE: Developed from data provided by the NSF SBIR program.

TABLE 4.2-5 Aggregate Proposals and Grants, Woman-owned and All
Businesses, 1995–2005

Businesses	Proposals				Grants			
	Phase I	Phase II	Phase IIB	Total	Phase I	Phase II	Phase IIB	Total
A. Woman-owned Businesses	2,069	203	15	2,262	268	79	15	362
B. All Businesses	16,897	2,299	161	19,357	2,833	1,059	161	4,053
Percent (A/B)	12.2	8.8	9.3	11.7	9.5	7.5	9.3	8.9

SOURCE: Developed from data provided by the NSF SBIR program.

NOTE: Phase IIB results are for the period 1998–2005, while data for Phase I and II are for
1995–2005. Also, note that the inclusion of woman-owned firms in all businesses tends to cause the
comparative results to look more favorable for the woman-owned firms than they would be if woman-
owned firms were compared to all *other* businesses.

funded. Woman-owned businesses contributed 9 percent of all Phase II applications submitted from 1995 to 2003 (203/2,299) and received 7.5 percent of
all Phase II grants (79/1059). Overall, from 1995 through 2005 woman-owned
businesses had 39 percent of their Phase II proposals accepted and all businesses
had 46 percent of their Phase II proposals accepted. Statistical tests of significance, however, showed these differences to be less robust than those at Phase
I. The null hypothesis—that there is no statistical difference in success rates at
Phase II—could not be rejected.[12]

Table 4.2-5 summarizes applications by and grants to woman-owned businesses from 1995 through 2005. Overall, woman-owned businesses submitted 11.7
percent of all applications and received 8.9 percent of all grants. Overall, woman-
owned businesses had 16 percent of their proposals approved (362/2,262), and
businesses overall had 21 percent of their proposals approved (4,053/19,357).

Minority-Owned Businesses. Minority-owned businesses—businesses
owned by a member of a U.S. minority group, such as Asian Americans, American Indians, African Americans, or Hispanic Americans[13]—generally fared less
well than businesses overall, both in their share of Phase I and Phase II grants and
in their application approval rates. However, disadvantaged or minority-owned

---

[12]The results of a paired t-test was $t = -1.3$ with 10 degrees of freedom, which was not less than
$t = -1.812$ at a level of significance of 0.05. However, the probability of getting this result if there
were no difference in success rates is only 0.224.

[13]In comparing minority- and woman-owned companies, it should be noted that there is double
counting in the data: Namely, some minority applicants and grantees are also women. These firms
are identified in the underlying database, but they are separately identified here only if specifically
stated as such.

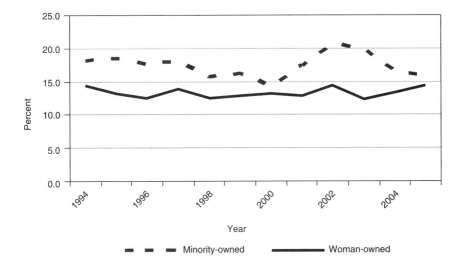

FIGURE 4.2-14 Share of Total Phase I Proposals from Minority- versus Woman-owned Companies, 1994–2005.
SOURCE: Developed from data provided by the NSF SBIR program.

businesses had a larger share of proposals (see Figure 4.2-14), a larger share of grants (see Figure 4.2-15), and a higher application approval rate than did woman-owned businesses (see Figure 4.2-16).

The shares of Phase I, Phase II, and Phase IIB grants that have gone to minority-owned businesses (shown in Figure 4.2-17) are slightly higher than the shares for woman-owned firms. The patterns are similar, as may be seen by comparing Figure 4.2-11 with Figure 4.2-17.

Figure 4.2-18 shows that minority-owned businesses, like woman-owned businesses, had a consistently lower application approval rate for Phase I proposals than businesses overall. This difference was found to be strongly statistically significant.[14] The annual success rate was also lower in most years for these companies in the Phase II grant process, but statistical tests of significance showed these differences to be less robust than those at Phase I. The null hypothesis—that there is no statistical difference in success rates at Phase II—could not be rejected.[15]

---

[14]The results of a paired t-test was t=−6.16 with 11 degrees of freedom, which was significantly less than t=−1.795 at a level of significance of 0.05.

[15]The results of a paired t-test was t=−1.43 with 10 degrees of freedom, which was not less than t=−1.812 at a level of significance of 0.05. However, the probability of getting this result if there were no difference in success rates is only 0.184.

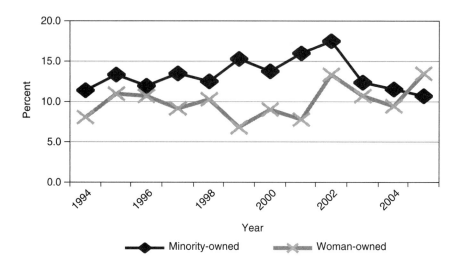

FIGURE 4.2-15 Share of Phase I Grants to Minority- versus Woman-owned Companies, 1994–2005.
SOURCE: Developed from data provided by the NSF SBIR program.

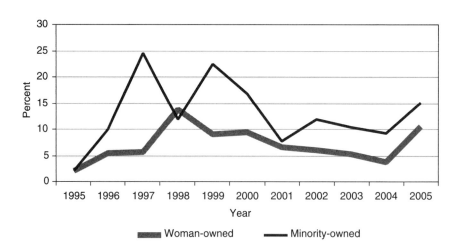

FIGURE 4.2-16 Share of Phase II Grants to Minority- versus Woman-owned Companies, 1995–2005.
SOURCE: Developed from data provided by the NSF SBIR program.

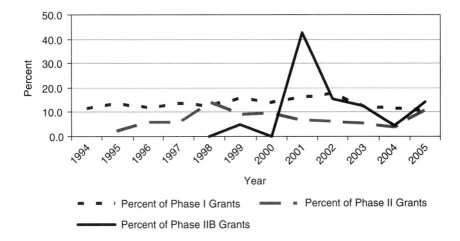

FIGURE 4.2-17 Shares of Phase I, Phase II, and Phase IIB NSF SBIR Grants to Minority-owned Businesses, 1994–2004.
SOURCE: Developed from data provided by the NSF SBIR program.

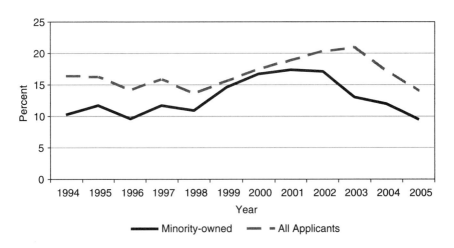

FIGURE 4.2-18 Phase I Success Rates for Minority-owned Businesses and for All Applicants, 1994–2005.
SOURCE: Developed from data provided by the NSF SBIR program.

**Box A
T/J Technologies, Inc.:
Bringing Nanoenergy Technologies to Market**

An innovative materials research company facing a long path to commercialization, T/J Technologies has used the SBIR program and other federal grants to get started and to build the firm's technological capacity and commercial potential.

Maria Thompson started T/J Technologies in 1991 with her husband by taking over a Department of Defense SBIR Phase I grant from a company that was divesting itself of its nonessential contractual obligations. From its initial work in ultrahard coatings, this minority-owned company moved on to developing ultracapacitors with the aid of additional SBIR grants. These grants, in turn, allowed T/J Technologies to develop advanced materials and devices for electrochemical energy storage and conversion. The company now holds several patents in the area of nanomaterials for alternative energy technologies, and it is recognized for its technologically advanced batteries, ultracapacitors, and fuel cells.

These advanced energy technologies power automobiles and other equipment more efficiently and safely. Novel nanomaterials in their Li-ion batteries allow for higher rates of performance, longer life, lower cost, smaller size, and lighter weight than similar batteries. Moreover, because these Li-ion batteries do not contain cobalt, they are far safer environmentally.

SBIR has been an integral part of T/J Technologies' building block strategy, says Maria Thompson. SBIR grants enabled the company to start and build its initial capacity. This in turn enabled the company to pursue government research contracts, which in turn makes it possible for the company to form commercial partnerships with much larger firms to test and demonstrate the commercial potential of its materials.

Table 4.2-6 summarizes SBIR applications from and grants for minority-owned businesses from 1995 through 2005. Overall, minority-owned businesses submitted 15.3 percent of all proposals and received 11.9 percent of all grants. Minority-owned businesses had 18 percent of their proposals approved (549/2,971), while all businesses had 21 percent of their proposals approved (4,053/19,357).

Women and Minorities as Principal Investigators. The NRC Phase II Survey provided data showing the frequency with which women and minorities serve on projects as the principal investigator—essentially, its technical leader. The survey found that 79 percent of projects surveyed had white males as principal investigators, while 21 percent had women, minorities, or both. Of these, 12 percent were women, 13 percent were minorities, and 4 percent were overlapping, i.e., minority women.

TABLE 4.2-6 Aggregate Proposals and Grants, Minority-owned and All Businesses, 1995–2005

Businesses	Proposals				Grants			
	Phase I	Phase II	Phase IIB	Total	Phase I	Phase II	Phase IIB	Total
A. Minority Businesses	2,950	296	21	2,971	383	145	21	549
B. All Businesses	16,897	2,299	161	19,357	2,833	1,059	161	4,053
Percent (A/B)	16.9	12.9	13.0	15.3	13.5	13.7	13.0	11.9

SOURCE: Developed from data provided by the NSF SBIR program.

NOTE: In this study, the terms "minority" and "disadvantaged" are used interchangeably.

4.2.6 Multiple Grant Winners and New Grant Winners

A company may receive multiple grants to develop multiple technologies aimed at multiple application areas, using a variety of commercialization strategies. An example provided by Luna Innovations, Inc., illustrates how receiving multiple grants is not synonymous with being a "grant mill," i.e., simply existing off multiple SBIRs without pursing any path to commercialization.

Luna, for example, has received SBIR grants from the NSF, Air Force, Army, Navy, EPA, DOC/NIST, USFA/DOA, DNA/DSWA/DTRA, DOT, and OSD. It has received multiple Phase I, Phase II, and Phase IIB grants from the NSF's SBIR program. The company has used SBIR grants to help move innovative ideas along a pipeline from which Luna has launched products and businesses. For every $1 of SBIR funding awarded, Luna reportedly has generated $2 in non-SBIR investment funding.

Luna has spun off six companies pursuing diverse technologies including optical test instrumentation, oil and gas sensors, wireless sensing, proteomics and clinical diagnostics, cancer inhibitors, and carbon-based nanomaterials. Recognized for its technical and business acumen, the company received the SBA Tibbetts award three times, most recently in 2006. It was named one of the 500 fastest growing companies in the United States in 1998. Founded in the early 1990s, Luna was listed on the NASDAQ in June 2006.[16]

Now, let us consider NSF's multiple-grant winners from a statistical perspective. Limitations are that at the time of this study records had been digitized only back to the early 1990s; furthermore, the classifications did not take into account grants that companies may have received from the SBIR programs of other agencies. As the Luna example showed and as the case studies found, companies typically participated in the SBIR programs of multiple agencies. Thus, it should

---

[16]This account of Luna was condensed from a draft case analysis developed by David Finifter, Professor of Economics and Public Policy, The College of William & Mary.

**Luna Innovations, Inc., Demonstrates the Use of
Multiple Awards from Multiple Agencies for
Multiple Technologies taken to Market along Multiple Paths**

- Founded in 1990 as Fiber and Sensor Technologies and renamed in 1993 as Luna Innovations, Inc.
- Listed on the NASDAQ [LUNA] in June 2006.
- Market capitalization of $35 million.
- Employee growth from 5 to 185.
- Spun off six additional companies.
- Diverse technologies in optical devices, advanced materials, and life sciences.

---

be noted that companies that received only one SBIR grant from the NSF may have received grants from other agencies.

Table 4.2-7 indicates for each year, 1996–2003, the percentage of Phase I grants each year that went to companies who had already received a grant that year, that is, to each year's multiple winners. The percentage ranges from 12.4 to 17.9 percent.

Most Phase I grant winners over the period covered received only one Phase I grant from the NSF, and less than 6 percent received more than five Phase I grants. Figure 4.2-19 shows the distribution of 1,618 company grant recipients over this same period by the number of Phase I grants they received. Ninety-four percent received one to five grants. At the top end, less than 1 percent (0.3 percent) received 31 to 40 grants. A single company received 40 grants over the period.

Figure 4.2-20 shows how multiple-grant winners affected the distribution of Phase I grants from 1996 to 2003. The 94.4 percent of companies that received

TABLE 4.2-7 Percentage of Phase I Grants Going to Multiple-Grant Winners Within a Given Year, 1996–2003

Fiscal Year	Percent of Grants to Multiple-Grant Winners
1996	17.9
1997	17.6
1998	14.9
1999	17.8
2000	12.0
2001	14.6
2002	18.9
2003	12.4

SOURCE: NSF SBIR program.

NOTE: In this table, "multiple-grant winners" is defined as having received more than one grant within the designated year.

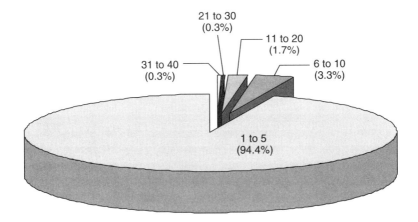

FIGURE 4.2-19 Distribution of Phase I Grantees by Number of Phase I Grants Received, 1996–2003.
SOURCE: Developed from data provided by the NSF SBIR program.

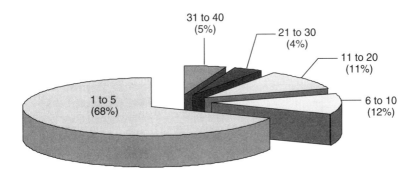

FIGURE 4.2-20 Distribution of Phase I Grants to Companies by Number of Grants Received, 1996–2003.
SOURCE: Developed from data provided by the NSF SBIR program.

between one and five grants as a group received only 68 percent of Phase I grants. Meanwhile, less than 6 percent of the multiple-grant winners received 32 percent of Phase I grants.

One company with six Phase I grants had all six followed up by Phase II grants. But all other multiple Phase I winners were less than 100 percent successful in having their Phase I grants followed by Phase II grants. The top nine companies in terms of Phase I grants received had 50 percent or more of their Phase I grants followed by Phase II grants. Eight of the top twenty Phase I grant-winners were not among the top 20 Phase II winners. Thus, there were multiple

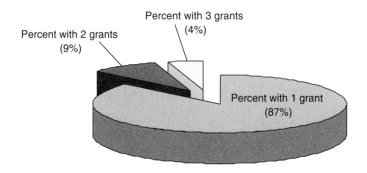

FIGURE 4.2-21 Distribution of Grantees by Number of Phase IIB Grants, 1998–2005.
SOURCE: Developed from data provided by the NSF SBIR program.

TABLE 4.2-8 Percentage of Phase I Winning Companies Who Had Not
Previously Received an SBIR Grant from the NSF, 1996–2003

Fiscal Year	Previous Winners	New Winners	Percent New Winners
1996	98	109	52.7
1997	99	116	54.0
1998	83	100	54.6
1999	113	81	41.8
2000	97	108	52.7
2001	92	95	50.8
2002	99	133	57.3
2003	138	236	63.1

SOURCE: NSF SBIR program.

Phase II grants to certain companies, but the concentration was much less than
at the Phase I level. For example, the maximum number of Phase II grants given
to a single company over the period 1996–2003 was twelve, which went to the
company that received forty Phase I grants.

As Figure 4.2-21 shows, 87 percent of Phase IIB grantees received only one
Phase IIB grant from the time the program started in 1998 through 2005. Nine
percent had received two grants; 4 percent had received three grants. Thus, the
majority of Phase IIB grants went to grantees that had never won one before.

New Grant Winners. In recent years, 42 to 63 percent of grant recipients
had never received an SBIR grant before.[17] As may be seen in Table 4.2-8, in most
years more than half the grant recipients were new to the program.

---

[17]Because the NSF SBIR computerized database is limited, these data on new winners do not cover
the entire period of the program's operation.

4.3 PROGRAM ORGANIZATION AND STRUCTURE

4.3.1 Organization

The organization chart in Figure 4.3-1 shows how the SBIR program fits within the broader structure of the NSF. The SBIR program is housed within the Directorate for Engineering, one of seven directorates and three offices comprising NSF. The SBIR program is grouped with the Small Business Technology Transfer (STTR) program, the Innovation and Organizational Change (IOC) program, and the Grant Opportunities for Academic Liaison with Industry (GOALI) program, under the designation "Industrial Innovation Programs" within the Office of Industrial Innovation (OII). Until recently, the OII reported to the Directorate for Engineering's Division of Design, Manufacture and Industrial Innovation, as shown by the solid black arrows in the figure. Now, the OII reports to the assistant director of the Directorate for Engineering, as indicated by the dotted red arrow.

NSF SBIR program management is centralized both organizationally and physically. The program's team is contained within one unit, the staff is relatively small, and team members are all on the fifth floor of NSF headquarters in Arlington, Virginia. There appears to be a strong degree of homogeneity within the program; indeed, it is a single program, unlike the SBIR programs at some of the other agencies in this study. According to program administrators, "Centralized management allows consistency within the program, efficient follow-up and tracking of proposal and grant management, efficient administration and technical resource management, better resource accountability, and consistent and timely information exchange."[18]

From the mid-to-late 1990s, the current centralized management replaced an earlier structure in which the program was largely distributed across the various NSF directorates and offices. Under the decentralized structure, the various NSF divisions had more of a stake in the program and could advocate for SBIR solicitations and grants that helped their fields. With the centralization, NSF program managers outside the program have much less involvement in the program. A potential effect may be that they are less supportive of the program.[19]

4.3.2 Staffing

Table 4.3-1 lists the SBIR/STTR management team, effective as of early September 2005. (Subsequent changes are not reflected.) In addition to the director and senior advisor, there were a total of seven program managers at that time. When program managers and assigned technical areas were examined

---

[18]"NSF SBIR Response to NRC Questions," January 2004.
[19]A potential topic of study would be the impacts of centralizing NSF's SBIR management.

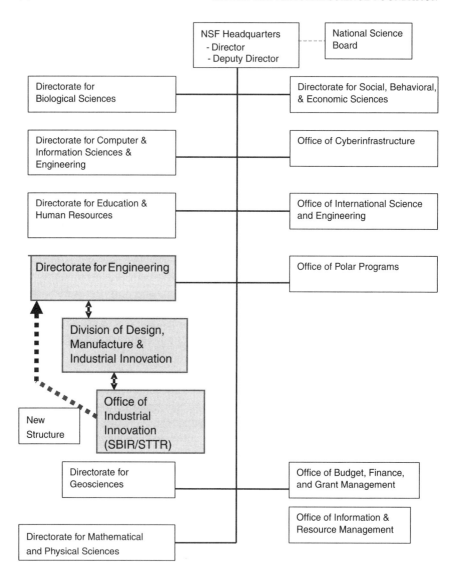

FIGURE 4.3-1 Organization Chart.
NOTE: The up-down black arrows represent the previous reporting path for the OII. The dotted red line represents the reporting path for the OII under the NSF's reorganization. The diagram may not accurately represent all of NSF postreorganization; the focus here is only on the change for OII.

TABLE 4.3-1 The NSF SBIR/STTR Team as of September 2005

Director, OII	Kesh Narayanan
Senior Advisor	Joseph Hennessey
Program Managers	Rosemarie Wesson
	Errol Arkilic
	Juan Figueroa
	Muralidharan Nair
	Sally Nerlove
	James Rudd
	Om Shai
BT Expert	George Vermont
Administrative Officer	Sandra Loving
Program Assistants	Caroline Hayer
	Jacqueline Moore
Contract Program Manager	Sonya Lucas
Contract Records Manager	Carla Lucas

SOURCE: The NSF's Web site (as of September 2005): <*http://www.nsf.gov*>.

NOTE: The current staff directory shows a number of changes, among them the listing of seven "Program Directors," six "Program Managers," and three "Experts." See <*http://www.nsf.gov/staff/sub_div.jsp?org=IIP&orgID=115*>.

,

initially in 2003, several additional program managers were listed,[20] and program managers were more clearly identified with specified technical areas than in 2004 or 2005. For example, in 2003, three program managers were listed for Advanced Materials, Manufacturing, and Chemical Processes; two were listed for Information-based Technologies and Electronics; and one was listed for Biotechnology. More recently, the program managers are posted on the NSF's SBIR program Web site as a group without technical area affiliation. Posted topic solicitation areas do identify program managers as points of contact.[21]

4.4 DESCRIPTIVE OVERVIEW OF THE NSF'S SBIR PROGRAM

4.4.1 Primary Program Objectives

According to the NSF, the primary objective of its SBIR/STTR program is "to increase the incentive and opportunity for small firms to undertake cutting-

---

[20]One program manager recently retired; another recently took another position within the NSF.

[21]In 2004 a change in posted program managers associated with Phase I areas occurred, with Electronics closed and Security newly opened. OII's strategic plan of June 2005 made further changes. (Program topics are discussed in more detail in Chapter 8.)

edge, high-risk, high-quality scientific, engineering, or science/engineering education research that would have a high potential economic payoff if the research is successful."[22] A recent strategic planning exercise has revisited and elaborated the OII's vision, mission, and goals.[23] Statements in the strategic plan emphasize accelerating industrial innovation by leveraging scientific R&D by small businesses so as to contribute to the U.S. economy. The NSF's new strategic plan acknowledges that to date "very few efforts have been made to foster and encourage participation by minorities and disadvantaged persons in technological innovation" and expresses the need to take action toward this objective.[24]

4.4.2 Program Phases

The NSF's SBIR program has the following phases: Phase I, Phase II, Phase IIB, and Phase III. Participation in Phase I is requisite to participation in Phase II, and participation in Phase II is requisite to obtaining a Phase IIB grant. Phase III, a phase without funding from the NSF, centers on commercialization and can start at any time following Phase I.

To reiterate, Phase I grants give up to $100,000 for an experimental or theoretical investigation of a proposed innovative research topic or activity over a six-month period. Phase II grants build on the Phase I research by funding additional research for up to two years. Phase II incorporates a reassessment of scientific, technical, and commercial merit and feasibility and provides funding up to $500,000. Phase IIB provides supplementary funding ranging from $50,000 to $250,000 for an extra year of research added to the existing Phase II period. A "supersized" version of Phase IIB, begun recently, provides up to $250,000 more and another year of funding. For every one dollar of Phase IIB funding sought, the applicant must show that it has obtained two dollars of financing from another source.

4.4.3 Use of Topics

The NSF's SBIR program solicits proposals in broad topic areas. From its beginning until the mid-1990s, the program operated in a decentralized manner, and its solicitation topics were closely aligned with the scientific and engineering disciplines of the NSF's scientific and engineering directorates. During this time there were 27 topics and over 100 subtopics reflecting the scientific interests of the directorates. After it was centralized, the program reduced the number of topics and subtopics and defined them in ways it thought to be more in tune with interests of the business sector. Until 2004, the five major topic areas were

---

[22]From the NSF's synopsis of its SBIR/STTR program found on its Web site, <*http://www.nsf. gov/funding/pgm_summ.jsp?Phase Ims_id=13371&org=DMII*>, in 2005.

[23]OII's strategic plan, pp. 7–8.

[24]Ibid, p.18.

(1) Advanced Materials and Manufacturing, (2) Biotechnology, (3) Chemical-based Technologies, (4) Electronics, and (5) Information-based Technologies. In 2004 a sixth topic, Security, was added.

OII's 2005 strategic plan explains that the solicitation topics now fit into three broad areas: (A) investment business focused topics, (B) industrial market driven topics, and (C) technology in response to national needs. (A, B, and C below). The topics list given in the strategic plan includes seven topics:

A. Investment Business Focused Topics
    1. Biotechnology (BT)
    2. Electronics-Technology (EL)
    3. Information-Based Technology (IT)
B. Industrial Market Driven Topics
    1. Advanced Materials and Manufacturing (AM)
    2. Chemical-Based Technology (CT)
C. Technology in Response to National Needs
    1. Security-Based Technology (ST)
    2. Manufacturing Innovation (MI)

It appears that areas A and B above are expected to be relatively stable and that area C is expected to change more frequently on short notice as national needs change.

In 2006 the Phase I solicitation (number 06-553) included one topic from each of categories A, B, and C: Advanced Materials, Information Technology, and Manufacturing Innovation. It included an additional topic, "Emerging Technologies" (also called "Emerging Opportunities"[25]) not included in the above list. This additional topic had a focus on near-term commercialization and was limited to a list of specific software and hardware areas.[26] The FY2007 Phase I solicitation, which began accepting proposals on November 4, 2006, and closed on June 13, 2007, includes Advanced Materials, Chemical Technology, and Manufacturing Innovation, Biotechnology, Electronics, Information Technology, and Emerging Opportunities. This inclusion covers all the topics listed in the strategic plan, with the exception of Security-Based Technology, possibly indicating a broadening of topic coverage. Phase II submission deadlines are tied to the expiration dates of the Phase I grants.[27]

Topics are used in the Phase I solicitations. Once Phase I grants are selected, the topics of these grants determine the "topic pool" from which Phase II grants

---

[25]SF SBIR Program Manager Errol Arkilic discussed the NSF's use of the Emerging Opportunities topic, reportedly with the use of a higher percentage of reviewers with strong commercial credentials, in an interview with Robert Jaffe, "Words from a Winner," SBTDC/SBIR Newsletter, October 2005.

[26]The topic area "Emerging Technologies" and the subtopics it includes were described in 2006 at the following NSF Web site: <*http://www.nsf.gov/eng/sbir/eo.jsp*>.

[27]Information on the FY2007 Phase I solicitation is provided at <*http://www.nsf.gov/eng/iip/sbir*>.

Box B
Nanosys: High-Performance Inorganic
Nanostructures for Solar Cells

Founded in 2001, Nanosys develops products based on a technology platform incorporating high performance inorganic nanostructures. Applications of this patented technology extend to a variety of industries, including energy, defense, electronics, computing, and the life sciences. One product with potential for significant commercial application is a new type of solar cell that is lightweight, flexible, and can be produced in large volumes using roll-to-roll mass manufacturing techniques. A practical solar cell technology that is low cost and widely available can contribute to the world's growing energy needs without impacting the environment.

In 2003 Nanosys applied for and received a grant from the NSF's SBIR program to further their efforts to develop and commercialize a "novel nanocomposite solar cell technology that combines self-assembled inorganic semiconductor nanocrystals into a plastic composite to produce lightweight, flexible solar cells of almost any size and shape." As the Nanosys leadership recognizes, the SBIR award confers the "stamp of approval" of the nation's leading science agency that acknowledges both the commercial value as well as the scientific importance of the work. This validation is helping Nanosys develop additional funding from private venture capital, further accelerating the development and commercialization of this technology.

---

are selected. In turn, the selection of Phase II grants determines the topic pool from which Phase IIB grants are made. If a topic is not included in Phase I, it is also not included in the subsequent Phase II and IIB solicitations.

The distribution by topic area of Phase IIB grants from 1998 to 2005 is shown in Figure 4.4-1. It shows that these supplemental grants funded projects in just four topic areas: Advanced Materials, Information Technology, and, to a lesser extent, Electronics and Biotechnology. (For additional discussion of topic development, see Section 8.1.)

4.4.4 Proposal Selection Criteria and Process

Proposal selection is driven by the following two criteria:

* Intellectual merit
* Broader impacts

These criteria are explained in program guidance and will be addressed in more detail in Section 8.3. Peer review is used to assess each proposal against the criteria. These reviews provide advice to SBIR management in selecting grantees.

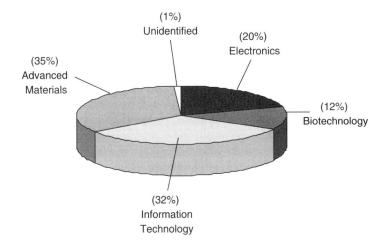

FIGURE 4.4-1 Phase IIB Grants by Topic, 1998–2005.
SOURCE: Developed from data provided by the NSF SBIR program.

Additional, less well-specified criteria are reportedly applied by program staff in the final selection decisions. (For more on selection criteria and process, see Section 8.3.)

4.4.5 Solicitations and Proposal Submissions

NSF solicitation schedules are published on its Web site. In 2004, for example, a Phase I solicitation was released on March 1, 2004, for two topics— Information-Based Technologies and Security Technologies. Proposals were due on June 9, 2004. A round of Phase II proposals for those topics for which Phase I projects were under way was due on July 27, 2005. Phase IIB proposals are due in 2007.

All proposals must be submitted via FastLane, an interactive real-time system used to conduct NSF business over the Internet.[28] The program reportedly adheres to a six-month processing time for all proposals.[29] (For more on the timing of solicitations, proposal submissions, and grant awards, see Section 8.6.)

4.4.6 Support for Commercialization

Because the NSF is not generally a procurer of the technologies it funds, it requires grantee companies to identify other routes to commercialization. The

---

[28]For more about FastLane, see <*http://www.fastlane.nsf.gov/fastlane.jsp*>.
[29]Draft of OII's strategic plan (June 2005), p. 18.

program emphasizes commercialization and begins educational support for commercialization during Phase I. All grantees are required to attend a business development and training workshop during Phase I. Phase II grantees meet annually, where they are briefed on Phase IIB opportunities and requirements. The annual conference provides workshops intended to assist with commercialization and opportunities to network with potential funding sources and partners. The Phase IIB program provides additional grant funding predicated on a grantee's ability to obtain third-party funding.

The NSF's SBIR program operates a Matchmaker program, which seeks to bring together SBIR recipient companies and potential third-party funders such as venture capitalists. Just recently, the NSF has begun to support participation of its grantees in SBIR Opportunity Forums, run jointly with other agencies, to bring grantees together with potential investors and partners.

Despite these efforts, commercialization is difficult and the program faces challenges in achieving commercial results. (Commercialization efforts and results are discussed in more detail in Chapter 5.)

4.5 NSF SBIR "SUCCESS STORIES"

Like many federal agencies, the NSF regularly publishes "success studies" (also called "nuggets" by NSF). Exhibits 4.5-1 and 4.5-2 provide SBIR success stories from NSF files. Note that these are success stories as defined by the NSF; the actual longer-term success of these projects has not been verified by this study.

Exhibit 4.5-1
SBIR Success Story from the NSF's
Office of Legislative and Public Affairs.
News Tip; March 20, 2003
Web's "Best Meta-Search Engine" Organizes
Documents from Anywhere in Any Language

Industry experts at Search Engine Watch recently named Vivisimo the Web's Best Meta-Search Engine for its ability to organize search results instantly into a computer-generated "index." The software behind Vivisimo's search engine can also be applied to any collection of documents, in languages ranging from English and German to Arabic and Korean.

A success story from the National Science Foundation's (NSF) Small Business Innovation Research and computer science programs, Vivisimo's Web site was recognized for the second consecutive year for its "outstanding performance in helping Internet users gather results from many Web search engines by using a single service."

The Vivisimo Web site demonstrates how the technology filters and automatically categorizes responses from search requests. The results resemble a human-generated index that can help guide researchers in the right direction.

"The clustering service on our Web site and underlying software technology show how users can comfortably explore much more information in an organized way, rather than being bombarded with disorganized information dumps," said Raul Valdes-Perez, president of Vivisimo, Inc. "Our Web site shows our business customers—whether Web, government, corporate, or publishing—what they can expect by installing our software products for their uses."

Getting answers to broad, exploratory questions can leave searchers, especially those with little knowledge about a topic, slogging through a morass of information. For example, searching for "Iraq" among the news stories on any Web news source will result in a long list of articles on global politics. Searching the entire Web can produce a similar, mostly undifferentiated list of sites about Iraq.

This is where Vivisimo steps in. Its Clustering Engine does a quick statistical, linguistic, and knowledge-based analysis of the search results which it then clusters into themes, thereby helping to identify trends or fine-tune searches without requiring users to know the correct terminology. For example, using Vivisimo to search news sites for "Iraq" might produce clusters of news articles under categories such as "weapons Inspectors," "Bush," "missiles," and so on. (The categories will change depending on the latest developments in the news.)

Vivisimo is supported by NSF's SBIR program, which emphasizes high-risk, high-payback innovations that are tied to NSF's mission of advancements in science, engineering, and education. All proposals are evaluated on the technical merit of their research and development, as well as their technology impact. NSF was the first of 10 federal agencies required to reserve a portion of their research and development funds for the SBIR program, which is coordinated by the U.S. Small Business Administration.

Related Links: Vivisimo: *<http://www.vivisimo.com>*, Vivisimo News demo: *<http://vivisimo.com/news/>*, NSF SBIR Program: *<http://www.eng.nsf.gov/sbir>*.

NSF Program Officers: Juan E. Figueroa, SBIR (703) 555-1212, *jfiguero@nsf.gov*, Ephraim P. Glinert, CISE (703) 555-1212, *eglinert@nsf.gov*, Principal Investigator: Raul Valdes-Perez, *valdes@vivisimo.com*, 412-422-2496

Exhibit 4.5-2
GPRA Fiscal Year 2004 "Nugget"
Retrospective Nugget—AuxiGro Crop Yield Enhancers

Emerald BioAgriculture Corporation (formally Auxein Corporation) of Lansing, MI, more than doubled sales of its commercial crop yield enhancer, AuxiGro®, over the past year. In research conducted under NSF's Small Business Innovation Research (SBIR) program, Emerald Bio developed a crop yield enhancer based on GABA (gamma aminobutyric acid). Currently, the product is used on field crops such as potatoes and sugar beets, high-value vegetables, grapes, snap beans, fruit trees, almonds, and grass for seed production.

Primary benefits of AuxiGro include increased yield on crops such as almonds and tomatoes, increased size and hence value on onions and potatoes, and increased sugar (Brix) content in crops such as grapes and sugar beets. Stress-protective benefits have also been realized with use of AuxiGro. When applied to cabbage, the plant can withstand extended exposure to freezing temperatures after treatment. When applied to grapes, the plant is better able to resist infection from powdery mildew disease. Since the active ingredient in AuxiGro is an amino acid that is naturally present in all living systems, AuxiGro can be considered a natural alternative to synthetic plant growth regulating compounds and in certain cases to synthetic chemical fungicides.

Environmental stresses such as drought, temperature extremes, and disease are important factors limiting agricultural productivity. Because crops that have been treated with AuxiGrow are better able to withstand a variety of stresses, they show improvements in many aspects of plant growth including flowering, fruit set, and root growth. The beneficial effects on crop growth provide growers with substantially improved yields. For every $1 that growers spend on AuxiGro they typically receive on average $35 in additional yield value from onions and $15 and $16 in extra yield value from snap beans and almonds, respectively.

AuxiGro is now being sold in the USA and countries in Europe, Asia, and Latin America. More than $8 million in private equity investment in Emerald Bio has followed the SBIR research support to date.

This work is notable because:
This research has developed GABA technology for large increases in agricultural productivity.

Program Officer: Ritchie Coryell
PI Name: Alan Kinnersley
Institution Name: Auxein Corporation
Grant Title: SBIR Phase II: "The Role of Gamma Aminobutyric Acid (GABA) in Plant Growth and Productivity"
Entered by Joann L. Alquisa

Approved for ENG by Joanne D. Culbertson

5

Commercialization

5.1 COMMERCIALIZATION STRATEGIES

The National Science Foundation (NSF) appears to have recognized early on that addressing commercial and financial issues throughout the initial stages of technology development—concurrently gathering information about markets, potential customers, competitors, strategic direction, and finance—is critical to effective and timely commercialization. This view has been "long recognized" by American industry.[1] It is a view also strongly embraced by other, more recently created, public-private partnership programs such as the Advanced Technology Program (ATP).[2]

The NSF's relative aggressiveness in encouraging early attention to business issues is particularly noteworthy. A likely reason for this early attention is the fact that, unlike the Department of Defense and other agencies, the NSF itself does not generally provide a market for the technologies it funds through the Small Business Innovation Research (SBIR) program. It likely also reflects the business training of the founders of the NSF program. Early on, the NSF's SBIR program specified a proposal selection criterion that relates to commercialization. It set forth the six-part "Commercialization Plan Guidelines" and in 2003 began requiring that grantees' commercialization plans have more fully developed financial projections at the beginning of Phase II. It has required all SBIR Phase I grantees

---

[1]Dawnbreaker, Inc., makes this point, citing the work of Robert G. Cooper—*Winning at New Products: Accelerating the Process from Idea to Launch* (Basic Books, 2001)—in Dawnbreaker, Inc., "SBIR: The Phase III Challenge," white paper, 2005, pp. 10–12.

[2]The ATP is currently being phased out. It has been adopted as a model public-private partnership program by other countries.

to attend a commercialization planning workshop. Since 2001, it has engaged the services of Dawnbreaker, Inc., a company that specializes in providing commercialization assistance to small advanced technology firms, to help Phase I grantees to prepare NSF's required commercialization plan for Phase II submissions.[3] In addition, the NSF's SBIR program has encouraged commercialization at the Phase II stage by offering its Phase IIB supplemental option, conditioned on the grantee obtaining third-party financing. The Phase IIB supplement stands out as an innovative method of encouraging companies to attract funding from other sources—a critical step toward commercialization. The development and enhancement of the NSF's Web-based Matchmaker program to promote partnering may also encourage commercialization.

The NSF's SBIR program is using Phase IIB Supplemental Grants as an innovative method of encouraging companies to attract funding from other sources, which
  • **provides an incentive for firms to partner with the investment community,**
  • **can identify technologies with greater market potential, and**
  • **helps bridge the funding gap to commercialization.**

Management of NSF's SBIR, the Office of Industrial Innovation (OII), has taken another recent step to provide more commercialization assistance to its Phase II grantees. The NSF's SBIR program entered into a partnership with the Department of Energy's (DoE) SBIR program to jointly sponsor the 2005 DoE Opportunity Forum.[4] This forum brought selected SBIR grantees face-to-face with prospective investors and allowed them to present their commercialization opportunities to the investors.

In addition to the 2005 DoE Opportunity Forum, OII brings together Phase II grantees on a regular basis at its annual grantee conferences. For example, OII cosponsored a Phase II Grantee Conference on May 18–20, 2006, in Louisville, Kentucky, with the Ewing Marion Kauffman Foundation, a Kansas City-based

---

[3]According to Dawnbreaker, Inc., it is difficult to give extensive attention to business planning during Phase I, despite the fact that early business planning is desirable. Factors limiting what can be done include the following: the short time duration of Phase I; the fact that small-company resources are generally tight during this phase; the difficulty of attracting external funding during this early phase; and limits on available funding for commercialization assistance through the SBIR program.

[4]The 2005 DoE Opportunity Forum brochure, "Partnering and Investment Opportunities for the Future." The forum took place on October 24–25, 2005, in Tysons Corner, Virginia.

organization whose goal is to catalyze an entrepreneurial society.[5] The conference showcased nearly 300 grantees to potential industrial strategic partners and venture capital investors. It also enabled one-on-one meetings of OII program officers with grantees, compensating at least in part for the general inability of program managers to conduct site visits to grantee companies due to a shortage of program administrative funding. At this conference, business and financial panels met, firms had poster presentations and maintained tables or booths, and the grantees made presentations grouped by industry sector.[6]

The NSF funds many very small, scientist-led firms that find commercialization quite challenging. OII's strategic plan includes a number of additional initiatives aimed at firms needing assistance to encourage commercialization. OII hopes to: develop plans to work with incubators, business schools, and other resources; provide innovation management courses to grantees; revise Phase I requirements to include more commercial information upfront; bring more business reviewers into the Phase I process; and bring investors and corporate partners to grantee conferences and workshops. Thus, it appears that the NSF is formulating plans to continue and intensify strategies aimed at fostering commercialization.[7]

5.2 COMMERCIAL RESULTS

It seems clear that the NSF's SBIR program intends for its grants to result in commercial goods and services. But how well is it doing in achieving this goal? In this section we examine evidence of commercial results drawn from five sources: (1) a survey of grantee firms that focuses on firm characteristics; (2) a survey of Phase II projects to focus on projects that went forward; (3) a survey of Phase I projects to find out why some projects did not continue into Phase II; (4) case studies of ten companies that received NSF SBIR grants; and (5) NSF-initiated data and analysis on commercialization. While each section following focuses on each of these sources in turn, to some extent findings from the various sources are interwoven.

5.2.1 Characteristics of SBIR-Funded Firms as Indicated by NRC Firm Survey Data

Influence of the SBIR program on Company Founding. The NRC Firm Survey found that 20 percent of 137 respondents attributed the founding of their

---

[5]"About the Foundation," Ewing Marion Kauffman Foundation Web site, <*http://www.kauffman.org/foundation.cfm*>.

[6]National Science Foundation, Office of Industrial Innovation, "SBIR/STTR Phase II Grantee Conference, Book of Abstracts," May 18–20, 2006, Louisville, Kentucky.

[7]In 2003 the NSF expanded the Phase I commercialization assistance program and awarded three-year contracts to Dawnbreaker, Inc., and Foresight Science & Technology. In a new contract competition in 2006, the NSF awarded three-year contracts to Dawnbreaker and Development Capital Networks (DCN).

companies to the SBIR program in full or in part. This finding was supported by the case studies in which several companies attested to the important role of the SBIR program in either the creation of their company (e.g., Language Weaver) or the restarting of their company after they encountered setbacks (e.g., ISCA Technologies).

Previous Business Experience. The survey found that founders of SBIR-awardee firms tended to have previous business experience. Fifty-six percent of the founders of respondent firms had started one or more other companies, and 52 percent of the founders had a business background. Furthermore, most of the founders had been employed with another private company prior to founding the survey firm.

Grants Received. Most of the respondent firms (86 percent) had received more than one Phase I grant from federal agencies. The reported range per firm was 1 to 462 Phase I grants, and the average number received was 31. Fourteen percent had received only one grant; 32 percent had received from two to five grants; 8 percent had received more than 100 Phase I grants. For most of the firms, these grants were spread over a number of years.

Most of these firms also had received more than one Phase II grant from federal agencies, but fewer Phase II than Phase I grants. The reported range was 1 to 182 Phase II grants, and the average number received was 14—less than half the average number of Phase I grants. Twenty-three percent received only one Phase II grant; 44 percent received from two to five grants; 7 percent received more than 50 Phase II grants.

Company Growth. A large proportion of the firms surveyed attributed a considerable part of their companies' post-SBIR grant growth to the SBIR program. Forty-six percent attributed more than 50 percent of their growth to the SBIR program. Table 5.2-1 summarizes the responses to the relevant survey question.

Surveyed firms reported employment growth from the time they received their first Phase II grant to the time of the survey. Figure 5.2-1 shows how the surveyed firms tended to increase their number of employees from the time of the firms' first SBIR grants to the time of the survey. At the time they received their first Phase II award, 60 percent of the firms had 5 or fewer employees, whereas by the time of the survey, this percentage had been nearly cut in half. The average number of employees increased from 13 to 36, and the range increased from 1 to 175 employees to 1 to 750 employees. Three percent of the firms had more than 200 employees at the later time.

Firm Revenue. Figure 5.2-2 shows the distribution of surveyed firms by their total revenue. At the low end, 13 percent of firms had annual revenues under $100,000. At the high end, 4 percent of the firms reported $100 million or more

TABLE 5.2-1 Company Growth Attributed to the SBIR Program

Percentage Growth	Percentage of Companies Attributing Growth to the SBIR Program
< 25	27
25 to 50	28
51 to 75	24
>75	22

SOURCE: NRC Firm Survey.

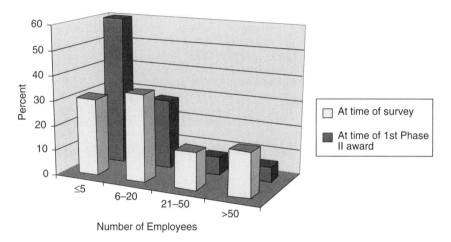

FIGURE 5.2-1 Firm Employment after First Phase II Award and at the Time of the Survey.
SOURCE: NRC Firm Survey.

in annual revenue. The largest group reported annual revenue of $1 million to $5 million.

Thirty-six percent of all the surveyed firms reported that more than half of their company's revenue during its last fiscal year was comprised of SBIR and/or STTR funding. Forty-two percent reported that SBIR and/or STTR funding comprised 10 percent or less of revenue in the company's last fiscal year, and 28 percent reported that it comprised 0 percent of company revenue.

Business Activity. Four of the firms in the survey sample had made an initial public stock offering—one in 2004, one in 2000, one in 1994, and one in 1983. Two more planned an initial public stock offering in 2005/2006. Eighteen percent of the surveyed firms had established one or more spin-off companies, for a total of 49 new spin-off companies.

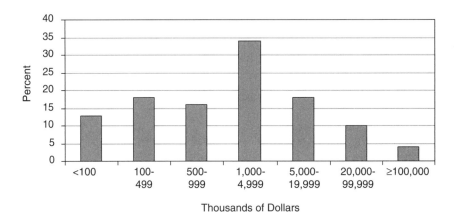

FIGURE 5.2-2 Distribution of Surveyed Firms by Annual Revenue.
SOURCE: NRC Firm Survey.

NSF-surveyed firms tended to be proactive in assessing the potential commercial markets for their SBIR products, processes, or services—slightly more so than the total of surveyed firms for all five of the agencies examined. Sixty-nine percent reported that they first determined potential commercial markets prior to submitting their Phase I proposal, and another 22 percent prior to submitting the Phase II proposal. Only 1 percent reported waiting until after Phase II.

The survey results revealed that market research/analysis in these firms is carried out by a variety of people, including the director of marketing, employees for whom marketing is their primary job, employees who take on marketing in addition to their regular duties, consultants, the principal investigator, and the company president or CEO. However, the company president or CEO was most often responsible for market research and analysis.

Similarly, sales were handled in various ways by the firms, as illustrated by Table 5.2-2. Corporate officers were most often responsible for sales, but a substantial percentage of firms surveyed reported having an in-house sales force. Firms designated as NSF grant recipients were more likely than firms in the total survey to achieve sales through the use of licensing, independent distributors or other companies with which they had formed marketing alliances, other companies that incorporate the product into their own, and spin-off companies.

Firm R&D Devoted to SBIR Activities. Thirty-six percent of the firms reported that more than half of their total R&D effort was devoted to SBIR activities during the most recent fiscal year, and 16 percent reported that more than 75 percent was devoted to SBIR activities. Thus, the SBIR funded a substantial share of the R&D activity of many grant-recipient firms.

TABLE 5.2-2 Methods of Accomplishing Sales of Product, Process, or Service

Method of Accomplishing Sales	Percentage of 130 Survey Respondents Using This Method
Corporate officers	50
An in-house sales force	39
Licensing to another company	39
Independent distributors or other company with which they have marketing alliances	36
Other company(ies) which incorporate product into their own	34
Other employees (other than corporate offices and sales force)	28
Spin-off company(ies)	9
None of the above	8

SOURCE: NRC Firm Survey.

NOTE: More than one method could be selected; thus, percentages do not add up to 100 percent.

Protection of Intellectual Property. The 137 firms in the survey reported 842 patents resulting at least in part from their SBIR and/or STTR awards, for an average of 6 patents per firm related at least in part to SBIR awards. This was double the average reported for the total firm survey sample covering all five agencies. The range for the firms in the survey identified as recipients of NSF SBIR awards was from 0 to 66 patents per firm, with 26 percent reporting no patents from SBIR/STTR and 28 percent reporting more than six patents each. This result is distinct from later patent results (from the Phase I and Phase II Surveys) that relate patents to the technology developed with funding from an individual award. The average number of patents attributed to individual SBIR projects is substantially lower than the number reported for the firm as a whole as resulting "at least in part to SBIR awards."

5.2.2 Commercialization Progress Indicated by NRC Phase II Survey Data

The NRC Phase II Survey provides recent evidence on the extent to which SBIR grant recipients have achieved commercialization and/or progress toward commercialization. The survey provides information on sales, on modes of commercialization, and on steps important to achieving commercialization, including marketing activities, interactions with other companies and investors, and attraction of funding from non-SBIR sources. It also provides information on employment effects, including the extent to which women and minorities are involved in the projects as principal investigators. Finally, it explores the extent to which the reported effects are believed by survey respondents to be attributable to impacts

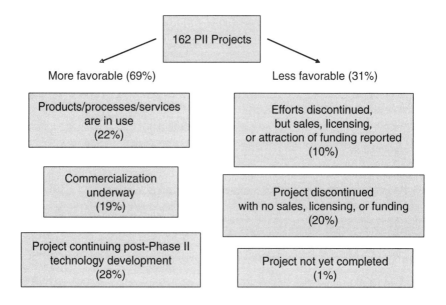

FIGURE 5.2-3 Commercialization Status of 162 Surveyed Phase II NSF SBIR Projects. SOURCE: NRC Phase II Survey.

to the SBIR program. Figure 5.2-3 summarizes the commercialization status of 162 Phase II NSF SBIR projects.

Projects Achieving Commercialization and Continuing Progress. Twenty-two percent of 162 respondents indicated that the referenced projects had resulted in products, processes, or services that were in use and still active. Fifty-one percent of the 162 respondents in the survey indicated some form of commercialization or progress toward commercialization of the technologies resulting from the referenced projects. Included in this 51 percent are those for which commercialization was under way (19 percent); those that had already achieved commercial results (22 percent); and those that had achieved sales, licensing, or additional funding before discontinuing the projects (10 percent). Another 28 percent of respondents indicated that the project was continuing technology development in the post-Phase II period. Altogether, 69 percent of respondents indicated some form of continuing progress—either in terms of commercial development (41 percent) or technology development (28 percent).

Average Sales Data. Reported sales incorporating the technology developed in the referenced projects were compiled from the 162 survey respondents. Average sales across the 162 projects were nearly $400,000, and average sales of associated licensees of the technologies exceeded $1.4 million. When respondents were asked to identify the types of sales either they or their licensees had experienced, 38 percent indicated sales of product, 22 percent, sales of services, and 3 percent, sales that involved processes.[8]

For projects with sales and those without sales who expected future sales from the technology developed in the referenced projects, respondents were asked to estimate the amount of sales expected for their company and its licensees between now (the time of the survey in mid-2005) and the end of 2006. Average expected sales across all 162 projects exceeded $800,000 for this time interval appropriate for the referenced projects.

These estimates of expected sales were based largely on educated guesses, ongoing negotiations, projections derived from current sales, and past experience. When asked the basis of the expected sales estimates, respondents said market research figured in only 26 percent of the future sales estimates. "Educated guess" was reported by 40 percent and "ongoing negotiations" was reported by 38 percent as the basis of the sales estimates. "Projection from current sales" was reported by 34 percent and "past experience" by 32 percent. In contrast, "consultant estimates" figured in 5 percent of the estimates.[9] For novel/disruptive technologies, educated guesses may be all that is available for market estimates. At the same time, program officials should be skeptical when Phase I applicants initially are very positive in their educated guesses of expected market size, closely followed by an educated guess that there is insufficient commercial promise to warrant a follow-on proposal or other effort.

As has been observed in other technology programs, a few projects often account for the majority of sales revenue.[10] This tendency was also observed in the SBIR survey results, where just eight of the projects—each of which had $2.3 million or more in sales—accounted for over half the total reported sales dollars for the surveyed projects. The project with the highest reported amount had $4.8 million in sales. Similarly, the results for sales by licensees of those survey projects' technologies were highly skewed by a single licensee that accounted

[8]These results are consistent with findings from a study of small companies funded by the Advanced Technology Program, which found that only 12 percent of applications of small firms in single-firm projects involved processes. ATP's inquiry cites the work of Utterback (1994) who noted that advances during early stages of innovation tend to be focused on a new product area, and later stages are more oriented to increased process efficiencies. According to the ATP report, "Small firms with limited, or no, existing product lines, such as is the case for the start-ups . . . are more likely to be addressing the earliest stages of innovation (i.e., those involving new products, or products with new capabilities) rather than process efficiencies." See Jeanne W. Powell, *Business Planning and Progress of Small Firms Engaged in Technology Development through the Advanced Technology Program*, NISTIR 6375, October 1999, p. 15.

[9]The source of these data is the NRC Phase II Survey.

[10]Venture capitalists report that, typically, a very few projects account for the bulk of their investment return. Similarly, the Advanced Technology Program reported a distribution of project successes that found a smaller percentage in the top than in the middle.

TABLE 5.2-3 Customers for Sales Derived for the Referenced Grants

Types of Customers	Percentage of Total Sales to Each Type of Customer
Domestic private sector	57
Department of Defense (DoD)	11
Export markets	11
Prime contractors for DoD or NASA	5
State or local governments	4
Other federal agencies	2
NSF	1
NASA	1
Support to universities and other institutions	9

SOURCE: NRC Phase II Survey.

for over half the total licensee sales dollars, amounting to $200 million or more in licensee sales.

Customers. Respondents reporting sales from the referenced NSF SBIR projects were able to identify their customers. As shown in Table 5.2-3, the majority of these sales (57 percent) went to the domestic private (civilian) sector. A total of 20 percent went to federal agencies or their prime contractors. Of this, 11 percent went to DoD and 5 percent to DoD or NASA prime contractors. Responses to a separate question revealed that 4 percent reported that the resulting technology was used in a federal system or acquisition program.

As would be expected given that the NSF is, for the most part, not a procurement agency, only 1 percent of total sales went to the NSF. Sales to export markets accounted for 11 percent of the total, and state and local governments accounted for 4 percent. Customers other than those listed accounted for the remaining 9 percent.

Mode of Commercialization. Respondents with sales or expecting sales were asked about their mode of commercializing technologies resulting from the referenced projects. Table 5.2-4 summarizes the responses. The most frequent answer was "as hardware" (54 percent), followed by "as process technology" (32 percent)[11] and "as software" (32 percent). The next most frequent responses were "as new or improved service capability" (24 percent) and "as a research tool" (21 percent). Twelve percent indicated the mode of commercialization to

---

[11]The previously noted low incidence of sales revenue reported from process technology (3 percent) seems in contrast to the reported 32 percent of respondents who reported their mode of commercializing to be process technology.

TABLE 5.2-4 Mode of Commercializing

Mode	Percentage of Respondents Choosing This Mode
As hardware (final product, component, or intermediate hardware product)	54
As process technology	32
As software	32
As new or improved service capability	24
As a research tool	21
As educational materials	12
As a biologic	3
No commercial product, process, or service was/is planned	2
As a drug	0
Other	8

SOURCE: NRC Phase II Survey.

be "as educational materials." "As a drug" (0 percent) was not an avenue for commercialization, but "as a biologic" (3 percent) was occasionally indicated. Eight percent of the modes used or expected to be used apparently fell outside the modes listed, and 2 percent responded that no commercial product, process, or service was/is planned even though they previously had indicated the contrary.

Marketing Activities. Asked about their marketing activities, respondents who had not discontinued their efforts without sales or additional funding indicated an emphasis on market planning, with approximately three-quarters reporting they had planned, had under way, or had completed preparation of a marketing plan. Their attention was also on market research and publicity and advertising, with about two-thirds of respondents indicating they had planned, had under way, or had completed these activities. About half the respondents also had planned, had under way, or had completed the hiring of marketing staff and test marketing, but a near equal percentage indicated these marketing activities were not needed.

Interactions with Other Companies and Investors. Asked about their activities with other companies and investors in the United States and abroad, respondents for 121 Phase II projects indicated the frequency with which they either had ongoing negotiations or had finalized agreements with other companies or investors. Overall, interactions with domestic firms and investors far exceeded interactions with foreign firms and investors.

> **Prevalent forms of interaction with other companies and investors, both domestic and foreign, were** licensing agreements, R&D agreements, marketing/distribution agreements, and customer alliances.

For many small companies, forming licensing agreements with other companies offers a path to commercialization that might not otherwise be possible due to large capital requirements and the need for an established market presence. Thus, it is not surprising that NSF SBIR grantees often form licensing agreements with other companies. The NRC Phase II Survey found that 20 percent of respondents had finalized licensing agreements with U.S.-based companies and investors and 21 percent had ongoing negotiations for licensing agreements with U.S.-based companies and investors. Ten percent had finalized licensing agreements with foreign companies and investors, and 7 percent had ongoing negotiations for licensing agreements.

That part of commercialization that is accomplished through licensing agreements tends to make the assessment of commercial progress more complicated than if the grantee firms commercialize solely by manufacturing and selling their own stand-alone products. Grantee innovations that are incorporated into the products of licensee companies may become difficult to assess separately. While royalty rates and fees paid to the grantee can be used to estimate the sales value of the product, sometimes licensing information is reportedly overlooked in agency assessments of commercial outcomes.

The next most frequently mentioned form of interactions of grantee respondents with other companies and investors—in decreasing order of frequency— were R&D agreements, marketing/distribution agreements, and customer alliances. Although few domestic sales of technology were reported finalized, a number of negotiations were under way. Foreign interaction centered on licensing, R&D agreements, and marketing/distribution agreements. Reports of interactions to sell technology rights or to either partially or totally sell the company to foreign companies and investors were few.

Attraction of Additional Funding Prior to Phase II. The NRC Phase II Survey identified sources of funding that preceded the referenced Phase II grants. Table 5.2-5 shows the percentages of respondents who indicated the referenced technologies received funding from each source, including prior SBIR grants. Thirty-five percent received prior funding primarily from internal company investment, including borrowed funds. Twenty percent received prior SBIR grants. Fifteen percent received prior funding from private companies other than venture capital, and 14 percent received prior funding from private investors. At

TABLE 5.2-5 Funding Sources for Research and Development of the Technology Prior to the Referenced SBIR Grant

Funding Source	Respondents Indicating Funding from Source Prior to Referenced SBIR Phase II Grant (%)
Internal company investment (including borrowed money)	35
Prior SBIR (excluding the Phase I which proceeded this Phase II)	20
Other private company (not venture capital)	15
Private investor	14
Prior non-SBIR federal R&D	13
Other	9
State or local government	5
College or university	4
Venture capital	2

SOURCE: NRC Phase II Survey.

13 percent, prior non-SBIR federal R&D funding was also notable. Relatively few received funding from state and local governments and colleges and universities. And, notably, only 2 percent received prior funding from venture capitalists.

Attraction of Funding from Non-SBIR Sources for Development and/or Commercialization. Bringing a technology from the research stage to commercialization requires considerable investment beyond that provided by the SBIR program. Obtaining follow-on funding is requisite to commercialization. Thus, survey data on the attraction of non-SBIR funding for the development and/or commercialization of SBIR projects is quite relevant to this section on commercialization.

First, the survey investigated additional funding received in the Phase II stage for matching or cost-sharing. In the case of the NSF, it is likely that this funding was to meet the third-party match for Phase IIB grants. Nearly one-third of the respondents reported that there were matching funds or other types of cost-sharing in conjunction with their NSF Phase II proposals. Table 5.2-6 shows the sources of the matching or co-investment funding that were proposed for Phase II. Most of the projects that reported matching funds got the match from other companies and internal sources. Angel or other private investment sources were also a relatively important source. By contrast, venture capital sources provided few matching funds.

In addition to investigating matching or co-investment funding, the survey asked respondents about the receipt of additional development funding for the project. The majority of the respondents (63 percent) indicated there was additional developmental funding put into the project.

TABLE 5.2-6 Sources of Matching or Co-investment Funding Proposed
for Phase II

Source of Matching or Co-investment Funding	Percentage of Projects for Which Respondents Reported Matching Funds Who Reported Each Source
Another company provided funding	58
Our own company provided funding, includes borrowed funds	38
An angel or other private investment source provided funding	21
A federal agency provided non-SBIR	4
Venture capital provided funding	2

SOURCE: NRC Phase II Survey.

TABLE 5.2-7 Sources and Amounts of Developmental Funding Reported
for Phase II NSF SBIR Projects

Funding Source	Average Amount per Project of Developmental Funding ($)
Non-SBIR federal funds	248,077
Private investment:	
(1) U.S. venture capital	39,450
(2) Foreign investment	19,290
(3) Other private equity	196,141
(4) Other domestic private company	57,925
Other sources	
(1) State or local governments	19,938
(2) College or universities	617
Not previously reported	
(1) Your own company (including borrowed money)	54,617
(2) Personal funds	22,154

SOURCE: NRC Phase II Survey.

Companies were asked to identify how much development funding from each of a number of potential sources went to the further development of the technology in each referenced project. Of 162 respondents, slightly more than half (93) reported additional funding by non-SBIR sources. Table 5.2-7 lists the reported developmental funding amounts by funding source, averaged over the 162 respondents to the question.[12]

The largest funding amounts for development came from two sources. Non-SBIR federal funding sources provided the largest amount. Other private equity

---

[12]It is assumed that those who did not report development funding received none.

sources provided the second largest amount. On average, U.S. venture capital provided substantially less than either the grant-recipient company or other domestic private companies.

Employment. Employment growth is relevant as an indicator of firm expansion often associated with commercialization. Describing employment changes in small companies associated with given project developments is more feasible than attempting the same in larger companies, yet ascribing causality remains difficult. While the data are descriptive rather than indicative of causality, the survey provides information on employment of respondent companies at the time the referenced proposals were submitted, and again at the time of the survey. On average, the companies had 21 employees at the time the referenced proposals were submitted and an average of 38 employees at the time of the survey, showing that the grant recipient firms grew on average over the period of the referenced projects.

To get at the employment effects directly associated with the referenced projects, respondents were asked how many employees were hired and how many were retained as a result of the technology developed by the project. On average, they reported 1.5 employees hired and 2 employees retained as a direct result of the technologies developed during these projects. Slightly less than half of the respondents reported no employees hired (48 percent) or retained (42 percent) as a result of the project. **Women and minorities received 21 percent of the project principal investigator positions.** These results emphasized employment in the grant-recipient firm, not in the larger economy where longer-term effects may be larger and measurement more difficult.

Attribution of Effects to the SBIR Grant. A challenge of evaluation is establishing attribution. Did the program cause the observed effect or would it have happened anyway? Asking counterfactual questions of participants about what they would have done without the SBIR grant is one approach to attempting to establish attribution of effect to the program.[13]

When asked if their companies would have undertaken the projects had there been no SBIR grant, approximately two-thirds of respondents responded either probably not (43 percent) or definitely not (24 percent). Nineteen percent were uncertain. Fourteen percent responded either probably yes (10 percent) or

---

[13]Using counterfactual questions does not entail experimental or quasi-experimental design, and, therefore, is not the strongest possible research design approach. A stronger approach is to use control or comparison groups to compare the actions of those who do and do not receive an SBIR grant. This approach has been applied to study the Advanced Technology Program (ATP), which compared the actions of a group of applicants who did not receive an award with a group who did, with a finding that the group that did not receive an ATP award was much less likely to be successful in attracting other sources of funding and much less likely to proceed with their projects. (Feldman and Kelley, *Winning an Award from the Advanced Technology Program: Pursuing R&D Strategies in the Public Interest and Benefiting from a Halo Effect*, NISTIR 6577, 2001.)

definitely yes (4 percent). The 14 percent who said they definitely or probably would have undertaken the project in the absence of the SBIR grant were then asked about the project's scope, timing, and progress in achieving similar goals and milestones. The majority (84 percent) of those who thought they would have continued without the SBIR grant said the project would have been narrower in scope, while a small number (13 percent) said it would be broader or similar in scope. Likewise, most of these companies (79 percent) reported that the start of the project would have been delayed without the SBIR grant, with the average delay being 15 months. Most of the responding companies (87 percent) reported that without the grant, the project would be behind in achieving similar goals and milestones. Table 5.2-8 shows how, in the opinion of the respondents, things would have been different had there not been the referenced SBIR grant.

Projects Discontinued. It is instructive to examine the 30 percent of survey respondents who indicated that efforts at the company related to the referenced Phase II grants had been discontinued. Why did these projects fail to take their innovations further?

Twenty percent of these projects were discontinued with no sales or additional funding resulting. Another 10 percent indicated that the project had resulted in sales or licensing or additional funding but subsequent efforts had been discontinued. Note that another 1 percent indicated that their projects had not yet completed Phase II.[14]

The 30 percent of projects that were discontinued were stopped for the various reasons given in Table 5.2-9. The table shows that usually the projects were discontinued for more than a single reason. It may be seen that the most commonly cited "primary reason" for discontinuing projects was "technical failure or difficulties," followed by "market demand that was too small," followed in turn by "insufficient funding," "licensing to another company," and "shifted priorities." Of lesser importance were "departure of the principal investigator," "lack of competitiveness of the product, process, or service," and "achievement of the project goal"—all listed with equal frequency.

"Not enough funding," "market demand too small," and "shifted company priorities" were frequently listed as among the secondary reasons for discontinuing the referenced projects. "Inadequate sales capability," "departure of the principal investigator," and "too high technical risk" were least often given as secondary reasons for discontinuing the projects.

Thirty percent of Phase II projects not continuing into further development or commercialization would appear to represent a substantial loss to the program. Therefore, further analysis of project failure may be warranted. Learning how to avoid or abate those factors causing projects to be discontinued may be possible

---

[14]Given the time period covered by the survey and the length of Phase II, all projects surveyed would normally have been completed by the time of the survey; reasons for the failure to complete the research were not ascertained.

TABLE 5.2-8 What Would Have Happened Without the SBIR Grant?

Alternatives	Percentage Reporting the Effect
Without the SBIR grant:	
• definitely would not have undertaken the project	24
• probably would not have undertaken the project	43
• uncertain	19
• probably would have undertaken the project	10
• definitely would have undertaken the project	4
Without the SBIR grant, the project would have been:	
• narrower in scope	84
• similar in scope	5
• broader in scope	11
Without the SBIR grant:	an average of 13 months
• the start of the project would have been delayed by	
Without the SBIR grant, the expected time to completion would have been:	
• longer	63
• the same	21
• shorter	0
• no response	16
In achieving similar goals and milestones, the project would be:	
• behind	68
• at the same place	5
• ahead	0
• no response	26

SOURCE: NRC Phase II Survey.

NOTE: These and the results below are only for the respondents who said they definitely or probably would have undertaken the project in the absence of SBIR.

in some cases, providing an effective way for the program to get more out of its funded projects. At the same time, it should be realized that much can be learned from technical failure. Technical failure can reveal which technical paths not to follow. And, technical failures can become hurdles that others find new ways to overcome.

5.2.3 Projects Not Continuing into Phase II as Revealed by NRC Phase I Survey Data

The NRC Phase I Survey focused on Phase I projects that were not followed by a Phase II project. The survey data provide insight about the commercial and

TABLE 5.2-9 Reasons Given by Companies for Discontinuing Phase II Projects

Reasons	Yes (%)	No (%)	Primary Reason (%)
Technical failure or difficulties	33	67	19
Market demand too small	42	58	17
Not enough funding	48	52	13
Licensed to another company	21	79	10
Company shifted priorities	40	60	8
Principal investigator left	17	83	6
Product, process, or service not competitive	29	71	6
Project goal was achieved (e.g., prototype delivered for federal agency use)	31	69	6
Level of technical risk too high	21	79	0
Inadequate sales capability	15	85	0
Other	19	81	15

SOURCE: NRC Phase II Survey.

NOTE: The OII reportedly does not stop projects for cause once a grant is made. This may be contrasted with the Advanced Technology Program, whose practice was to stop projects for cause and to track the cause of "terminated" projects.

noncommercial outcomes of Phase I grants. The data also provide insight about factors that contributed to the reported outcomes.

Of 248 projects for which survey responses were received, 46 percent did receive a follow-on Phase II project and were asked only several questions about assistance in preparing the Phase I proposal. The remaining 54 percent (135 respondents) did not receive a follow-on Phase II project and were questioned more extensively.

Figure 5.2-4 summarizes results for the 135 surveyed projects for which no Phase II follow-on grant was received. The branch to the left shows more favorable results for commercialization, the right less favorable results.

Why the Phase I Grant Was Not Followed by a Phase II. Nearly 60 percent of the Phase I grants were not followed by a Phase II grant because, even though the company applied, it was not selected by NSF. Another third of the companies did not apply for a Phase II grant. Of those that did not apply, slightly more than half did not apply because the Phase I did not demonstrate sufficient technical promise, and slightly more than a third did not apply because of lack of sufficient commercial promise. Another 16 percent indicated that the NSF was not interested in a Phase II, likely indicative of a lack of technical or commercial promise. Some other reasons given by the companies for not pursuing a Phase II

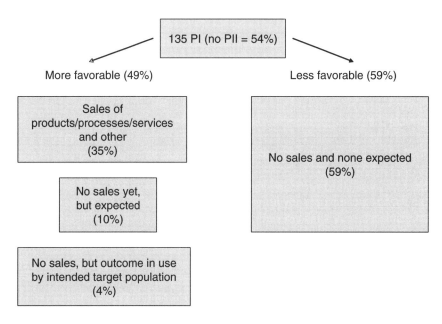

FIGURE 5.2-4 Commercialization Status of 135 NSF SBIR Phase I Projects.
SOURCE: NRC Phase I Survey.

were that proposal preparation was too difficult to be cost-effective, the company shifted priorities, the research goals were met by Phase I such that no Phase II was needed, and the principal investigator (PI) was no longer available.

Noncommercial Benefits from Phase I Grants. Of the Phase I grants not followed by Phase II grants, most respondents reported noncommercial benefits—primarily knowledge gains. The 135 Phase I grants resulted in 49 scientific publications of 52 submitted. The 135 Phase I grants also resulted in a total of 35 patents granted of 48 patent applications filed; 16 copyrights issued of 16 applications; and 14 trademarks issued of 15 applications filed. On an average basis, 0.3 patents were reported granted per Phase I grant.[15] Other benefits reported by the responding firms were: enabling the founding of the firm or keeping the firm in business; employees hired and retained; and various benefits to the public, including environmental benefits, energy supply advantages, training, and public

[15]As noted in Section 5.2.1, the average number of patents per grant is much less than the average number of patents per firm reported as being at least in part related to SBIR grants. While Phase I grants surveyed resulted in an average of 0.3 patents granted, firms surveyed reported an average of 6 patents related at least in part to SBIR grants.

service benefits. Only 8 percent of these projects were reported to have no non-commercial benefits.

Post-Phase I Developments. Forty-five percent of the companies reported that they dropped the project after not getting a Phase II award. Forty-two percent went on to receive either another Phase I or a Phase II grant in the technology (though not as a direct follow-on). Ten percent went on to receive subsequent federal non-SBIR contracts or grants in the technology. Thirteen percent pursued the technology after Phase I, but the effort did not result in any subsequent grants, contracts, licensing, or sales.

Reported sales averaged close to $100,000 for the Phase I-only projects (compared to average sales of the Phase II survey projects of nearly $400,000). Licensing fees averaged under $1,000. The top five Phase I-only grants for which sales were reported accounted for 62 percent of total sales for this group and ranged from $700,000 to $2,033,589.

While most respondents reported that they did not attract additional funding for the Phase I technology development, 27 percent did report such funding. Additional development funding was obtained from a number of sources, with the largest amount coming from the company itself, followed by funding from other federal sources and from other domestic private companies. Less than 10 percent of the companies reported receiving developmental funding from U.S. venture capital sources.

Without the Phase I Grant: Had there been no Phase I grant, only 13 percent of the respondents thought they probably or definitely would have undertaken the research in approximately the same way. Forty-five percent thought they probably or definitely would not have. Another 21 percent thought they probably would have undertaken the research, but most of these thought either the scope of the project would have been reduced or its completion slowed if there had been no Phase I grant. Fifteen percent were uncertain.

5.2.4 Commercialization as Illustrated by Selected Case Study Data

Case studies tell the stories of actual companies who have received NSF SBIR grants—adding flesh to the bones of the survey data. Table 5.2.10 provides an overview of 10 cases.[16] Although they differ in many respects, the ten companies in the case study set were found to be similar in at least three respects.

- All of the companies were receiving revenue at the time of the study.

---

[16]The case studies were developed by interview and supporting research. The set of case studies is presented in full in Appendix D of this report, together with an account of how the cases were selected.

TABLE 5.2-10 Case Study Company Demographics

Company	Location	Founding Date	Technology	Annual Revenue ($)	No. Employees	Gov't Grants and Contracts (%)	Originations
Faraday Technology, Inc.	Clayton, OH	1991	Electrochemical technologies	2 million	10 FT 9 PT	48	Woman-owned (inactive)
Immersion Corporation	San Jose, CA	1993	Sense of touch added to computer interfaces	23.8 million	141	4	Stanford U spin-off
ISCA Technologies	Riverside, CA	1996	Pest monitoring and control	2.4 million	12	40	Founded by U of CA-Riverside researcher
Language Weaver, Inc.	Marina Del Rey, CA	2002	Software for statistical machine language translation	12 million	28	60-70	Founded by 2 professors at USC
Mer Corporation	Tucson, AZ	1985	Rapid manufacturing prototyping in metal and composites; other	7.9 million	75	60	Arco Chem spin-off
MicroStrain, Inc.	Williston, VT	1987	Micro sensors and wireless sensor networks	5 million	20	>26	Founder started company upon leaving graduate school
National Recovery Technologies (NRT), Inc.	Nashville, TN	1983	Sorting plastics; electronics-based metals recycling	4 million	14	NA	Older company changing technology focus
NVE Corporation	Eden-Prairie, MN	1989	Electronics based on electron spin	12 million	70	35	Founded by Honeywell former executive
T/J Technologies	Ann Arbor, MI	1991	Nanomaterials for batteries and fuel cells	2.5 million	24	15-20	Minority-woman-owned (active) and operated
WaveBand	Irvine, CA (now owned by a company in Nevada)	1996	Electronically steerable smart antenna	5.5 million	24	>50	Spin-off of Physical Optics; acquired in 2005 by Sierra Nevada Corp (large co)

NOTE: See Appendix D for details.

- The companies expressed a belief that SBIR grants were critical to their ability either to get started at all or to develop capabilities critical to their existence and continued strength.
- Without exception, the companies sought and received grants not only from the NSF but from SBIR programs operated by other agencies and, in a number of cases, from other kinds of government funding programs as well—notably the Defense Advanced Research Projects Agency (DARPA) and the Advanced Technology Program (ATP).

Given the limited number of case studies performed, and the diversity that characterizes them, the cases are illustrative only. They cannot be taken as necessarily representative of the universe of SBIR grant recipients. Yet, a number of common themes were found to run through this set regardless of their diverse locations, technology fields, year of founding, and other particulars.

Table 5.2-11 lists five broad and overlapping common themes that emerge from the cases, each of which has implications for commercialization, and each of which is discussed in turn in the following paragraphs.

No Linear Path from Phase I, Phase, II, and Phase IIB, into Phase III. The company interviewees made it clear that a simple linear model of Phase I, Phase II, and Phase IIB grants is woefully inadequate to mount the kind of

TABLE 5.2-11 Broad Common Themes Running Through the Cases Studies

Theme
There is no linear path from NSF Phase I, Phase II, and Phase IIB into Phase III
R&D finance is a major challenge to firms
SBIR enables firms to   • Start a company   • Survive in hard times   • Test creative ideas, develop new capabilities, and build capacity   • Develop, extend, and renew technology platforms   • Grow an intellectual property portfolio essential to commercial success   • Leverage other sources of funding   • Maintain control until technical progress enables more effective partnerships
Firms pursue multiple paths to commercialization, including contract research as an ongoing line of business
Commercializing is typically costly and challenging
Multiple benefits result from commercialized technologies

SOURCE: Analysis of 10 case studies.

technology research and development program necessary to bring an advanced technology to the point of attracting private sources of funding on favorable terms. Some described how they might get a concept going with a Phase I SBIR grant from one agency, and later take it forward with an SBIR grant from another agency. Some spoke of using SBIR grants to get an idea to a certain point and then taking the idea forward with an ATP award. Some explained how they used DARPA funding to get an idea to the point that it could be taken forward with a combination of SBIR and ATP funding. Some leveraged government research grants into government research contracts, with the goal of that leading into commercial deals in the private sector. Others got a jump start in university research labs, often with government support, and then spun out a business. In none of the cases was there a simple sequence of NSF Phase I, Phase II, and Phase IIB grants that provided adequate preparation for a launch into Phase III commercialization.

> *"The SBIR grants served as building blocks for us,"* explained Maria Thompson, President of an award-winning African-American, woman-owned firm, T/J Technologies, Inc. *"Without the SBIR, we couldn't have won the ATP. And, without the ATP and SBIR, we may not have had the technology with which to earn the larger contracts and joint development agreement. . . . They are all linked."*

In all the cases, the story involved multiple funding sources and a complex path to Phase III. The brief examples that follow, drawn from the case studies, illustrate the lack of a simple, linear path leading from SBIR financing to commercialization.

Ongoing Search for Research Funding. A common theme among the case study companies was that to stay in the high-tech game, continuous research, and, hence, ongoing funding of research, is needed. Although several of the companies appeared to be shifting emphasis from research to commercial product development, all the firms referred to the need for continuous innovation to keep their edge. In no case was the story a one-way path of conducting research, developing a product, going to market, and then enjoying sales. These companies all seemed to define themselves as current—not past—innovators, and all appeared to be looking ahead to the next technical challenge.

Box A
There Is No Linear Path from NSF Phase I, Phase II,
and Phase IIB into Phase III

Faraday Technologies, Inc. made the point that a "failed" Phase I grant provided the seed for later electronics work that now provides 35 percent of the company's business and accounts for 8 of its patents. "It is not a tidy path; it is a cumulative process."

Immersion Corporation's technology—adding sense of touch to computer applications—got its start in a university laboratory. SBIR funding from NIH, DoE, DoD, Navy, Army, and the NSF, as well as an ATP award, figured in its financing of technology development. Now the company has a number of products in the market and is relying principally on private equity funding.

Language Weaver, which offers a seemingly straight path of NSF grants leading to commercialization, actually had about 20 years of university research heavily funded by DARPA underlying its technology development. ATP also provided funding for further development of the technology following a series of NSF SBIR grants.

NVE Corporation initially licensed rights to civilian applications of its MRAM computer memory technology from Honeywell, where its development had received military support. With a combination of SBIR grants from the NSF and other agencies, at least two ATP awards, funding from work on agency BAAs, and other sources of revenue, NVE developed sensors based on the MRAM technology and has continued to develop MRAM.

T/J Technologies, Inc. started by taking over a DoD SBIR Phase I grant from a company that was divesting contractual obligations. This grant served as a building block for T/J to obtain a series of SBIR grants from multiple agencies, which in turn served as a building block for it to obtain larger ATP awards. As it developed its technology in advanced materials for energy storage and conversion devices, T/J was positioned to pursue research contracts with the Army under BAAs. From there, the company was able to develop partnerships with global firms.

See Appendix D for more details.

The emphasis on research to support ongoing innovation means that funding for research is a constant rather than a one-time issue for these companies.

Because all these companies are in various stages of commercialization, usually involving more than one product, they have a continual need for financial support to get products, services, and processes to market. In most cases these companies were receiving sufficient revenue from sales, licensing fees, and contract research to supplement their laboratory research and were no longer threatened by financing gaps in government grant programs. But in all the cases, the companies seemed to be wrestling with the issue of how to get the funding needed to sustain technological strength and make advances needed for survival and growth. Supplying research funding appeared to be an ongoing and constant concern for these innovative companies even though they all had successfully developed technologies and had made substantial progress in commercializing them.

As one company's president explained its ongoing research needs, "We need a family of materials for multiple applications . . . to demonstrate a whole system, and that takes time and money." Another company president described how his company had used SBIRs to rejuvenate its technical base as demand for existing products fell off. Even the company that most seemed to be shifting emphasis from research to commercialization indicated that it would continue its internal research and reliance on the fruits of ongoing university research.

Difficulty in Obtaining Private Funding. Most of the companies spoke of difficulties in obtaining private funding to develop their technology base and take it to the point of commercialization. One of the difficulties centered on the unfavorable negotiating power of a small firm that may be pressured to seek financial assistance from venture capitalists or other companies before the firm is ready. As one company president put it, "They are going to try to get all of your IP [intellectual property], for only a fraction of its value." A representative of a company that has been successful in obtaining private funding also noted the difficult path in obtaining it, explaining that the company founders were turned down multiple times before they received confirmation of the technology by receiving an NSF SBIR grant and became able to demonstrate the technology's potential.

Another difficulty in finding private funding centered on the type of technology. The two software companies in the set of case studies, both located in California, were more successful in obtaining venture funding than most of the other case study companies. As a company president of a materials research company put it: "Venture capitalists don't fund manufacturing. They don't fund materials research. The development time horizon is generally too long."

It is also possible that companies located in venture capital hotbeds like the San Francisco Bay Area or Boston may have an easier time raising private funds than companies based in regions without such a large entrepreneurial community devoted to high technology.

... the company founders were turned down multiple times [by private funding sources] until they received confirmation of the technology by receiving an NSF SBIR grant and becoming able to demonstrate its potential.

The SBIR Program as an Enabler and a Lifeline. All the companies that were interviewed described at least one important role played by the SBIR program, which varied depending on the needs of the company. In a number of the cases, an SBIR grant figured in the start-up of the company. In other cases, an SBIR grant helped the company survive when it had encountered hard times. In all the cases, SBIR grants were used to further the companies' technical capabilities in ways that enabled them to bring new and improved products and services to customers. Furthermore, all the companies appeared to use the SBIR grants as a means to leverage other sources of funding. Several companies saw SBIR grants as a way to retain control of their companies. Several saw SBIR grants as a way to build capability in order to improve negotiating strength so as to attract partners and investors on more favorable terms. The following examples from the case studies illustrate various roles the SBIR program has played for these companies.

Intellectual Property as a Competitive Advantage. As a group, the case study companies appeared to pursue aggressive patenting and trade secret strategies, and they regarded intellectual capital as their core strength and the source of their competitive advantage. In all the cases, SBIR grants figured prominently in the development, extension, and renewal of the core technologies that comprised the companies' intellectual property base and enabled the firm to file for patents.

Faraday Technology, for instance, has, since its founding in 1991, had 23 patents issued in the U.S. and three issued abroad, which historically amounts to 1.4 issued patents per employee. Patents and the fees they generate are the central focus of Faraday's business strategy, and the company investigates citations by other companies of its patents to obtain knowledge about potential customers. (See Appendix D.)

Immersion Corporation has more than 270 patents issued in the United States, and another 280 patents are pending in the United States and abroad. According to Immersion, its patent portfolio is at the heart of its wealth-generation capacity. (See Appendix D.)

ISCA Technologies has received a trademark on its most recent insect lure technology—a technology that is expected to generate substantial growth for the company, as well as benefits for the environment.

Box B
The SBIR Program as an Enabler and a
Lifeline for High-Tech Companies

Faraday Technologies, Inc.: The SBIR grant enabled the company to undertake research that it otherwise would not have done. It sped the development of proof of concepts and pilot-scale prototypes, opened new market opportunities for new applications, led to the formation of new business units in the company, and enabled the hiring of key professional and technical staff. The SBIR [program] "is well structured to allow taking on higher risk. . . ."

Immersion Corporation: SBIR grants gave Immersion the ability to grow its intellectual property portfolio, the core of its commercial success. The company leveraged the government funding to attract investment funding from private sources. It has grown to a capitalized value of $173 million.

ISCA Technologies: The SBIR program was essential to the survival of the company after it hit a major financial setback on its initial path. "The NSF SBIR gave us lots of prestige; it gave us credibility." The company used SBIR funding to upgrade its technology and find new markets.

MER Corporation: The SBIR program allowed the company to steadily improve and advance its R&D capabilities. It also enabled the owners to not lose control of the company.

Language Weaver: "The STTR/SBIR from NSF created Language Weaver and what we are today. Without that we would have shelved the technology."

MicroStrain, Inc.: The company found the NSF SBIR program, with its "more open topics," particularly helpful in the early stages when the company was building capacity.

National Recovery Technologies, Inc.: "Without the SBIR program, NRT wouldn't have a business. We couldn't have done the necessary technical development and achieved the internal intellectual growth. . . . SBIR saved our bacon."

NVE Corporation: The SBIR program is "the mother of invention." SBIR and other government R&D funding programs are essential to NVE being able to perform the advanced R&D that has allowed the company to produce products for sale and to license intellectual property.

T/J Technologies, Inc.: "The SBIR grants served as building blocks. . . . Without the SBIR, we couldn't have won the ATP. And, without the ATP and SBIR, we may not have had the technology with which to earn larger contracts and joint development agreements. So they are all linked. . . . The cutting edge intellectual property that we have developed through the SBIR and ATP programs has attracted multiple players to us. Small companies have a stronger negotiating position when more than one company competes for their technology."

WaveBand Corporation: Initially, the company (which is a spin-out of another company) was dependent on SBIR grants. After focusing on military objectives, the company went through a cycle of Phase I, Phase II, and Phase IIB NSF SBIR grants to develop not only technical prowess but also commercial strength for the company. "The technologies the company developed under Phase II SBIR research are vital to its commercialization success."

See Appendix D for more details.

Language Weaver has more than 50 patents pending worldwide; these patents underpin its commercialization approach. (See Appendix D.)

The other companies in the set of case studies reported similar essential roles for their intellectual property. (See Appendix D.)

> . . . intellectual capital—the core strength and source of competitive advantage

Multiple Paths to Commercialization. The case study companies were found to be pursuing a mix of approaches to commercialization, including licensing, partnering, providing contract services, and producing and selling product (see Appendix D). Licensing was an important route to commercialization for about half the case study companies. This finding is consistent with NRC Phase II Survey results, which also showed the importance of licensing as a path to commercialization. For example, the major route to commercialization for Faraday Technology has been to license "fields of use" to interested customers. As another example, Language Weaver describes itself as "a core technology house based on licensing its software" directly to customers and indirectly through partners who license Language Weaver's technology and incorporate it into their own products. And, as noted in the section that follows, licensing as a commercialization strategy was also emphasized by Immersion, MER Corporation, and NVE Corporation.

> . . . agencies who collect information about SBIR impacts typically ask only about product sales, whereas, in fact, most SBIR grantees are not OEM suppliers of product.

Most of these companies generally had not built, and did not plan to build, large, commercial-scale production facilities. However, it was not unusual for the companies to maintain small-scale production capabilities or to arrange for small-scale contract production and to sell directly to customers on a limited basis. Some maintained pilot-scale production facilities for making prototypes or limited production facilities to produce a single line of product. In keeping with this finding, interviewees frequently commented that the strength of the company was research, not manufacturing. Another comment heard several times was that the scale of a production facility needed for competitiveness was huge and the capital cost requirements were enormous—far beyond the capacity of a small company. One company's marketing director noted that agencies who collect information about SBIR impacts typically ask only about product sales, whereas in fact, most SBIR grantees are not OEM suppliers of product.

T/J Technologies, a materials development company facing a relatively long product development cycle and what were described as "prohibitive" costs to build production facilities, is emphasizing partnerships with large global companies to reach markets. MER Corporation is commercializing through a mix of strategic alliances, joint ventures, licensing, and production and sale of product. NVE is also pursuing a mixed strategy—commercializing its MRAM technology primarily through "an intellectual property business model," while it continues to design, fabricate, and sell directly a variety of sensors and signal coupler devices for both commercial and defense applications. Although it has the largest annual sales revenue to date and the largest revenue from direct product sales among the companies in the case study set, Immersion Corporation has limited manufacturing operations, arranges for some contract manufacturing, and "far and away, depends on licensing fees as its major source of revenue." NRT has maintained a steady annual revenue stream on the order of $2 million–4 million for a number of years from the sale of equipment and now is seeking larger markets through partnerships both to operate and to sell equipment.

Among the case study companies, those that appeared most focused on direct product sales as the major path to commercialization were ISCA Technologies and MicroStrain. Currently, ISCA's annual revenue from sale of product is approximately $1.5 million. It has multiple product lines and is anticipating a dramatic increase in sales in the near future from a new trademarked product. Likewise, MicroStrain's main path to commercialization has been the sale of sensors and systems of networked sensors. In the past, WaveBand has mainly focused on sales of antenna systems to defense agencies. More recently, it has also worked with suppliers in the auto industry on adaptive cruise control for cars and with suppliers in the avionics industry on guidance and landing systems. In any case, WaveBand's recent acquisition by a large systems integrator may alter its commercialization strategies.

Contract Research as an Ongoing Line of Business. Another recurring theme from the case studies was that contract research is often used as a bridge to commercialization, and is also seen by some as a way of life. One interviewee, for example, characterized his company as "an innovation house for a number of companies that are not well positioned to innovate themselves." Other interviewees said the large number of innovative small companies performing contract research had, in aggregate, provided a practical replacement for the large corporate research labs of the past that have been reduced in size or shut down. More often, the case study companies appeared to pursue contract research as a business sideline to generate revenue. For example, MER Corporation relies heavily on government engineering contracts as a source of revenue, and describes itself as operating as an engineering services company. MicroStrain performs contract research as a source of revenue, but it reportedly focuses on product sales. T/J Technologies currently obtains most of its revenue from contract research, but its longer-term strategy is reportedly to develop partnerships for commercializing its material technologies.

Contract research appears to be important to most of these companies either as an interim or an ongoing commercialization strategy.

Challenges of Commercializing. These 10 cases emphasize that even under the best of circumstances getting to market is difficult. Even those companies that were relatively successful in commercializing spoke of the difficult challenges faced by small companies trying to develop and commercialize a technology. "I think commercialization is very hard for people," said an interviewee from one of the most rapid commercializers, commenting on the challenge of finding additional funding sources and partners as early as needed. Another interviewee noted the trepidation of entering into partnerships and negotiating arrangements with large, powerful companies early in development. But the difficulties did not end with start-up. Several of the companies that appeared to have achieved commercialization had then come close to folding as a result of events outside of their control, such as a default on product orders from a foreign buyer (see the ISCA Technologies case study in Appendix D) and a collapse of markets due to an adverse Supreme Court decision (see the National Recovery Technologies case study in Appendix D).

"... commercialization is very hard ..."

Multiple Benefits from Commercialized Technologies. All of these case study companies linked commercialization of their technologies to SBIR grants and, in turn, linked commercialization of their technologies to the generation of multiple benefits, including direct economic benefits. Though net income or profit data are confidential and not available, revenue data are available and summarized in Table 5.2-10. Annual revenue ranged from $2 million to nearly $24 million and averaged $7.7 million across the 10 companies.

Beyond the return to company owners, there may be employment benefits. Table 5.2-10 gives employment data as of the time the 10 case studies were conducted. The number of employees ranged from 14.5 to 141 and averaged 42 employees per company. Although attribution of employment to the SBIR grant was not attempted, we can conclude that all these companies have grown since their founding and, from statements of the company founders, CEOs, and presidents, we have evidence that SBIR grants played an important role in the development of all of these companies.

Each of the companies identified a number of additional benefits conveyed by their products or services. For example, Faraday's process technologies result in lower cost manufacturing and higher quality output for its customers, as well as potential beneficial environmental effects. Immersion's technology can boost the productivity of software users, enhance online shopping experiences, enhance entertainment from computer-based games, improve skills of medical professionals, increase auto safety, enable industry to experience prototypes virtually before building costly physical prototypes, capture 3-D measurement from physical objects, and assist visually impaired computer users. ISCA's technology offers cost savings, quality improvements, and increased profitability for its customers, as well as environmental, health, and safety benefits. Language Weaver's technology offers a significantly higher rate of accuracy in language translation than counterpart rule-based machine translation systems and greater speed than human translators—this could have important military and civilian applications. MicroStain's sensors and networks of sensors offer the benefits of alerting managers to emerging problems in time to take preventative action, conserve resources, improve performance, and increase safety.

> Potential benefits from the resulting products and services—beyond those accruing to the companies—include lower costs and higher quality for customers, reduced threats to the environment, improved safety, improved outcomes for medical patients, alerts to emerging structural problems, and faster translations in military and civilian situations.

5.2.5 Commercial Progress as Indicated by Agency-Initiated Data and Analysis

The NSF's approaches to analysis and data compilation are discussed in Section 8.8. The results of program-initiated analysis of commercialization are summarized in this section. In addition to the routine publication of "nuggets" and "success stories" (see Exhibits 4.5-1 and 4.5-2), several ad hoc survey studies have been conducted by the program since 1995.

The survey studies have relied primarily on telephone interviews using a structured interview guide. The studies collected data to show measures of performance for a selected group of companies. One such study was completed in 2004. It is referred to here either as the "2004 NSF SBIR Commercialization Survey" or the "Coryell Study," after the NSF SBIR program manager who conducted it.[17] Another study, conducted in 1996, will be called the "Tibbetts Study," after another NSF SBIR program manager.[18] A related study carried out by a contractor, but never completed and released, will be called the "Dawnbreaker Study," after the contractor. All three studies were internal studies, not published in the open literature. A new internal effort to collect commercialization data was begun in the summer of 2005, and provides limited data.

The Tibbetts Study, according to the NSF's SBIR program office, was the first agency-initiated study to produce program performance metrics. Study results were included in an earlier NRC report.[19] The study covered a group of 50 companies, all of which had commercialized results of their SBIR grants.

The Dawnbreaker Study (intended to extend the Tibbetts Study but never completed) conducted interviews with 30 companies and developed approximately 20 "success stories" based on the interviews. The study also reported quantitative indicators of success for the 20 "successful companies," including cumulative sales dollars, total investment, number of new jobs, number of patents and copyrights, use of trade secrets and trademarks, and number of collaborators from industry and universities. The results of this study were much less favorable for NSF commercialization than the previous Tibbetts Study.

Between 2000 and 2004, the Coryell Study surveyed 34 companies. Among the companies surveyed were 17 "stars" selected by program managers as companies for which they had high expectations of outstanding accomplishments.

Among the findings of the Tibbetts Study are the following:

• 100 percent of the 50 selected firms had commercialized their SBIR-funded innovation.
• Sixteen of 50 firms said that the SBIR projects were key to starting the company.
• Forty-five of the 50 companies said the SBIR projects were critical to their growth and/or survival.
• $2.2 billion in sales were reported to be directly related to NSF SBIR.
• The 50 companies were granted an estimated 377 U.S. and 732 foreign patents that related directly or indirectly to SBIR program research or funding.[20]

---

[17]Alan Baker, "Commercialization Support at NSF" (draft), p. 2.

[18]The two former program managers are Ritchie Coryell and Roland Tibbetts, both identified earlier as playing key roles in the founding of the program.

[19]National Research Council, *The Small Business Innovation Research Program: Challenges and Opportunities*, Charles W. Wessner, ed. Washington, DC: National Academy Press, 1999.

[20]Thus, the Tibbetts Study found an average of seven and a half U.S. patents per company in the survey that were directly or indirectly related to SBIR research or funding. This figure is comparable

• Private follow-on investment was $963 million, of which $527 million was considered directly related to NSF SBIR projects.
• The 50 companies had 959 research collaborations: 404 with industrial firms, 394 with universities, and 111 with national laboratories (not attributed to specific projects).
• The 50 companies achieved specific technical breakthroughs and innovations.
• The 50 companies achieved specific commercial successes.
• A table of performance indicators was developed from data compiled for the 50 companies.

Among the results of the Dawnbreaker Study (based on a draft report provided by the NSF's SBIR program office) are the following:

• Cumulative sales directly or indirectly attributable to the selected 20 NSF SBIR projects totaled $31.8 million—much lower than the amount reported in the Tibbetts Study even when adjusted for differences in the number of companies.
• A conclusion that the 50 companies included in the Tibbetts Study represented the "cream of the crop," including three of the most successful commercializers.
• A conclusion that companies in the Tibbetts Study had been funded for a longer time prior to the interviews than those included in the Dawnbreaker Study, thus contributing to the larger revenues found by the Tibbetts Study.
• Twenty draft "success stories."
• A table of indicator data for 20 companies.

The Coryell Study includes survey data for approximately 300 projects; those data are summarized in Table 5.2-12. Using a criterion of "fully successful"—defined as having achieved a "first sale"—the study (based on preliminary results) concluded that 40 percent of the companies surveyed (FY96–98) were "fully successful."

The Coryell Study provided more extensive results for 34 grant winners: 10 with no Phase IIB grants and 24 with Phase IIB grants (fifteen of which had more than one). Figure 5.2-5 shows commercial results for 33 of the 34 companies reporting commercialization. Product sales attributed to NSF SBIR grants averaged $3.5 million for the 33 companies. (NOTE: Questions used to develop the findings given in Figure 5.2-5 are shown in Appendix E.)

The Postproject Annual Commercialization Report. A postproject annual

---

to the average of six per firm reported by the NRC Firm Survey, as reported in Section 5.2.1—that is, assuming that the NRC Firm Survey patents reported were U.S. patents. It would be expected that the Tibbetts group of firms would show a higher average than the NRC Firm Survey, because the Tibbetts firms were drawn from the highest achievers.

TABLE 5.2-12 Survey Results of the "Coryell Study"

	NSF Initial Survey Results, 2003							
	Fiscal Year 1998		Fiscal Year 1997		Fiscal Year 1996		Fiscal Years 1996–1998	
	Number	Percent	Number	Percent	Number	Percent	Number	Percent
Full success	36	29	55	46	31	53	122	40
Likely success	20	16	25	21	2	3	47	16
Commercial failure	35	28	22	18	16	27	73	24
Technical failure	21	17	11	9	7	12	39	13
Other	12	10	7	6	3	5	22	7

SOURCE: National Research Council Symposium, "The Small Business Innovation Research Program: Identifying Best Practice," Washington, DC, May 2003.

Results for 33 Companies Reporting

- 27 companies had product sales by NSF SBIR
- 70 product lines are due to NSF SBIR
- 42 of these would not have been developed
- NSF SBIR product sales have totaled $116 million
- Average time to first sale is 2.5 years

FIGURE 5.2-5 Commercial Findings from the Coryell Study.
SOURCE: NSF.

commercialization report was long required by the program of all Phase II grantees. While it would appear to be a potential source of evaluative information, reporting compliance by grantees has been low. Moreover, the reports that were filed have not been used to assess commercialization. Hence, there are no findings available from this effort.

New Monthly Postcompletion Telephone Interview of Grantees. A new interview survey was implemented in July 2005. Thirty companies reaching the third-, fifth-, or eighth-year anniversary after project start were interviewed by an NSF program manager in this telephone survey. Using the OII's definition of commercial success,[21] the first survey report found that half the 24 companies

---

[21]The "minimum requirement for success" for each group is defined by the NSF's SBIR program office as follows: (1) For the third-anniversary group, success is defined as having sales (of any amount)

responding were "fully successful," and overall, the 24 companies had achieved in sales an amount that equaled at least the value of their grants.[22] Of the half that were deemed not successful, the reasons given were technical failure, poor cost competitiveness, and insufficient demand. The postproject Annual Commercialization Report and the new monthly interview are discussed further in Section 8.8.

5.2.6 Commercialization Insights Provided by a Committee of Visitors

Expert review of the SBIR program is provided every three years by the NSF SBIR Committee of Visitors (COV). Findings from its 2004 review concerning commercialization are timely and relevant in a number of ways to the current NRC assessment of program performance and its efforts to provide recommendations for program improvements.

The COV recommended that "more consideration be given to commercial potential in evaluating Phase I proposals. . . ." Specifically, it recommended that the review panels for Phase I proposals have more well-qualified representatives from the business sector. (The COV's mode of assessment is discussed in Section 8.8.1.)

5.3 CONCLUSIONS ON COMMERCIALIZATION

Frequency of Commercialization. The NRC surveys, the case studies, the three surveys conducted by the NSF, and the new postcompletion telephone survey all show a range of commercialization results. The Tibbetts survey, which provides the largest revenue estimates, had a concentration of the most successful companies. The NRC Phase II Survey results showed that 22 percent of the referenced projects reported products/processes/services in use. An additional 10 percent of the referenced projects had produced sales, had licensing fees, or had attracted additional funding, but were discontinued.[23] Because the 10 percent figure includes attracting additional funding, we cannot conclude that 32 percent (22 percent + 10 percent) achieved sales. Furthermore, we do not know to what extent "sales" includes samples put out to customers for trial or testing rather than commercial sales. Hence, we can conclude only that between 22 percent

---

one year after completion; (2) for the fifth-anniversary group, success is defined as having sales on the order of $600,000 three years after completion; and (3) for the eighth-anniversary group, success is defined as having more than $1.5 million in cumulative sales 6 years after completion. (Based on a description provided by NSF SBIR program staff.)

[22]Note that this statement should not be confused with the more generally accepted cost-effectiveness criterion that project benefits equal or exceed project costs. This is because sales revenue does not equal net income or net private benefits accruing to companies.

[23]For projects funded by procurement agencies, producing product and then discontinuing may mean the intended goal of the project was met, but for NSF-funded projects, discontinuing may more realistically be taken to mean that the project was not fully successful commercially.

and 32 percent of projects in the Phase II survey had achieved some degree of sales revenue.

Of course, the percentage of projects resulting in sales revenue may rise in the future, as indicated by the 19 percent of respondents reporting commercialization under way and the 28 percent of projects still active in the developmental stage. Yet, these could be offset in the future by some of those projects now active becoming inactive, and some of the prospective projects not achieving sales. If half those expecting commercialization achieve it and remain active, and if a quarter of those still in development achieve commercialization and remain active, this would raise the upper bound of projects achieving sales revenue and remaining active to as high as 38.5 percent. This projected upper-bounds figure approaches the finding of the larger Coryell Study, which found that 40 percent of the projects surveyed had achieved a first sale.

Commercialization as Signaled by Sales Revenue. The NRC Phase II Survey showed that relatively few of the projects had achieved significant sales revenue. When the total reported sales revenue was averaged across all survey projects, sales revenue averaged about $0.4 million per project and licensee sales averaged about $1.4 million per project. Furthermore, just eight of the projects—each of which had $2.3 million or more in sales—accounted for over half the total sales dollars reported by survey projects. The project with the highest reported annual sales amount had $4.8 million in sales. Similarly, a single licensee accounted for over half the total licensee sales dollars.

The set of 10 case study companies reported annual company revenue ranging from $2 million to nearly $24 million. The average annual company revenue per company was $7.7 million. This included revenue from all company projects—not just a single NSF project.

The Tibbetts Study, which included some of the most successful NSF SBIR projects and was taken over a longer period of time than did the other studies, found much larger sales revenue. It was reported cumulatively by the study rather than annually; hence, the data from the various sources are not comparable as reported. Cumulative direct sales attributable to NSF SBIR projects ranged from $1 million to $500 million. Cumulative direct and indirect sales ranged from $2 million to $2.6 billion.

To the extent that the past is a predictor of the future, more than half the projects funded by SBIR likely will not achieve commercialization. Somewhere between 20 percent and 40 percent will probably achieve some level of commercialization. For most of these projects, sales revenue will likely remain relatively small, with approximately 10 percent having more robust sales in the range of $2 million to $10 million annually—again, if the past is a predictor of the future. A smaller percentage of projects will likely achieve large growth and huge sales revenues, i.e., be commercial "home runs." These patterns are similar to those found in other private and public investments in high-risk advanced technology investments, where many research projects must be seeded to yield a few commercial home runs. Meanwhile, many small successes together will continue to comprise a potentially important component of the nation's innovative capability.

Based on the counterfactual data, the NSF's SBIR program can take credit for most, but not all, of the observed effects. A reported 14 percent of the surveyed firms believe that they definitely or probably would have done the referenced projects anyway, although most of these projects reportedly would have been narrower in scope, delayed in starting, and slower in progressing without the NSF SBIR program.

6

Support to Agency Mission and to Small Business

6.1 AGENCY DIFFERENCES: CONTRACTING VERSUS GRANT AGENCIES

Hearings surrounding passage of the SBIR-authorizing Act emphasized procurement barriers to small companies, with attention focusing on the large procurement agencies. In the then-existing environment, procurement agency proposals were said to be often too long and large for small businesses to undertake; projects were often bundled into packages too large for small businesses; and small businesses lacked the close tie-ins with project managers, who tended to prefer established institutions. Large "mission" agencies reportedly made many noncompetitive grants.[1]

In the defense world, the lines between the agencies and the companies they engaged were often blurred. This blurring of lines may have stemmed from the historical need of the military to control its supply of armaments and related goods. In contrast, from the beginning of its interactions with small business, there were very clear lines of distinction between the National Science Foundation (NSF) and the companies it engaged. The newly established Small Business Innovation Research (SBIR) program's emphasis on agency procurement did not fit the NSF model, which is perhaps paradoxical given that the NSF program is considered the first SBIR program.

The focus on innovation by the new, multiagency SBIR program did, however, closely match the attention to innovation of the NSF's predecessor program, though its motivation differed. The NSF's emphasis was on adhering to scientific

---

[1]U.S. Small Business Administration, *The State of Small Business: A Report to the President*, Washington, DC: U.S. Government Printing Office, 1994, pp. 125–126.

excellence while countering an institutional bias in favor of funding academic research. In contrast, the new multiagency SBIR program was focused on rectifying the large disparity in research capability of large versus small firms, stemming in part from federal procurement practices.

Federal procurement of research and development carries with it the "independent research and development and bid and proposal expense" (IR&D/B&P) that provides firms with funds that they can use for independent research. "The 100 or so major defense contractors accounted for an estimated 97 percent of all IR&D."[2] The IR&D/B&P funds had primarily assisted large defense contractors in doing research that might lead to additional government contracts and, in the process, lessened the relative competitiveness of smaller firms lacking these funds for research. The SBIR program was seen as providing a way for the major procurement agencies to increase the capability of small businesses to also conduct research that might lead to federal procurement. Increasing the research capabilities of small businesses was seen as a way to decrease barriers to their entry into the defense contracting world and thereby to increase competition among agency vendors. In contrast, it was seen as a way for the NSF to fund innovative research in small firms for the purpose of generating technology that increases national economic prosperity, one of the NSF's mission goals.

6.2 NSF—A NONPROCURING AGENCY

The NSF is an independent federal agency created "to promote the progress of science; to advance the national health, prosperity, and welfare; [and] to secure the national defense. . . ."[3] It is an important funding source for basic research conducted by the nation's colleges and universities. In a number of disciplines, such as mathematics, computer science, and the social sciences, it is the major source of federal funding. The agency carries out its mission largely by issuing grants to fund specific research proposals selected through meritorious peer review. "NSF's goal is to support the people, ideas and tools that together make discovery possible."[4]

The NSF's SBIR program follows the lead of the agency, emphasizing discovery and innovation that are put forward through a bottom-up process. Though the program defines topics, it defines them in ways that leave room for individual firms to decide what approach they will take. Its emphasis is on stimulating small firms to innovate, not on procuring goods and services that the agency needs.

The program encourages commercialization through the marketplace rather than through procurement channels. Unlike the defense agencies, the NSF would seldom be a customer for the results of its funded research. It was reportedly

---

[2]Ibid., p. 127.

[3]"NSF at a Glance," description of the agency provided at its Web site: <*http://www.nsf.gov/about*>.

[4]Ibid.

clear to the program's early developers from the outset that if the research results of funded projects were to be widely used, the use would have to come through avenues other than the agency's procurement channels.[5] Believing as they did that requiring companies to give attention to marketplace commercialization need not compromise the quality of research, the program designers were free to emphasize marketplace commercialization as a way to increase the economic impact of the agency's new initiative.[6]

In summary, the driving force for the NSF's SBIR program was to channel funding for innovative research to small firms for the purpose of generating a broad economic payoff, as a departure from funding only academic research. The driving force for the "mission" agencies' SBIR programs was both to rectify the large disparity in research capability of large versus small firms, stemming in part from past federal procurement practices, and to increase procurement options for the agencies. However, the resulting program parameters were closely similar: agency grants were to be made to small businesses for innovation, and commercialization—defined broadly to encompass sales into the marketplace, agency procurement, and subcontracts and sales to prime contractors—was to be encouraged.

6.3 NSF SUPPORT OF SMALL BUSINESS

6.3.1 Basic Demographics of NSF Support for Small Business

The NSF's funding for universities dwarfs that to small businesses. Its support for small business is centered in its SBIR/STTR [Small Business Technology Transfer] programs within the Office of Industrial Innovation (OII), and the SBIR program is much larger than the STTR program. Therefore, the demographics of the SBIR program, as summarized in Section 4.2, provide the principal demographics of NSF support for small business. In recent years, the NSF's SBIR program has provided close to 300 awards annually, totaling nearly $100 million to small firms.

6.3.2 NSF Small Business Research Funding as a Share of NSF R&D Spending

While the SBIR program directs R&D funding to small businesses, it remains a small percentage of the total budget of the NSF. The congressionally directed

---

[5]In contrast, some agencies—particularly DoD and NASA—are customers for the results of some of their SBIR grants. In fact, from the outset, a funded project might clearly have a single customer—the funding agency—and there might be a strong symbiotic relationship between the grantor and the grantee.

[6]Interview with Ritchie Coryell, NSF SBIR program dtaff, on October 23, 2003. (Note that Mr. Coryell has since retired.)

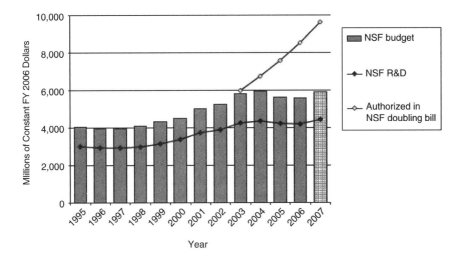

FIGURE 6.3-1 National Science Foundation Budget, FY1995–2007 (Budget authority in millions of constant FY2006 dollars). SOURCE: AAAS. NOTE: Authorized levels are authorizations in Public Law 107-368 (Dec. 2002). The solid line shows the portion of the NSF's annual budget authority for the NSF's R&D.

expenditure on the SBIR program is 2.5 percent of the NSF's extramural research budget. The largest part of the NSF's budget by far goes to research and development by entities other than small businesses.

In FY2004, the NSF's total budget for R&D was $4.1 billion, consisting of $3.8 billion for the conduct of research and $0.3 billion for R&D facilities.[7] Figure 6.3-1 provides a look at recent trends in R&D funding at the NSF, which will be reflected in turn in funding available to the foundation's SBIR program. In 2005 and 2006, the total current R&D budget is essentially flat, declining slightly in real terms, as shown by the solid line across the bars. In constant FY2006 dollars, the 2004 budget represented a peak, followed by a decline in 2005 and a further small decline in 2006, followed by an increase in FY2007.

Despite the fact that the NSF's SBIR program is not procuring for federal requirements, the results of the NRC Phase II Survey showed customers of NSF-funded technologies to include federal agencies and primes. While slightly more than half of sales of NSF Phase II-funded technologies were reported to be going

---

[7]American Association for the Advancement of Science, AAAS R&D Funding Update on NSF in the FY2007 Budget, available online at <*http://www.aaas.org/spp/rd/nsf07hf1.pdf*>.

TABLE 6.3-1 Sales of NSF PII–Funded Technologies Go to Multiple Markets with a Broad and Diversified Customer Base

New and improved products to domestic civilian sector	57%
Meeting federal agency needs	20%
International competitiveness	11%
Support to universities and other institutions	9%
Meeting needs of state and local governments	4%

SOURCE: NRC Phase II Survey.

to domestic civilian markets, as Table 6.3-1 shows, 20 percent were reported to be meeting federal agency procurement needs.

Other survey results show that the NSF's SBIR program is enabling small firms in ways supportive of NSF's mission. It is achieving a high level of participation by firms new to the NSF's SBIR program each year. In 2003, the most recent year examined, more than 60 percent of grantees were new to the SBIR program. As described in more detail elsewhere in this report, it appears that the program is changing small firms' conduct of R&D, enabling them to undertake projects they otherwise would not have, to broaden the scope of research, and to accelerate results.

7

Contributions to Knowledge

7.1 RESEARCH PROGRAM PERSPECTIVES

By nature of their principal activity, Federal research programs share the goal of generating scientific and technical knowledge. They differ, however, in the importance they place on pushing the technology envelope, overcoming high-risk technical barriers, and performing high quality research. They also differ in the importance they attach to disseminating knowledge widely outside the innovator. In addition, they differ in the importance placed on developing performance measures for research quality, knowledge creation, and knowledge dissemination.

7.1.1 Attention to Research Quality and Knowledge Creation

In its statement of program purpose, the Small Business Innovation Research (SBIR) program of the National Science Foundation (NSF) emphasizes that it intends "to increase the incentive and opportunity for small firms to undertake cutting-edge, high-risk, high quality research." In its statement of goals, the NSF's SBIR program states firstly that it aims to promote development of intellectual capital, making awards for research which builds on recent discoveries in basic sciences and engineering. As would be expected for consistency, the NSF's SBIR program emphasizes intellectual merit as one of two merit review criteria for selecting proposals for funding. Thus, the NSF's SBIR program in its statements of purpose, goals, and criteria—hence, intentions—consistently assigns importance to research quality and knowledge creation. In light of its stated purpose, goals, and criteria, and given that the peer-review selection process appears to have integrity, this study found nothing that would suggest the program is falling down on research quality or knowledge creation. At the same time, it is recognized that research quality is difficult to measure and that the value of knowledge created is difficult to predict, takes time to be realized, and is highly variable among projects.

7.1.2 Attention to Knowledge Dissemination and Spillover Effects

Technology development programs, in contrast to basic science programs, generally view knowledge gains as a means to a desired end, not the end point itself. Knowledge gains are seen as capacity building, as providing answers to questions impeding innovation, and in some cases as a means of broadening the scope of program benefits beyond those accruing directly to funding recipients and their customers. In the longer run, the social impacts of technology development programs come to reflect both the direct effects and the spillover effects of public R&D investment.[1]

Agencies that emphasize the value of research to accomplish their own mission-driven goals, such as the Department of Defense (DoD), tend to place less emphasis on measures of knowledge dissemination as an important program output. Programs that emphasize the generation of broad-based economic benefits, such as the Advanced Technology Program (ATP), emphasize both the value of their programs achieved through knowledge dissemination to others and the direct economic impact achieved by the innovator.

The NSF's SBIR program would appear at first glance to be close to the ATP in emphasizing knowledge dissemination and spillovers, in that its second major merit review criterion is "broader impacts." However, the details of the NSF's broader-impacts criterion appear to put less emphasis on broadening economic benefits of the proposed activity to society than ATP's. More important, in practice the NSF's SBIR program has largely ignored potential spillover benefits from knowledge dissemination. "Broader impacts" appears to be defined by the program largely as commercial results.[2] Patenting was the only knowledge-related measure found in NSF evaluation studies of its SBIR program,[3] and it appears that patent data were collected to signal commercial activity.[4] No effective pro-

---

[1] In contrast, basic research programs view knowledge creation as their primary goal. For these programs, most of the agency's output/outcome/impact measures will likely focus on knowledge creation and dissemination.

[2] NRC Program Manager Survey and discussions with program managers.

[3] It should be acknowledged that the required interim, final, and post-grant annual commercialization reports (see Appendix F) each include a request for the reporting, if applicable, of publications, including the reporting of "scientific articles or papers appearing in scientific, technical, or professional journals . . . and any publication that will be published in a proceedings of a scientific society, a conference, or the like." However, no program reporting or use of this information on knowledge creation and dissemination has thus far been discovered in either program management or performance metrics. And the requirement for postgrant annual reports has recently been dropped.

[4] Patents preserve ownership rights to innovations that may be critical to being able to exploit them commercially.

gram reporting or use of publication information to indicate knowledge creation and dissemination was uncovered in either program management or performance metrics.

In fact, it was concluded that the NSF SBIR program management sends mixed messages to grantees about the importance of knowledge dissemination, telling them at conferences to "forget about publishing and focus on commercializing." At the same time, grantees appear in fact to be producing knowledge outputs, including publications, presentations, networking, and patents. There appear to be opportunities for the program to provide a more consistent message that may stimulate knowledge spillovers and to compile a more comprehensive set of indicators of knowledge creation and dissemination for evaluation purposes.

7.2 NRC STUDY FINDINGS ON KNOWLEDGE CREATION AND DISSEMINATION BY THE NSF'S SBIR PROGRAM

The National Research Council's study, through its surveys and case studies, investigated the outputs commonly used to assess knowledge creation and dissemination. The following sections present results for patents, copyrights, trademarks, and scientific publications; licensing agreements; sales of equity; partnering relationships with other companies and investors; and relationships of grantees with universities. The result of another type of analysis that would have been potentially useful, citation analysis, was not performed. In addition, survey results pertaining to the risk profile of projects funded are given.

7.2.1 Patents, Copyrights, Trademarks, and Scientific Publications

Patents, copyrights, trademarks, and scientific publications are important indicators that knowledge has been both created and disseminated. The NRC Phase II Survey grants provided information on intellectual property, including patents, copyright, trademarks, and scientific publications. The responses for the 151 grants (or projects which they represent) reported the knowledge outputs shown in Table 7.2-1 as of the time of the survey. Note that the average number of patents granted (or received) per Phase II SBIR grants in the survey is 0.67. This is more than double the average number reported in the NRC Phase I Survey per Phase I SBIR grant not followed by a Phase II grant.[5]

7.2.2 Licensing

Licensing agreements depend on the protection of the intellectual property. They are another indicator of the creation and dissemination of knowledge.

---

[5]As noted in Section 5.2.1, the reported average number of patents per SBIR grant in the survey is much lower than the reported average number of patents per firm resulting at least in part to SBIR awards.

TABLE 7.2-1 Phase II Survey Results on Intellectual Property

Type	Number Applied for/ Submitted	Average	Number Received/ Published	Average
Patents	159	1.05	101	0.67
Copyrights	49	0.32	42	0.28
Trademarks	42	0.28	33	0.22
Scientific publications	266	1.76	250	1.66

SOURCE: NRC Phase II Survey.

TABLE 7.2-2 Licensing Activities of Phase II Surveyed Grantees with U.S. and Foreign Companies and Investors

Focus of Interactions	Finalized Agreements (%)	Ongoing Negotiations (%)
Interactions with U.S. companies and investors	20	21
Interactions with foreign companies and investors	10	7

SOURCE: NRC Phase II Survey.

Respondents reported licensing as the predominant activity they engaged in with other companies and investors both in the United States and abroad. Table 7.2-2 shows the frequency with which respondents said they had finalized or were negotiating licensing agreements to commercialize technologies resulting from the referenced grants. Respondents appear to form licensing agreements with foreign companies and investors approximately half as often as they form them with domestic companies and investors.

The intense use of licensing signals the underlying importance of intellectual property protection to high-tech small businesses. Case study results also highlight the importance of intellectual protection and licensing activities as a major commercialization strategy of the small businesses. For example, consider the case study of Language Weaver. The company describes itself as "a core technology house based on licensing its software" directly to customers and indirectly through partners who license Language Weaver's technology and incorporate it into their own products. Licensing activities tend to increase the diffusion of a technology's effect, and as noted by Jaffe, licensing tends to increase spillover effects, particularly market spillovers.[6]

---

[6]Adam Jaffe, *Economic Analysis of Research Spillovers: Implications for the Advanced Technology Program*, NIST GCR 97-708, pp. 42–44.

7.2.3 Tracking Knowledge Dissemination by Citation Analysis

Citation studies have been used extensively to show the transfer of knowledge from federally funded projects to others outside the walls of the funded projects, thereby demonstrating the wider potential impact of the federal funds. In the case of paper-to-patent citations, this is done by examining references to scientific and engineering papers on the front pages of U.S. patents. References are also made to previously issued patents. Both sets of patent and nonpatent references comprise the "prior art" of patents.

Citation analysis has been used at various times by the U.S. Department of Energy, the National Institute of Standards and Technology, the Agricultural Research Service, the National Science Foundation,[7] and other federal agencies to show the movement of knowledge from scientific research programs—where impacts are difficult to measure—to industrial technology—where impact measurement is more easily tracked.[8] Patent citation trees are routinely used by ATP, for example, to show the dissemination of technical knowledge via patents from completed projects to other companies and other organizations.[9]

No evidence was found, however, of publication or patent citation analysis by the NSF's SBIR program. Further, no evidence was found of the systematic collection by OII of the detailed publication and patent data from SBIR projects needed to support citation studies. With regard to publications, evidence was found of NSF SBIR program managers sending mixed signals to grantees regarding the importance, or lack thereof, of publications as a project output.[10]

Yet, as indicated by the results of the NRC Phase II Survey conducted for this study, patents and scientific publications are being produced by the NSF's SBIR program. Hence, opportunities exist to encourage program participants to publish when it will not compromise their ability to commercialize. Both publication and patent citation analysis could be used to demonstrate and track knowledge dissemination from NSF SBIR projects to others.

---

[7]The referenced use of citation analysis by the NSF lies outside the NSF's SBIR program. The NSF supported extensive work by CHI Research, Inc., to develop and "clean" databases needed to perform publication citation analysis.

[8]For an example of a citation study performed for a federal R&D program, see J. S. Perko and Francis Narin, CHI Research, Inc., "The Transfer of Public Science to Patented Technology: A Case Study in Agricultural Science," *Journal of Technology Transfer* 22, no. 3 (1997):65–72.

[9]Advanced Technology Program, *Performance of 50 Completed ATP Projects: Status Report Number 2,* NIST Special Publication 950-2, Gaithersburg, MD: National Institute of Standards and Technology, pp. 266–270.

[10]This statement is based on observation of program manager oral responses to proposed questions at an SBIR conference regarding the importance of publications.

TABLE 7.2-3 Equity Sales of Phase II Grantees to U.S. and Foreign Companies and Investors

Focus of Interactions	Company Merger Final (%)	Company Merger Ongoing (%)	Sale of Technology Rights Final (%)	Sale of Technology Rights Ongoing (%)	Partial Sale of Company Final (%)	Partial Sale of Company Ongoing (%)	Sale of Company Final (%)	Sale of Company Ongoing (%)
Interactions with U.S. companies and investors	0	4	5	16	2	6	2	3
Interactions with foreign companies and investors	2	1	4	3	2	2	0	1

SOURCE: NRC Phase II Survey.

7.2.4 Equity Sales

Sales of equity by NSF SBIR grantees to others represent transfers of knowledge. While licensing leaves ownership of the technology with the SBIR firm at the same time that it allows the licensee use of it, a sale of equity transfers ownership to a new owner.

Among the NRC Phase II Survey respondents, activities to transfer equity centered on sales of technology rights to other domestic companies and investors rather than sales abroad. Table 7.2-3. shows that much of this activity was still in process at the time of the survey. Although at the time of the survey, none of the grantee companies had been sold to foreign companies or investors, there was indication that this activity was under way to some extent.

Equity sales are sometimes an essential element in a commercialization strategy. In some cases, companies with the ability to commercialize are located outside the United States, and they may require ownership as a condition for commercializing.[11]

7.2.5 Partnerships of Small Firms with Other Companies and Investors

Partnering with other organizations and people also accomplishes knowledge transfer. For small companies, the formation of partnerships with other compa-

---

[11]For example, according to Brodd there are no volume lithium-ion battery manufacturers in the United States and this may influence commercialization strategies of small companies performing R&D in lithium-ion batteries. Ralph J. Brodd, *Factors Affecting U.S. Production Decisions: Why Are There No Volume Lithium-Ion Battery Manufacturers in the United States?* ATP Working Paper Series, no. 05-01, June 2005.

TABLE 7.2-4 Percentage of Phase II Surveyed Grantees Forming Partnerships with U.S. and Foreign Companies and Investors

Partnering for:	With U.S. Companies and Investors		With Foreign Companies and Investors	
	Finalized (%)	Ongoing Negotiations (%)	Finalized (%)	Ongoing Negotiations (%)
Licensing Agreement(s)[a]	20	21	10	7
R&D Agreement(s)	17	17	5	7
Marketing/Distribution Agreement(s)	16	12	8	2
Customer Alliance(s)	12	18	3	4
Manufacturing Agreement(s)	8	10	3	2
Joint Venture Agreement(s)	3	10	1	2

SOURCE: NRC Phase II Survey.

NOTE: (a) Licensing agreements may or may not entail close partnering, whereas the other listed forms generally do require close alliances and partnering.

nies is often an essential strategy for commercializing a technology. The larger companies they partner with often have manufacturing capacity, marketing know-how, and distribution paths in place. Grantees whose technology is far upstream of consumer goods may need to: partner with other companies for the additional research needed to integrate their technologies into larger systems; partner with Original Equipment Manufacturers (OEM) who purchase the grantees' output as intermediate goods; and form alliances with customers to more effectively reach markets.

The NRC Phase II Survey provided insight about the kinds of partnerships being formed by SBIR recipients. As shown in Table 7.2.4, partnerships for R&D, for marketing and distribution, with customers, and for manufacturing were found to be formed by these grantees.

The case studies also illustrate how partnering as a business strategy transfers knowledge among partnering companies and their researchers. For example, NRT developed a metals processing technology in a strategic collaboration with another company, wTe Corporation, which has an automobile shredder division and specialized knowledge in this application. The joint objective of the companies was to develop an optoelectronic, ultra-high-speed process for sorting metals into pure metals and alloys and apply it to automobile shredding.

7.2.6 Small Firms and Universities

Many of the companies also have a variety of relationships with universities by which knowledge is created and disseminated. Many funded projects have involvement by university faculty, graduate students, and/or university-developed technologies. University faculty and students establish small businesses. Faculty members serve as proposal reviewers. Universities assist firms with proposal preparation and serve as consultants or subcontractors on projects. They also sometimes provide facilities and equipment to assist projects. Even high-school teachers and students sometimes work on projects through a supplemental program provided by the NSF to encourage such opportunities. Sixty-eight percent of companies responding to the NRC Firm Survey had at least one founder with an academic background. Of these, 40 percent of company founders had been employed by a college or university prior to founding the company.

The NRC Phase II Survey showed that nearly half (47 percent) of the referenced projects involved some form of university involvement. The survey data show the prime mode of involvement to be faculty members or adjunct faculty members working on the referenced project in a role other than as principal investigator—as a consultant, for example. The next most frequent modes of involvement were those of graduate students working on the project and university or college facilities or equipment being used on the project. To a lesser extent, universities/colleges were subcontractors on the projects. In some instances, project technologies were originally developed in universities or colleges by one of the participants in the referenced projects. On occasion, the technologies for the referenced projects were licensed from a university or college. Table 7.2-5 indicates the extent to which each type of university involvement occurred in the sample Phase II projects.

The NRC Phase II Survey results show that the NSF's SBIR program helps move research concepts out of the university. Of the Phase II projects surveyed, 14 percent involved technology that was originally developed at a university by a project participant. Five percent of the technologies within the Phase II survey projects were licensed from a university.

Some of the case study firms were found to have ongoing affiliations with universities. For example, NVE Corporation—formerly known as Nonvolatile Electronics and, now, simply as NVE—has maintained ongoing affiliations with the University of Minnesota, Iowa State University, and the University of Alabama. Established by a retired Honeywell executive, NVE developed substantial intellectual property in MRAM technology, nonvolatile computer memory. As NVE pursued MRAM development, it saw related potential applications, such as magnetic field sensors, and has developed several business lines in magnetic field sensors.

7.2.7 Risk Profile

Survey results for funded projects suggest that they have high technical risk—an indicator that they are taking on challenging scientific problems. The NRC Phase I Survey reported technical difficulties as the chief reason for dis-

---

Box A
Language Weaver: Validating University Research

Like a newly born gazelle, Language Weaver found its legs early. It has rapidly developed a fully functional commercial software that uses a statistics-based translation technology from a prototype developed by the company's founders—professors and researchers at the University of Southern California's Information Sciences Institute. While the company founders still have their university posts, the company, under professional management, has grown to 35 employees, with several millions of dollars in revenues.

The founders' insight that they had a marketable product came just as the Internet bubble burst, making private venture capital difficult to secure. "When we were trying to start the company," related Mr. Wong, a founder, "it was a little before 9/11 and no one cared about languages. There were no Senate hearings about languages. Then 9/11 happened, and at the time, we had already submitted a proposal to the NSF. But we didn't hear back until November, and by then the NSF was able to bootstrap us to get us working quickly, moving code from the university to the company. . . . It was after the SBIR grant that everything happened. We started getting government interest as it became apparent that we had something interesting. But we would not have been positioned to move quickly to respond to the need if it hadn't been for that first small amount of NSF funding and the confirmation of the technology."

The Language Weaver translation technology uses a code-breaking approach rather than the standard rule-based approach of existing translation software. The result is very high quality, near simultaneous translation of languages, including Arabic, Farsi, and Somali where, given the United States' current national security requirements, the demand for translation has outstripped the supply of available translators.

---

continuing Phase I projects. The survey of Phase I projects also suggested high technical risk. Of the Phase I projects that did not get a follow-on Phase II grant, a leading reason was technical barriers.

An example of a technical barrier that one of the case study companies, MicroStrain, Inc., is successfully addressing is how to make microminiature, digital, wireless sensors that can autonomously and automatically collect and report data in a variety of applications. These applications range from protecting the Liberty Bell when it was moved, to determining the safety of a high-traffic bridge, to tracking damage in Navy aircraft.

Another example of a technical barrier overcome is provided by the efforts of Language Weaver, Inc., which has boot-strapped a statistical machine translation technology with national security and economic potential out of a university into use on a fast-track basis. It is being used now to translate Arabic, Farsi, Chinese, and other languages into English, and English into other languages with reportedly far greater speed and accuracy than other techniques.

TABLE 7.2-5 Involvement by Universities and Colleges in Phase II
Survey Projects

Type of Relationship Between Referenced Project and Universities/Colleges	Respondents Reporting the Relationship (%)
Faculty members or adjunct faculty member worked on the project in a role other than principal investigator	37
Graduate students worked on the project	27
University/College facilities and/or equipment were used on the project	25
A university or college was a subcontractor on the project	17
The technology for this project was originally developed at a university or college by one of the participants in the referenced project	14
The technology for the project was licensed from a university or college	5
The principal investigator for the project was at the time of the project an adjunct faculty member	5
The principal investigator for the project was at the time of the project a faculty member	1

SOURCE: NRC Phase II Survey.

Box B
MicroStrain Inc.: Real-Time Monitoring and the Liberty Bell

How do you move the Liberty Bell without turning its famous crack into an infamous one? That was the dilemma that National Park Service curators faced in 2003 when they moved the delicate American icon from its longtime home at the Bicentennial Pavilion in Philadelphia to a new display space at the Liberty Bell Center. Casting impurities make the Liberty Bell prone to cracking, but thanks to new microsensor technology developed by MicroStrain Inc., with SBIR awards from the NSF, the bell reached its destination without incident.

The Liberty Bell was moved safely, thanks in part to a technology developed by MicoStrain that uses a network of wireless sensors to autonomously detect and report motion as small as one-hundredth the width of a human hair. Such networks of wireless sensors can be used to monitor in real time the structural health of bridges, roads, trains, dams, buildings, ground vehicles, aircraft, and ships, alerting those responsible to potential failures before they become disasters. Both private manufacturers and government customers—from local municipalities to the Navy—are drawing on this technology to reduce the costs of equipment maintenance and replacement while improving both the safety and the reliability of the nation's infrastructure.

Steve Arms started MicroStrain after completing his graduate degree in engineering. He found the NSF's SBIR program to be particularly helpful in the early stages when he was building the company's technological and commercial capacity. The NSF's relatively more "open topics allowed the company to pursue the technical development that best fit its know-how," he explained, adding that as a result, the company is better able to respond to the needs of its commercial and government customers.

7.3 INDICATORS, NOT MEASURES OF BENEFIT

Economic benefits from knowledge efforts become apparent when they are actually used by others to develop new and improved products, processes, and services. This means that collecting data about knowledge generation and dissemination activities provides only an indirect measure of impacts. Such data can, however, help us construct indicators of potential economic impacts. Examples of possible indicators include the number of patents per research dollar, characteristics of collaborative networks formed, and sales of commercialized goods and services. Trends in these and other indicators may indicate that developments are occurring along an indirect path—as would be expected for projects that are progressing toward the generation of broad impacts.

It is apparent from the NRC Phase II Survey results that it would be possible to compile multiple indicators of knowledge generation and dissemination and early commercialization achievements from NSF SBIR projects, as well as to track them over time. Thus far, however, it appears that such indicators have been developed only partially and on an ad hoc basis. It also appears that more could be done to systematically compile and track indicators of knowledge generation and dissemination if desired.

8

Program Management

The discussion of program management includes developing topics, planning and implementing the grant cycle, conducting outreach to stimulate grant-worthy proposals, selecting proposals for grant, encouraging firms to take the next steps needed to develop and commercialize their technologies, bridging across funding gaps, implementing reporting by grantees, and evaluating program results. The treatment of program management also covers the degree of program flexibility, size of grants, online capabilities, and administration of the program.

8.1 TOPIC DEVELOPMENT AND SELECTION

8.1.1 Topics

As the Small Business Innovation Research (SBIR) program evolved from control by the National Science Foundation's (NSF's) research divisions to centralized management in the mid-to-late 1990s, solicitation topics were reduced in number and oriented less toward scientific disciplines and more toward broad technology areas that would better mesh with business sectors. A purpose was to orient the topics in a way that would increase private-sector commercialization of the innovations derived from the grants. "The NSF SBIR/STTR [Small Business Technology Transfer] program aligned the solicitation topics with external investment and market opportunities and simultaneously preserved the science and engineering alliances with the NSF directorates."[1]

---

[1]Office of Industrial Innovation (OII), Strategic Plan (draft, June 2, 2005), p. 13.

In 2005, the program identified the following seven broad topic areas:

Biotechnology (BT)
Electronics Technology (EL)
Information Based Technology (IT)
Advanced Materials and Manufacturing (AM)
Chemical Based Technology (CT)
Security Based Technology (ST)
Manufacturing Innovation (MI)

When a solicitation is held in a topic area, it is fleshed out with subtopics. The subtopics add specificity to the solicitation. (For additional discussion, see Chapter 4.)

For example, Table 8.1-1 shows the 2004 "Advanced Materials" (AM) second- and third-level subtopics. Note that the third level serves to eliminate, as well as to define, areas of inquiry. AM included "Manufacturing" and "Chemical Processes," but these are shown here only to the second level.

In its first solicitation, the Securities Technologies (ST) topic area was defined as cross-disciplinary, and proposals submitted under this topic had to represent the convergence of at least two of the following three technologies: nanotechnology, biotechnology, and information technology (both hardware and software). Proposals also had to "be responsive to a subtopic within the solicita-tion," listed in Table 8.1-2, effective April 2004.

8.1.2 Sources for Topic Ideas

According to NSF SBIR program management:

> . . . topics are rooted in the agency's vision and strategic goals. In particular, SBIR and STTR are uniquely positioned to emphasize NSF vision of innovation. Since NSF is not the final customer for the SBIR/STTR grantees, it is impera-tive that our grantees are positioned to tap into private sector capital, which is essential for commercializing the technology developed under the SBIR grant. Therefore, NSF topics reflect the market opportunity and are aligned with the broad investment business. At the same time the topics also resonate with the science and engineering disciplines that NSF supports within its Directorates and Divisions. (SOURCE: "NSF SBIR Response to NRC Questions," January 2004)

This study found no formal process for soliciting outside input in the genera-tion of NSFs topic ideas—such as a white-paper process used to develop thematic ideas in concert with industry. Rather, the sources for topic and subtopic ideas were said to come from NSFs program managers as they interact with industry and others at conferences and workshops, or through an approach devised by a program manager and approved by the SBIR director.

TABLE 8.1-1 Advanced Materials, Second- and Third-Level Topics—
Example 1 from the 2004 Solicitation

A. Advanced Materials
 1. Environmentally benign technology
  • Improved techniques for recycling
  • Processing of recycled materials
  • Pollution prevention/avoidance processes
 2. High temperature materials
  • Metal, ceramic, and composite materials developed for high temperature applications (e.g., improved turbine blade materials/processing)
 3. Structural materials
  • Improved strength, toughness, fracture resistance, etc., materials
  • Processing and material improvements
 4. Corrosion-resistant coatings
  • Surface coatings and modifications which lead to improved corrosion resistance
  • Improvements in materials for corrosion resistance
 5. Tribological and wear-resistant coatings
  • Surface coatings and modifications which lead to improved wear resistance and/or reduced friction
  • Material improvements in tribology/wear
 6. Engineered materials
  • Improved processing and/or materials with engineering applications other than those listed above (No nanotechnology, biotechnology, or electronic materials)
  • Smart materials
  • Shape memory alloys
 7. Surface modification and thin film technology
  • Process improvements for modifying surfaces and applying thin films
  • Material improvements related to process modification that are not related to corrosion or tribology
B. Manufacturing (also developed to a third level—but only shown to level 2)
 1. Polymer processing and rheology
 2. Casting/molding processes
 3. Machining and material removal processes
 4. Deformation processes
 5. Powder material processing
 6. Composite manufacturing processes
 7. Additive manufacturing
 8. Manufacturing process control
 9. Machine design
 10. Joining and assembly processes
 11. Nontraditional material removal processes
 12. Manufacturing systems
C. Chemical Processes (also developed to a third level—but only shown to level 2)
 1. Separations applications
 2. Novel catalytic systems
 3. Photochemical or electrochemical applications
 4. Fluid flow applications
 5. Combustion-related processes
 6. Applications of plasma technology
 7. Thermal energy applications
 8. Reactor engineering applications
 9. Chemical technology

NOTE: The first-level topic is Advanced Materials; the second-level topics are those numbered; the third-level topics are those preceded by bullets.

TABLE 8.1-2 Security Technologies, with Subtopics Listed—Example 2 from the 2004 Solicitation

A. Prevention
   1. Tools and systems for smart buildings/structures for
      • Public resources (energy, water, air) monitoring and control
      • Human resources coordination during emergency situations
   2. Networked sensors and tools to provide real-time information on structural integrity
   3. Systems beyond optical recognition (finger, facial, or retinal) that provide quick (under two minutes), unambiguous identity authentication
B. Detection
   1. Terahertz sources and detectors
   2. Systems utilizing hardened RFID and other modalities for secure supply chain management, traceability, and counterfeit detection
   3. Multiscale integration tools including new-generation packaging to enable nano-micro-meso system integration
   4. Compact, cost-effective, environmentally friendly and long-lived power supplies (e.g., for widely disbursed wireless sensor networks)
      • Biomimetic
      • Energy-scavenging systems
      • Photovoltaic systems
      • Acoustic-voltaic systems
      • Other systems that provide energy densities exceeding 1000 Wh/kg
   5. Proteomic-based biometric systems (NOTE: NOT PCR-based)
C. Treatment
   1. Site-specific wireless and wireline data/information systems to empower responders and emergency managers
   2. Systems and tools for wide-area rapid treatment dissemination (including agricultural applications)
   3. Systems with information management capability for rapid susceptibility
D. Remediation
   1. Systems and approaches for chemical (including industrial), biological, or radiological event remediation
      • Homes
      • Workplaces
      • Reservoirs
   2. Stand-alone, single-use, widely dispersible sensors or detectors for environmental monitoring (NOTE: NOT lab-on-a-chip systems)
   3. Environmentally friendly agent-specific widely dispersible decontamination media
      • Organic-based
      • Inorganic-based
E. Attribution
   1. Taggants and anticounterfeit/product authentication systems with unique spectral and other signatures
   2. Field-deployable front-end sample preparation systems to extend the reach of laboratory-based analytic equipment

An example of the latter, an NSF SBIR program manager promoted revision of the manufacturing topic. According to the program manager,[2] the manufacturing topic area had, over time, become "stale" and "not enough connected to the real world of manufacturing, such that introducing high tech alone would not help manufacturing's competitive problem." The manufacturing subtopics were described as holdovers from the time when the SBIR had operated as a decentralized program with close affiliation to its research divisions and with an academic orientation to the topics.

To revitalize the manufacturing topic area, the program manager put together a panel, using volunteers from the topic area's review panels. As a result, "the topics became more open and more targeted to issues of established manufacturing. In addition, the experts sent a recommendation for more reviewers for the manufacturing area."[3]

Topics are posted as part of a solicitation announcement. During the time between the initial solicitation announcement and the proposal due date, it appears that modifications may be made at the subtopic level if developments or ideas from the outside suggest to the program manager(s) that modifications are warranted. These modifications reportedly are handled by addendums to the solicitation list. Letters may be sent out to companies to alert them to the additional ideas. Supplemental funds may be provided to support newly identified areas of special interest.

Through outreach activities of program managers and postings on the program's Web site, further information on acceptable subtopics is conveyed to the public. For example, feedback was given to potential proposers on the acceptability of homeland security as a research topic area prior to issuing the topic area and after release of the solicitation. Outreach activities by program managers can further delineate topic areas after release of the solicitation.

NSF SBIR program management provided the following statement regarding topic modification after announcement of a solicitation:

> NSF SBIR/STTR topics are not modified or changed once the solicitation is announced and published on the NSF SBIR Web site. The published solicitation includes submission instructions and proposal submission deadlines. However, NSF SBIR program management will make modifications to the topics areas prior to the solicitation announcement. NSF SBIR program management will stay aligned with NSF strategic goals, make changes based on current technological trends, and give careful consideration to the market and investment community. (SOURCE: "NSF SBIR Response to NRC Questions," January 2004)

Presumably, the apparent discrepancy between what this inquiry found and the NSF management statement lies in the distinction between topics and subtopics. Furthermore, it should be noted that what may appear at first glance to

---

[2]Based on interview with NSF Program Manager Cheryl Albus on January 7, 2004.
[3]Ibid.

represent a change at the subtopic level, may not always actually change the acceptability of a given research theme. For example, next-generation vehicles were spelled out in one solicitation's subtopic list, but not in the next. Yet, next-generation vehicles reportedly remained an acceptable topic of research proposed under other specified subtopics.

The program may need more transparency in defining its topics and more communication with its partnering communities. A useful model is provided by the Advanced Technology Program. (ATP). From 1994 through 1998, the ATP used a white-paper process whereby any organization or individual could propose a topic area for a competition. The purpose was to ensure maintenance of a bottom-up approach that would result in selection of topics in touch with industry and the marketplace. The program published guidelines for preparing and submitting the white papers. White papers were grouped by common themes, and when ideas gained momentum, public workshops would be held to assess the level of interest, to further develop the ideas, and, if merited, to prepare a topic description that would be widely circulated for comments. A review board periodically considered emergent ideas for topics, helped establish priority among competition topics, and recommended to the director those that might form a subprogram for a series of funding competitions. In this way, the bottom-up selection process was maintained as the program experimented with topic specification. At the same time, in parallel to the "topic competitions," the ATP held an "open competition" as part of each solicitation to provide an open door to all topics.

8.1.3 Agency-Driven versus Investigator-Driven Approach to Topics

In most cases, NSFs topic specification leaves open the approaches and techniques that an investigator can take to respond to the particular problem or opportunity in the topic. At the same time, the NSF's SBIR program information in the past stated that the applicant must propose within the announced topic areas or the proposal is rejected, and it currently states that the proposal may not be considered if it is not responsive to the program announcement/solicitation.[4]

While none of the companies interviewed in the case studies complained that the NSF overspecified topics, they did comment that the NSF SBIR program's broad definition of topic areas is unique compared with other agency programs. However, firms did comment on the length of time they had to wait until their topic areas come up for solicitation. According to one program official, the time to wait for a topic to be repeated has gone up, to as long as eighteen months.[5] Also, the larger eligible small firms were concerned that the NSF's SBIR program limits the number of proposals a firm can submit in response to a given solicitation to four.

---

[4]Reasons for returns are listed under "Frequently Asked Questions," found at <*http://www.nsf. gov/funding/preparing/faq/faq_r.jsp?org+IIP#returns*>.

[5]Telephone interview with Joseph Hennessey, NSF, March 3, 2006.

TABLE 8.1-3 Criteria Used to Guide "Topic" Development

Criteria Used to Guide Development of Topics[a]	Weight (%)
Cutting edge of the field	45
Likely commercial technologies	45
Other	10

[a]The response suggested that "topic" was interpreted broadly to include subtopics.

SOURCE: NRC Program Manager Survey.

8.1.4 Topic Decision Making

The broad topic areas that were in effect several years ago were reportedly developed by the SBIR program management staff acting as a working group.[6] The subtopic areas change, on average, about 35 percent each year.[7] The decisions are made by the program manager(s) responsible for the solicitations, with informal input from industry and with concurrence from other members of the program management staff who edit and adjust the topics/subtopics according to the criteria assigned the weights shown in Table 8.1-3.[8] The program director makes the final topic selection.[9]

The Office of Industrial Innovation's (OII's) 2005 Strategic Plan noted the "increased awareness of and the necessity for the SBIR/STTR program to be aligned with national needs."[10] It set forth an action plan to: (a) identify technologies with external investment/market focus; (b) exploit emerging discoveries from NSF-supported science, math, and engineering disciplines as subtopics; and (c) respond to national priorities set forth by the administration and other emerging or pressing societal needs. The perspective expressed in the 2005 Strategic Plan is that most of the currently identified major topics will stay relatively constant and that changes will likely be made in response to emerging national needs.

The NSF has also made decisions about topics it will not accept. Guidelines state that it will return proposals if they propose research in the following areas: (1) weapons research; (2) biomedical research (except bioengineering research); (3) any topics that fall in the area of classified research; and (4) any topics for which the primary purpose is demonstration, technical assistance, literature survey, or market research.

---

[6]Based on interview with NSF Program Manager Cheryl Albus on January 7, 2004.

[7]NRC Program Manager Survey completed by Joseph Hennessey, NSF. Note that the survey referred only to topics, not subtopics, but the nature of the responses suggested a broader interpretation than the major topics only.

[8]Ibid.

[9]Ibid.

[10]Office of Industrial Innovation, Strategic Plan, p. 13.

Applicants are instructed to be responsive to NSF topics if they wish their proposals to be considered. They are instructed to designate "one, and only one, of the topics, keeping in mind that a firm cannot submit more than 4 proposals per solicitation (which includes the parent company and any of its subsidiaries). The topic name and the appropriate subtopic letter MUST be identified on the cover sheet."

In connection with its 2003 solicitation, the NSF provided the following instructions for classifying proposals by topic, noting that the NSF had made substantive revisions to subtopics under its then-four technology areas:

> NSF has established a cascading decision-making procedure in selecting the fit of each sub-topic under the four broad solicitation topics. The hierarchy for the fit of sub-topic starts at the top with BT [*Biotechnology*], followed by EL [*Electronics*], followed by IT [*Information Technology*], and finally AM [*Advanced Materials*]. The following are presented as illustrative examples. If the research is biology-based, it is BT. If the research is electronics or information or materials-based for applications in biotechnology such as devices for medical or bioinformatics or biocompatible materials, it should be submitted to the BT topic. If the research is electronics or photonics or magnetism-based, it is EL. Most instrumentation outside the BT application area fits into EL. If research is information or materials-based, such as embedded software or nano carbon tubes for use as semiconductors for electronic applications, it should be submitted to the EL topic. If the research is computer science or cognitive science-based, it is IT. If the research is modeling and simulation of engineering applications with software as the resultant commercial product, it is IT. If the research is on structural materials or chemical processes, it is AM. If research is on mechanical parts or manufacturing processes, it is AM. These examples are not meant as a comprehensive list of research opportunities but to assist in finding the proper fit for research ideas under the four NSF solicitation topics."

> (SOURCE: NSF Web site, <*http://www.eng.nsf/gov/sbirspecs*>, as of November 2003.)

The NSF SBIR FY2006 Phase I solicitation lists four topics: Advanced Materials (AM), Information Technology (IT), Manufacturing Innovation (MI), and Emerging Opportunities (EO). Each topic is further developed under "Research Topics." Additional information on recent developments regarding topics lists is provided in the overview of Section 4.4.3 of this report.

Funding is not apportioned among the topics/subtopics in an attempt for equality among them. Rather, the strategy is to fund the projects with the most merit. Sometimes topics/subtopics are narrowed or eliminated to reduce the number of applications.[11] It was not determined when this narrowing or elimination of topics/subtopics occurs or what its effect is on applicants—both issues of possible concern.

---

[11]NRC Program Manager Survey, op. cit.

In conclusion, the NSF's SBIR program headquarters decides the topic areas for the program and the subtopics for a given competition with the assistance of program managers and, sometimes, other program-manager-devised sources, such as convened panels.[12] The topics are fixed once the solicitation is issued, but subtopics appear to evolve in response to ongoing developments.

8.2 OUTREACH

The NSF has an active outreach program and has sponsored the SBIR spring and fall national conferences since the inception of the program in 1982. These conferences provide information on upcoming competitions, include workshops that offer training, and allow face-to-face meeting opportunities for grantees and prospective applicants with program managers and each other. In addition to the prime audience of small businesses, these conferences are aimed at sales and marketing professionals, university researchers interested in business, scientists, prospective partners, and others.

Notifications of upcoming conferences and meetings are easy to find through an online search, which yields notices of meetings on several Web sites. For example, one site sponsored by the NSF gives substantial information about the SBIR and STTR programs. [13] It lists national and regional conferences and events, details upcoming solicitations, provides links to federal agencies, provides a guide to state resources, announces partnering opportunities, and enables companies and others to join an email list to receive notices of NSF-sponsored national SBIR/STTR conferences.

The Small Business Administration (SBA) provides outreach for the program as a whole.[14] State SBA-related sites, as well as other federal agency sites, are also rich sources of SBIR/STTR information for prospective applicants. The Pacific Northwest National Laboratory (PNNL) operates an agency-wide SBIR/STTR online alert service.[15]

Privately operated sites also provide SBIR information. The SBIR Resource Center™ operates one such privately operated site that purports to provide up-to-date information covering ten SBIR/STTR agencies in one place.[16] Another site claims to be "the most comprehensive and easy to use SBIR information site."[17]

---

[12]Based on a telephone interview with NSF Program Manager Rosemarie Wesson on December 1, 2003, and a face-to-face interview with NSF Program Manager Cheryl Albus on January 7, 2004.

[13]See <http://www.sbirworld.com>.

[14]See <http://www.sba.gov/sbir>.

[15]The PNNL-operated online alerting service may be found at <http://www.pnl.gov/edo/sbir.stm>.

[16]For more about the SBIR Resource Center, see <http://www.win-sbir.com>.

[17]See the Web site for "SBIR Gateway" at <http://www.zyn.com/sbir>.

8.2.1 Agency Outreach Objectives

A review of NSF-supported Web sites targeted at potential applicants suggests that the NSF has the objective of reaching areas that have submitted large numbers of applications as well as those that have not. Searches of NSFs outreach offerings identified many varied outreach opportunities. For example, a 2004 search of the NSF-supported *SBIRworld.com* revealed a national NSF SBIR outreach conference scheduled in Atlanta, Georgia, and a recently completed national conference in Boise, Idaho. Both Georgia and Idaho are states that supply a relatively low number of SBIR proposals. The search revealed a workshop, "How to Prepare Winning Proposals for SBIR and STTR," scheduled in Livermore, California—California being the state with the highest number of SBIR applications and grants. The search showed several regional meetings—including one in North Carolina, a relatively low-application state, and one in Ohio, a relatively high-application state. Similarly, a 2005 search of the same NSF-sponsored Web site showed multiple workshops aimed at prospective applicants in diverse parts of the country. It included an "SBIR Grant Writing Workshop" to be held at Florida State University and an "SBIR/STTR Phase I Proposal Preparation Workshop" to be held at the Moore School of Business, part of the University of South Carolina. Florida is a mid-tier state in terms of numbers of applications, and South Carolina has fewer SBIR applications than average. Thus, the outreach activities of the NSF are not limited to a single geographical region or to either low- or high-application states.[18]

Through its association with the "Experimental Program to Stimulate Competitive Research and Institutional Development" (EPSCoR), however, the NSF's SBIR program, like the other agency SBIR programs, has a special tie to low-application states. The EPSCoR program aims to increase the ability of states that receive a low proportion of federal research funds to become more successful in attracting such funds. It often partners with other state programs that have similar goals. For example, in conjunction with EPSCoR, Nevada's Small Business Development Center and the Nevada Commission on Economic Development joined forces to increase the number of SBIR grants going to firms in the state. They tried to raise funds for state-based "Phase 0" competitions and to provide assistance to companies developing SBIR proposals.[19] The NSF's SBIR program receives supplemental funding from EPSCoR that allows it to participate in SBIR events in EPSCoR states. Each year, the OII schedules one of its national conferences in an EPSCoR state and one in a non-EPSCoR state.[20] The OII works jointly with EPSCoR to develop a competitive research infrastructure within the

---

[18]Not all the state and regional effects would be NSF sponsored. In some cases, when resources permit, NSF staff members participate in state SBIR-related workshops and meetings.

[19]This outreach activity in the State of Nevada is described at <*http://www.nevada.edu/epsor/sbir.html*>.

[20]Telephone interview with Joseph Hennessey, NSF, October 18, 2005.

TABLE 8.2-1 NSF SBIR Outreach Activities and Their Relative
Importance

Type of Outreach Activity	Importance as a Share of NSF's Overall Outreach Program (%)
SBIR National Conferences	50
State conferences	15
NSF National Agency SBIR Meeting	10
Other agency conferences and outreach meetings	10
The SWIFT Bus Tour	10
Academic conferences	5
Total	100

SOURCE: NRC Program Manager Survey.

EPSCoR states and territories. According to program administrators, "Special
consideration is given by the SBIR/STTR program regarding funding decisions
for proposals received from EPSCoR states and territories in collaboration with
the EPSCoR program."[21]

8.2.2 Outreach Programs

Table 8.2-1 lists the major types of outreach activities undertaken by NSF
SBIR staff and indicates the relative importance program managers place on each
activity.

According to the NRC Program Manager Survey, the NSF views the SBIR
national conferences as the premiere outreach activity because they draw the largest number of applicants and are the most cost-effective. Next in importance are
the state conferences. The NSF assigns its own national conference a weight of 10
out of 100, indicating that it is of equal importance to other agencies' conferences
and outreach meetings and of equal importance to "the SWIFT Bus Tour"—a
bus tour of program managers from several federal agencies who publicize SBIR
grant opportunities by periodically traveling together to regional state-sponsored
small business conferences and meetings. Academic conferences are considered
by OII to be the least important of the outreach activities.

The SBIR office often engages in partnering to provide outreach services.
Table 8.2-2 lists the types of organizations that partner with the NSF for outreach.
As noted previously, the NSF also partners online with various sources to make
its outreach activities known.

Assistance programs that help companies prepare their SBIR proposals are
offered by universities, state agencies, regional associations, and mentor companies. While these programs are generally not sponsored or run by the NSF, the

---

[21]"NSF SBIR Response to NRC Questions," January 2004.

TABLE 8.2-2 Partnering to Provide Outreach Services

Partners of NSF to Provide Outreach Services
Business organizations
State and other nonfederal government agencies
Academic units
Private firms

SOURCE: NRC Program Manager Survey.

program often contributes its staff to serve as speakers.[22] In that sense, the events comprise a component of the NSF outreach program.

The NRC Phase II Survey provided information on the frequency with which respondents received assistance from these organizations in preparing their Phase II proposals. Most of the respondents (91 percent) received no assistance in preparing the Phase II proposals that led to the referenced grant; only 9 percent did. All of those who received assistance found it useful: Most (62 percent) found it "very useful," and the rest (38 percent) found it "somewhat useful." Those who did receive assistance received it from universities (5 percent), from state agencies (3 percent), and to a lesser extent from mentor companies (1 percent). None of the survey respondents received assistance from regional associations.

The case studies also provided examples of companies applying for SBIR grants with the help of these assistance programs. For example, MicroStrain tapped Vermont's EPSCoR Phase O grants to leverage its ability to gain federal SBIR grants. T/J Technologies learned how the SBIR worked from MERRA, a Michigan-based organization aimed at boosting the state's technology businesses. Even after MERRA was dissolved, T/J received assistance from its former staff members to obtain additional federal research funds.

It should also be noted that NSF SBIR program managers provide one-on-one counseling to individual potential applicants. Metrics for this activity are given in Section 8.2.3.

8.2.3 Agency Outreach Benchmarks and Metrics

The NRC Program Manager Survey elicited the following metrics for the NSF outreach:

- An estimated 20 percent of a program manager's work time is spent on outreach activities.
- The NSF SBIR program manager attends 8 conferences on average each year, spending an average of 20 total days in attendance.

---

[22]Telephone interview with Joseph Hennessey, NSF, October 18, 2005.

- The NSF SBIR program manager spends an average of about four hours per week providing one-on-one counseling to individual potential applicants (not counting existing grantees).

The NSF has achieved nationwide coverage in terms of the geographical dispersion of Phase I and Phase II applications. The program has also received applications from Puerto Rico and the U.S. Virgin Islands. By the start of 2005, Phase IIB had attracted applications from 36 states.

8.3 GRANT SELECTION

8.3.1 Description of Selection Processes for Phase I, Phase II, and Phase IIB Grants

At its Phase I and Phase II stages, the NSF's SBIR program uses a peer review process to identify proposals for potential selection. Individual reviewers rate proposals, then, meeting as a panel, provide a consensus funding recommendation that goes to the relevant program managers, who may accept or override the recommendation. The program managers then make recommendations to NSF headquarters as to whether to fund each proposal. These recommendations may be accepted or overturned on a variety of grounds. Examples of grounds for overturning a program manager's recommendation are that the firm had essentially the same proposal funded in another program, that there was insufficient funding to permit funding the proposal, or that one or more "Additional Factors" are triggered, such as the firm had received "an excessive number of grants."[23]

The selection of proposals for grants at each phase centers on the application of the same two formal selection criteria. However, the detailed guidance on considerations to be taken into account in applying these two criteria differs for each phase. Guidance from the NSF indicates that while proposals must address both merit review criteria, reviewers are asked to address only those considerations that are relevant to the proposal being considered and for which they are qualified to make judgments. Furthermore, the considerations are termed "suggestions and not all will apply to any given proposal." The application of "additional factors" may also enter the selection process, as discussed further in Section 8.3.3.

The two formal merit review criteria used for both Phase I and II proposals, including Phase IIB proposals, are the following:[24]

---

[23]These examples are based on an interview with NSF Program Manager Rosemarie Wesson on December 1, 2003, and a face-to-face interview with NSF Program Manager Cheryl Albus on January 7, 2004. The examples are consistent with later responses of OII staff to the study's Program Manager Survey, which identified additional criteria that are sometimes applied in making grants. The use of the additional criteria was confirmed by discussions with OII management on August 17, 2005.

[24]The National Science Board approved revised criteria for evaluating proposals at its meeting on March 28, 1997 (NSB 97-72).

(1) What is the intellectual merit of the proposed activity? (This criterion addresses the overall quality of the proposed activity to advance science and engineering through research and education.)

Underlying considerations in applying this criterion are the following:

- Is the proposed plan a sound approach for establishing technical and commercial feasibility?
- To what extent does the proposal suggest and explore unique or ingenious concepts or applications?
- How well qualified is the team (the principal investigator, other key staff, consultants, and subgrantees) to conduct the proposed activity?
- Is there sufficient access to resources (materials and supplies, analytical services, equipment, facilities, etc.)?
- Does the proposal reflect the state-of-the-art in the major research activities proposed? (Are advancements in state-of-the-art likely?)
- Added to the foregoing for Phase II proposals As a result of Phase I, did the firm succeed in providing a solid foundation for the proposed Phase II activity?

(2) What are the broader impacts of the proposed activity? (This criterion addresses the overall impact of the proposed activity.)

Underlying considerations in applying this criterion are the following:

- What may be the commercial and societal benefits of the proposed activity?
- Does the proposal lead to enabling technologies (instrumentation, software, etc.) for further discoveries?
- Does the outcome of the proposed activity lead to a marketable product or process?
- Evaluate the competitive advantage of this technology versus alternate technologies that can meet the same market needs.
- How well is the proposed activity positioned to attract further funding from non-SBIR sources once the SBIR project ends?
- Can the product or process developed in the project advance NSF goals in research and education?
- Does the proposed activity broaden the participation of underrepresented groups (e.g. gender, ethnicity, disability, geography, etc.)?
- Has the proposing firm successfully commercialized SBIR/STTR–supported technology where prior grants have been made?

Prior to listing the two main criteria listed above, the NSF Web site states, "Other factors that may enter into consideration include the following: the balance among NSF programs; past commercialization efforts by the firm where previous grants exist; excessive concentration of grants in one firm or with one principal investigator; participation by woman-owned and socially and economically disadvantaged small business concerns; distribution of grants across the States; importance to science or society; and critical technology areas."

After listing the two merit criteria, the NSF lists another set of "additional factors" to be addressed in proposals and taken into consideration by reviewers: "Integration of Research and Education," and "Integrating Diversity into NSF Programs, Projects, and Activities."

The NSF statement of selection criteria next indicates another criterion for Phase II proposals: "Review of the Proposal's Commercialization Plan." Considerations in this review include the following:

- Market Need, Expected Outcomes, and Impact: Does the company present a compelling value proposition for the Phase II Project? Does the discussion of need demonstrate that there is market-pull and breadth of potential commercial impact for the innovation? In addition, does the proposer make a solid case that there are potential societal, educational, and scientific benefits of this project? Does the noncommercial impact add to the overall significance of work being proposed?

- The Company: Does the company have focused objectives and the appropriate core competencies? Does the company have the appropriate resources to perform the tasks being proposed and to take the project through to commercialization? If the company has several years of experience, has it experienced growth? Does the company have a good record of commercializing prior SBIR/STTR projects or other research? Does it appear that the company can grow/maintain itself as a sustainable business entity?

- The Market, Customer, and Competition: Does the PI/company understand the market in which the product will be introduced? Is the customer adequately and correctly described? Are the benefits to the customer and the hurdles to acceptance of the innovation adequately described? Does the PI/company know and understand the competitive environment? How would you rate the proposer's ability to execute a marketing and sales program to bring the technology successfully to market in view of this competition (or competitive environment)? What are the strengths and weaknesses of the company's marketing and sales strategy?

- Intellectual Property (IP): Is intellectual property addressed and are there plans for sufficient protection to get the product to market and attain at least a temporal competitive advantage? What is the company's prior record in this area? Please comment on the company's strategy to build a sustainable business through protection of intellectual property.

- The Financing: Has the company properly estimated the amount of funding needed in Phase III? Does the company have a high probability of securing this funding? Has the PI/company identified specific companies for financial commitments, prototype purchase, and/or will they fund themselves? If there are no "hard" commitments for funding (that is, letters of interest or intent), does the company have a solid road map for pursuing the funding needed to commercialize?

- Revenue Stream: Are the plans for generating a revenue stream adequately described? Are the revenue projections and the assumptions behind the revenue projections realistic? Is the revenue stream sustainable? Will it lead to robust company growth or at least sustain the product (and/or the service) through its life cycle?

While Phase I proposals are evaluated on both of the main criteria, the focus has been on technical feasibility. Approximately 25 percent of reviewers have both technical and business backgrounds, such as being a CEO of a past grant recipient company. At its meeting in 2004, the Committee of Visitors (COV) recommended that more consideration be given to commercial potential in evaluating Phase I proposals and that the review panels for Phase I proposals have more well-qualified representatives from the business sector.

At the Phase II level, selection criteria since 1992 have focused on commercial potential and commercialization planning, in addition to the research proposed. In 2004 the COV recommended that Phase II reviewers give more attention to the societal impact considerations in the "broader impacts" criterion, implying that the current focus is too narrow. Thus, changes in Phase II would also result from OII's implementation of COV-recommended changes in considering the second merit selection criteria, "Broader Impacts."

The Phase IIB selection process has always focused on the ability of applicants to secure third-party financing. There is no indication of change in this practice. However, there has been a recent change in the selection process for "supersized" Phase IIB grants. Now the selection process requires oral presentations by applicants and their third-party financiers as part of the grant-decision process. According to a company participant who had gone through the selection process for a supersized Phase IIB grant, the focus of the oral presentation was on the business case of the company and its financing.[25] NSF management also reportedly would like to increase its due diligence by funding program manager staff visits to companies, particularly those who are applying for the supersized Phase IIB funding.

---

[25]See the case study of Language Weaver in Appendix D.

The review process concludes with a debriefing of unsuccessful applicants and notification of winners. The debriefing provides the following materials electronically to the principal investigator and the company officer/organization representative: (a) verbatim copies of reviews, excluding the names of reviewers; (b) summaries of review panel deliberations, if any; (c) a description of the process by which the proposal was reviewed; and (d) the context of the decisions (such as the number of proposals and grant recommendations, and information about budget availability).

8.3.2 Peer Review Panels—Membership, Selection, and Qualifications

As part of a proposal, SBIR applicants are asked to provide a list of prospective reviewers whom they regard as experts in the relevant field(s). Applicants are also asked to provide a list of individuals whom they do not wish to be considered as reviewers for their proposal.

The program materials state that "special efforts are made to recruit reviewers from nonacademic institutions, minority-serving institutions, or adjacent disciplines to that principally addressed in the proposal." Yet 60 percent of the reviewers continue to come from academia; 20 percent are industry scientists; 15 percent are other industry personnel; and 5 percent come from other sources.[26] Warnings and requests are made to all prospective reviewers about the need to avoid potential conflicts of interest. Per policy, the NSF will not use a reviewer in the peer review process if that individual has any affiliation with a company who has a proposal under review for funding in that funding cycle. This includes employees of the company or consultants to that company. The NSF has used past and potentially future applicants as reviewers if they determine that no conflict of interest exists.

According to discussions with several program managers, it is each NSF program manager's responsibility to find reviewers for his or her topic areas.[27] Reportedly, they give more attention to the designation by applicants of "who should not review their proposal" than the designation of "who should review it."[28] The program managers devise various "schemes for building their reviewer pools," in addition to obtaining suggestions from applicants. One example that was given as a way program managers build reviewer pools is to send out letters to deans of major university departments asking them to recommend to their new faculty members that they participate as reviewers in the SBIR program.[29]

According to program officials, "there are no restrictions prohibiting submitters in a current solicitation, or past or future applicants to serve as NSF

---

[26]NRC Program Manager Survey.

[27]Based on an interview with NSF Program Manager Rosemarie Wesson on December 1, 2003, and a face-to-face interview with NSF Program Manager Cheryl Albus on January 7, 2004.

[28]Interview with Cheryl Albus, NSF Program Manager, January 7, 2004.

[29]Ibid.

reviewers."[30] A case against the practice of using current proposers as reviewers, even if they are not used to review directly competing proposals to their own, is that when grant dollars are limited, a decision to fund one proposal may mean insufficient funding for other proposals. A potential real or apparent conflict of interest could arise when a current applicant reviews the proposals of others in the same budgetary cycle—a problem that would not be eliminated by requiring all reviewers to sign a conflict of interest (COI) form.[31]

All NSF SBIR reviewers are provided with instructions and guidance regarding the SBIR peer review process and are compensated for their time. NSF policy regarding the NSF review process and compensation can be found in the *NSF Proposal and Grant Manual*.[32]

Proposals are divided into appropriate technical topics, and review panels are assembled. Each panel at the Phase I stage typically deals with eight to ten proposals. Each reviewer is provided his or her proposals electronically 30 days prior to the panel meeting to prepare individual reviews, which are then submitted electronically to the NSF. The panel typically meets as a group to discuss each proposal with the program manager. The perceived strengths and weaknesses of each proposal are then presented in a panel summary.[33]

For Phase I proposals a minimum of three and a maximum of six reviewers are used. The average is four reviewers.[34] A typical panel makeup for Phase I reviews is three technical reviewers and one reviewer with a combined technical and business background. However, as has been noted previously, the SBIR COV concluded in 2004 that inadequate consideration is being given to commercial potential in evaluating Phase I proposals. The COV has recommended that more well-qualified representatives from the business sector be used to staff business panels for Phase I review.

Phase II proposals typically receive a minimum of three technical and three business reviews. Technical and business reviewers external to the NSF's SBIR program office perform the review. At Phase II, specific arrangements for proposal reviewers vary depending on the program manager. Some program managers, for example, mail their proposals to technical reviewers and convene a panel for business reviews. Some convene a combined panel of technical and business reviewers. Within this variability, however, the business reviewers are reportedly always convened as a panel.[35] The business reviewers include entre-

---

[30]"NSF SBIR Response to NRC Questions," op. cit.

[31]The conflict of interest form is based on NSF Manual 14, *NSF Conflicts of Interest and Standards of Ethical Conduct*. The form may be found on-ine at <*http://www.eng.nsf.gov/sbir/COI_Form. doc*>.

[32]NSF SBIR guidance for reviewers is provided online at <*http://www.eng.nsf.gov/sbir/peer_ review.htm*>. The *NSF Proposal and Grant Manual* is provided online at <*http://www.inside.nsf. gov/pubs/2002/pam/pamdec02.6html*>. See also "NSF Response to NRC Questions," op. cit.

[33]NRC Program Manager Survey, op cit., and "NSF SBIR Response to NRC Questions," op. cit.

[34]Ibid.

[35]Ibid.

preneurs, business school professors, professional investors, corporate managers and investors, and others with business and financial experience.

Reviewers at the Phase IIB review come from within the SBIR program office. They are SBIR program managers and other program administrators. These program managers and administrators typically have technical and business qualifications. However, it should also be noted that at this phase, primary reliance is placed on the ability of applicants to obtain third-party financing to signal that a proposal should be funded. Discussions with program officials suggest that few, if any, applicants who meet the third-party financing requirement are not approved. In fact, the practice is that would-be applicants who do not meet the financing requirement withdraw their applications; they are not rejected and are not recorded in the database as failed proposals. Thus, at this phase, internal program reviewers are asked primarily to exercise their judgment to decide if third-party financing requirements are adequately met.

It should be noted that the COV's recommendation in 2004 that the reviewers give more attention to societal impact considerations has implications for the future make-up of "business reviewers." It was the experience of the ATP, which has a similar criterion for broad benefits potential (interpreted as such), that entrepreneurs and other business experts tend to be strong in assessing potential for commercialization, but they are not necessarily strong in assessing potential for widely dispersed spillover benefits to society. To increase the likelihood that factors contributing to spillover effects are taken into account, the ATP engaged economists as reviewers, together with business reviewers, and provided briefings on what constituted broad societal benefits to help reviewers better understand this selection criterion.

8.3.3 Transparency of Selection Process

Although the selection process at first glance appears relatively straightforward, it appears less transparent on further examination. In some instances, the NSF will employ "additional factors" as described earlier. As indicated in Section 8.3.1, these "additional factors" that may enter into consideration during proposal selection include the following: the balance among NSF programs; past commercialization efforts by the firm; excessive concentration of grants in one firm or with one principal investigator; participation by woman-owned and socially and economically disadvantaged small business concerns; distribution of grants across the states; importance to science or society; and critical technology areas. However, there is no further explanation of how, or when, these factors might be applied.

The case studies found that several of the companies had come to the conclusion that the NSF's SBIR program selection process was unfair. Their concern arose from the reportedly uneven application of the factor "excessive concentration of grants in one firm." According to the case study interviewees, companies are unable to tell if and when this factor may be applied by NSF, presumably to reduce multiple awards to companies. The interviewees said they are not told in advance if they have an excessive concentration of grants, or what might constitute an excessive concentration. They observed that other companies, who have received more grants than they have received, continued to receive grants at a time when the relevant case study firms were told they had had enough. Thus, they say they proposed to the NSF's SBIR program believing they were eligible and their proposal received favorable reviews, but then they were turned down based on the NSF's application of the "excessive concentration" factor. The companies say that if they had known they were considered by NSF to have received too many grants, they would have avoided the costs of applying.[36]

Inspection of the program data does not reveal if and how the program is using these other factors. For example, given that woman- and minority-owned companies have a smaller approval rate than other companies, one might conclude either that their proposals are of poorer quality or that the "additional factor" in this case is being applied in reverse. Of course, it is also possible that the approval rate would be even lower without application of "the factor." Or, perhaps, this additional factor is not being applied. The point is that the selection process lacks transparency when it comes to application of "additional factors."

As long as the main selection criteria are applied without triggering "additional factors," the case study companies appeared to believe the selection process to be "fair."

8.3.4 Scoring Procedures

Reviewers are instructed to score all proposals against the technical merit and broader impact criteria. The NSF does not provide a weighting system for reviewers to use in scoring proposals against the criteria. The reviewers do not assign a numerical score to proposals and do not score the individual components of proposals, such as qualifications of the principal investigator, adequacy of facilities, qualifications of other staff, commercial potential, etc. Rather, they give the proposals an overall rating of "Excellent," "Very Good," "Good," "Fair," or "Poor."[37] Each reviewer submits a summary rating and an accompanying narrative.

The reviewers are then typically convened as a panel to make a consensus recommendation. The panel makes this consensus recommendation based on the merit of the proposal and whether it should be recommended for funding. The panel gives each proposal one of three designations: (1) "Highly Recommended," (2) "Recommended if Funding Permits," and (3) "Do Not Consider for Funding." Based on a recommendation by their advisory committee the NSF SBIR program added a recommendation category in the Phase II review panel process

---

[36]See case studies in Appendix D for the MER Corp and the NVE Corporation.
[37]NRC Program Manager Survey, op. cit.

called "Fund with Revision" in addition to "Highly Recommend" and Do not Fund." A typical proposal that receives a recommendation of "Fund with Revision" is one that is generally meritorious but is missing some key information in the technical and/or commercialization plan that, if available, might result in a "Highly Recommend" classification. The program manager contacts the Phase II proposer and typically provides the proposer with copies of the individual reviews and asks the proposer to address the issues in a short period of time. It is not a resubmission of the proposal but clarification of issues. If the program manager is satisfied with the response from the proposer, he or she can recommend the proposal for an award.

Indeed, all input from the individual reviewers and the panel of reviewers is advisory to the program manager. The program manager acts on the advice of the panel, taking into account the other proposals under consideration and presumably the "additional factors" which govern selection. The SBIR program manager—who is also the topic manager—makes a final recommendation. The final grant decision is made by the NSF Grants Office.

8.3.5 Role of Program Manager

The program manager plays a key role in grant selection. The program manager's responsibilities include contributing to topic development, providing elements of program outreach, selecting proposal reviewers, and making recommendations for funding decisions—taking into account the advice of reviewers. If the 2004 recommendations of the COV are followed, the program manager may also make site visits to grantees in the future.

8.3.6 Resubmission Procedures and Outcomes

The NSF's SBIR program has a limited resubmission policy, which operates at the discretion of OII. It affects Phase I, II, and IIB applications differently.

Phase I proposals that are declined by the NSF can be resubmitted to a new solicitation, but Phase II declined proposals can not be resubmitted, even by invitation. There is no appeal or reconsideration of a Phase I proposal that has been declined. A company whose Phase II proposal has been declined can request a reconsideration of the decline decision by a standard NSF process. This is not a rebuttal of reviews or a rereview of the proposal. The reconsideration process is carried on by an independent group (outside of the SBIR program) to determine if the correct procedures and policies were followed in the review and decision process.

For selected Phase II proposals that are considered contenders given improvements in their commercialization plan, applicants may be given two to five days to resubmit. They are provided the review panel's comments on the proposal's commercialization plan. An invitation to resubmit must be received by an applicant from the review panel chair; otherwise, Phase II proposals that have been declined are not eligible for resubmission. About a third of invited resubmissions reportedly receive Phase II funding.[38]

There is no resubmission for a Phase IIB application. Phase II grantees have a single attempt to gain a Phase IIB grant. Once that window passes, grantees are not permitted to reapply.[39]

As noted by several of the case study companies, there is no appeals process. Any invitation to resubmit must come from the panel chair; an appeal from an applicant to revise and resubmit is not permitted.

Several of the case study companies indicated frustration that they were not allowed to revise and resubmit under two conditions: (1) when the company judged from the reviewer comments that reviewer objections would be quick and easy to remedy, or (2) when the company deduced from split reviewer opinions and poorly explained rationales for the rejection decision that a reviewer failed to understand the proposal and that the fault likely lay with the reviewer rather than the proposal. That there is a basis for the expressed frustration of these companies is suggested by observations in the 2004 COV report. Commenting that there is room for improvement in the feedback given to small businesses from the review process, the COV observed that when there is "wider variation in individual reviews," better documentation of the basis for a consensus decision is needed. The case study results also suggest the importance of providing applicants with clear communication about a funding decision based on conflicting reviews. If a fault is found with a proposal that appears to be easily remedied, it is equally important either that the submitter be allowed to make quick changes and resubmit or that a debriefing make it clear why the decision has gone against the applicant, eliminating the chance of resubmission.[40]

8.4 TRAINING (AFTER SUCCESSFUL APPLICATION)

Small technology-driven companies seeking to commercialize a product, service, or process may need special assistance in any of the following areas: (1) identifying potential partners and investors; (2) understanding negotiation processes with a variety of other businesses; (3) valuating technologies for negotiating licensing, partnering, spin-offs, and sales of technology; (4) protecting intellectual property; (5) changing or augmenting internal cultures to reflect the importance of commercialization as well as research; (6) making arrangements for tests, trials, certifications, and demonstrations required for commer-

---

[38]Robert-Allen Baker, "Commercialization Support at NSF," undated draft, p. 11.

[39]Based on program information posted at the NSF Web site and an interview with Joseph Hennessey, NSF on March 3, 2006.

[40]In discussions with NSF SBIR program managers about the absence of an appeal process, they made a compelling argument that they would be unable to hold to their target selection schedule and also accommodate an appeals process, but they could accommodate resubmission by invitation.

cial success; (7) enhancing abilities to present business opportunities to potential investors; and (8) generally shaping strategic choices and developing more detailed business plans.

8.4.1 Training Programs for Agency Phase I and Phase II Grantees

Within the constraints imposed by legislation and feasibility, the NSF's SBIR program provides training support to small businesses. Specifically, the training is intended to foster commercialization by building grantees' business acumen. The commercialization assistance offered by the NSF in Phase I aims at assisting the companies "to attend to issues related to commercialization using an educational model."[41] The assistance, while administered in Phase I, has centered on enabling companies to develop a commercialization plan as required for Phase II. The NSF reportedly adopted this strategy because it believed that grantees would benefit from having business coaching and a commercialization plan whether or not they proceeded to Phase II.

In 2001 the NSF's SBIR program began the Commercialization Planning Program (CP^2), a commercialization assistance program run by Dawnbreaker, Inc. In a recent round of training, Dawnbreaker assisted 76 companies in developing commercialization strategies for their NSF-funded technologies.[42]

The training, which entails intervention at the Phase I stage, is aimed at assisting participating firms to begin (or continue) the development of a commercialization strategy for the NSF-funded technology. The Dawnbreaker training takes participating firms through preliminary strategic planning, interacting with customers, sizing markets, and examining the strengths and weaknesses of competitors. Based on their analyses, participants are asked to refine their plans for marketing, distribution, and financing. The process yields a Commercialization Plan, approximately 15 pages long, that conforms to NSF Phase II solicitation guidelines.

Until late in 2002, Dawnbreaker alone provided the program under contract with the NSF. That year, the contract was renegotiated and split with another vendor, Foresight Science & Technology, Inc. The training program provided by The Foresight Science and Technology, Inc., Training Group uses online training lessons and tutorials, a commercialization/marketing plan template, and supplemental resources, including lessons on deal making and regulations.[43] Both

---

[41]Dawnbreaker, Inc., *SBIR: The Phase III Challenge: Commercialization Assistance Programs* (1990–2005), July 15, 2005, p. 15.

[42]More about the NSF's Commercialization Planning Program (CP^2) may be found at the trainer's Web site, <*http://www.dawnbreaker.com/gov/nsf.html*>.

[43]Both of the Phase I commercialization assistance contractors meet with their assigned clients during the Phase I Grantees Workshop. Additional information about Foresight Science & Technology and its training offerings may be found at the company's Web site, <*http://www.seeport. com/training/*>.

Dawnbreaker and Foresight maintain Web sites that offer password-protected entry for SBIR companies.

Dawnbreaker's CP[2] has its kickoff meeting at a Grantees Workshop, a two-day event which all NSF Phase I grantees are required to attend. All Phase I grantees must include in their Phase I budget the costs of attending the workshop and an explicit statement acknowledging that attendance at the Grantees Workshop is required.[44]

The Grantees Workshop is held at the NSF halfway through the six-month grant period. During the conference, the contractor (Dawnbreaker) meets with each grantee company's principal investigator in a one-on-one meeting that lasts an hour. Thereafter, the grantee companies can continue in the CP[2] and work with the contractor to develop a commercialization plan if they wish.

In 2003, the NSF modified the guidelines for the CP[2]. The modification required more fully developed financial projections as a part of a company's business plan. To meet this requirement the training was simplified in order to focus on developing financial projections.[45] In keeping with this increased focus on financial projections, the two companies providing commercialization assistance have increased their attention to providing market research information to participating companies.[46]

The focus until 2005 had been on providing assistance in Phase I. A new NSF initiative in 2005 supported participation of a group of Phase II NSF grant recipients in an "Opportunity Forum™" that features networking with and presentations of business opportunities to potential investors. To this end, the NSF formed a partnership with the Department of Energy (DoE) that allowed the NSF to "piggyback" on a DoE-arranged Opportunity Forum™ run by Dawnbreaker.[47,48] The NSF identified about 20 Phase II companies with topics of interest to DoE, and approximately 12 to 14 of these participated in the forum.[49]

Participation of companies in the forum generally requires special coaching to assist them in meeting the informational needs of an audience comprised of potential partners and investors. The training associated with the forum typically includes a kickoff meeting, an advanced workshop, and the forum itself. The tools developed by companies include a PowerPoint presentation that emphasizes business opportunities rather than technical details.

The NSF's Matchmaker program also attempts to help grantees find suitable partners for commercialization. It encourages grantees to obtain additional

---

[44]NRC Program Manager Survey.

[45]Dawnbreaker, *SBIR: The Phase III Challenge*, op. cit., p. 15.

[46]Ibid., p. 18.

[47]DoE, the agency with the earliest Opportunity Forum, has reportedly long opened its Opportunity Forum to partnering with SBIR programs run by other agencies. See also Dawnbreaker, *SBIR: The Phase III Challenge*, op. cit. p. 5.

[48]DoE Opportunity Forum brochure, 2005, shows joint participation in theforum of both NSF and DoE SBIR Phase II grantees.

[49]Telephone interview with Joseph Hennessey, NSF, on October 18, 2005.

sources of funding by matching interested Phase II grantees with prospective partners and investors from the venture-capital, angel-investor, and strategic-partner communities.[50] As of late 2005, the activity had not met program expectations regarding business participation rates, and the OII was looking for ways to increase participation.[51]

It should be noted that because companies often receive SBIR grants from multiple agencies that offer commercialization assistance, some NSF grantees have received commercialization assistance from other agency programs prior to, during, or after the CP^2, and generally through the same contractors.[52]

8.4.2 Rating the Effectiveness of Various Training Efforts

In addition to providing training for the NSF's SBIR program, Dawnbreaker has provided training for participation of Phase II companies in Opportunity Forums for other agencies, including DoE, EPA, the National Institutes of Health (NIH), and the ATP of the National Institute of Standards and Technology (NIST). Although it is still too early to find metrics for the effectiveness of the training for the NSF, Dawnbreaker has compiled metrics for these other programs that may serve as benchmarks for evaluating the effectiveness of its Phase II training for the NSF.

The primary metrics have been the amount of funding received by Opportunity Forum participants within 18 months of the forum, and the percentage of participants who received funding within 18 months. The highest recorded success rates for participation in an Opportunity Forum to date have been for the NIST-ATP program, for which 70 percent of the 20 participating companies reportedly obtained a total of $60 million in investor funding within 18 months of the forum.[53] The benchmark performances by SBIR companies range from 40 percent to 68 percent of just over 30 participants receiving close to a total of $40 million within 18 months. No control group was used against which to compare these metrics.

The NRC Program Manager Survey results provide a comparative rating by NSF program managers of the usefulness of the various support functions provided to grantees.[54] Table 8.4-1 shows the rating of comparative usefulness. The

---

[50]SBIR Phase II grantees and interested prospective partners and investors are encouraged to sign up for the Matchmaker program by sending an email to *http://www.SBIRMatch@nsf.gov>*. See also "NSF SBIR Response to NRC Questions," op. cit.

[51]"NSF SBIR Response to NRC Questions," op. cit.

[52]An example of a company that has completed Dawnbreaker commercialization programs offered by the NSF and by other agencies is Materials and Electronchemical Research (MER) Corp, which is among the case study companies given in Appendix D. MER participated in DoE's commercialization assistance programs in 1989, 1994, and 1995, and in the NSF CP^2 in 2001 and 2003. Ibid, p. 9.

[53]Dawnbreaker, *SBIR: The Phase III Challenge*, op. cit., p. 18.

[54]As was noted earlier, rather than have all its program managers complete this survey (as did the other agencies), the NSF responded by having its senior advisor, Joseph Hennessey, complete the survey on behalf of all of its program managers.

TABLE 8.4-1 NSF Rating of Support Functions by Their Usefulness

Type of support function	Usefulness Rating (1= highest)
Business plan development	1
Commercialization planning assistance to Phase I grantees	2
Partnering	3
Matchmaking with VC and other funders	4
Information and planning	4
Government contracting guide	5

SOURCE: NRC Program Manager Survey.

most useful training functions were reported to be "business plan development" and "commercialization planning assistance to Phase I grantees."

8.4.3 Take-up Rates and Projections

All NSF Phase I grantees attend the Grantees Workshop. Thereafter, there is an attrition rate. Between 40 percent and 80 percent of eligible companies each year since the program began have reportedly continued to participate in the NSF's CP^2 after the Grantees Workshop.[55]

8.4.4 Constraints on Commercialization Assistance Training

Training for grantees is constrained and guided by the 1992 SBIR reauthorization legislation. The legislation allowed agencies to spend no more than $4,000 per company for commercialization assistance during Phase I.[56] This amount has remained unchanged since 1992, and, hence, has declined in constant dollars. It appears not to be based on a realistic assessment of the actual cost of providing such services.

The 1992 legislation also makes it possible for individual companies—authorized by an agency—to spend up to $4,000 of their Phase II grant for such services.[57] However, "the cost to a vendor of negotiating individual contracts and nondisclosure agreements with Phase II companies makes . . . this method of working untenable. . . . experience has demonstrated that Phase II services offered in a programmatic fashion and paid for by one or more agencies are . . . preferable."[58] Again, the amount set has declined in constant dollars. The bottom

---

[55]The rates are based on attrition rates after the mandatory Grantees Workshop, which reportedly ranged from 20 percent to 60 percent in recent years. See Robert-Allen Baker, Commercialization Support at NSF, op cit., Section 5.2.2, p. 5.

[56]Dawnbreaker, *SBIR: The Phase III Challenge*, op. cit., p. 2.

[57]Ibid.

[58]Ibid.

line appears to be that it has been easier under the existing legislation to provide individual company business planning under Phase I than Phase II, and that, at best, the ability to do so has been eroding over time.

8.5 THE NSF PHASE IIB PROGRAM

8.5.1 Description

Introduced as a pilot program in 1998, the NSF SBIR Phase IIB program is intended to help bridge the gap between technology research and commercialization by providing an incentive for SBIR grantees to seek partnerships with investors and to continue their research while securing the support of third-party investors. The Phase IIB option extends "the R&D efforts beyond the current Phase II grant to meet the product/process/software requirements of a third-party investor to accelerate the Phase II project to the commercialization stage and/or enhance the overall strength of the commercial potential of the Phase II project."[59] The NSF's SBIR program essentially holds back what could otherwise be part of its Phase II funding and grants it as a supplement to those Phase II grantees who show evidence of commercialization potential as indicated by their ability to attract third-party funding. The Phase IIB grant is provided in a partial match to the amount of third-party funding.

Prior to November 1, 2003, the maximum amount of a Phase IIB grant was $250,000. This supplement extended the Phase II grant for one year, and the combined initial Phase II and supplemental IIB grants typically would not exceed three years in duration.

After November 1, 2003, the maximum Phase IIB supplement was increased to $500,000. For a Phase IIB supplement in excess of $250,000, the initial Phase II grant could be extended for two years, with the combined initial Phase II and supplemental IIB grants not exceeding four years in duration. Thus, the total cumulative grant for the Phase II and Phase IIB supplement increased from $750,000 to $1 million in 2003.

To be eligible to apply for a Phase IIB grant, a company must have completed one year of work on the initial Phase II grant (or receive special permission from the NSF SBIR program officer). They must also meet the requirements for third-party funding, with a third-party investor providing at least $100,000. Furthermore, to be eligible, the applicant must apply for the Phase IIB grant during the original performance period of the relevant Phase II grant. The NSF's SBIR program office announces the deadlines for submission of Phase IIB proposals. (For example, a choice of two deadlines was recently offered: March 1 or September 1.)

---

[59]The source is the NSF SBIR Web site, which provides information on the Phase IIB option, *http://www.nsf.gov/eng/sbir/phase_IIB.jsp#ELIGIBILITY.*

As in the case of a Phase II grant, there are two proposal review criteria.

(1) What is the intellectual merit of the proposed activity?
This criterion is supported by the following additional questions:

- Will the completion of the proposed activity lead to a solid foundation of the scientific and engineering knowledge and understanding base?
- Has the firm progressed satisfactorily in the Phase II activity to justify a Phase IIB activity?
- Is the proposed plan a sound approach for establishing technical feasibility that could lead to commercialization?

(2) What are the broader impacts of the proposed activity?
This second criterion is supported by the following additional questions:

- Does the commercialization plan summary in the proposed activity show a clear path to commercial and societal benefits?
- Does the proposed activity reflect changes to the Phase II commercialization plan that further improves the chances of conversion of research in order to provide societal benefits?
- What are the expectations of the third party, and how effective will the third-party-funded activity be in leading to commercial and societal benefit?
- What are the competitive advantages of the subject technology?

8.5.2 Use of Matching Funds

The Phase IIB grant requires third-party funding twice that of the NSF grant. To put it another way, the NSF matches third-party investment with $0.50 on the dollar. The minimum size of a Phase IIB grant is $50,000, requiring $100,000 in third-party funding. The maximum Phase IIB grant is $500,000, requiring $1 million in third-party funding. The additional federal funds are to be used only for advancing the research of the project. The third-party investor funds can be used for research or for business-related efforts in order to accelerate the innovation to commercialization, including market research, advertising, patent applications, and refining business plans. The method by which the investor will provide the funding to the company must be identified. The third-party funding can be cash, liquid assets, or tangible financial instruments but not in-kind or other "intangible assets." Loans and investments with contingency clauses are not acceptable. Self-funding does not qualify for the Phase IIB option.

For Phase IIB grants up to and including $250,000, the third-party funding may consist of other government funding, such as other federal funding and state and local funding, as well as private-investor funding. A Phase IIB grant in excess of $250,000 is considered "supersized," or a "Phase IIB+ grant," and the third-party funding for the amount in excess of $250,000 must come from one or more nongovernmental private investors. For example, a Phase IIB grant of $400,000 would require third-party funding of $800,000. Of this amount, as much as $500,000 could include funding from other government sources (i.e., the third-party funding against the first $250,000 of the Phase IIB grant), and the remaining $300,000 must be from private investors (i.e., the third-party funding against the additional $150,000 of the Phase IIB grant).

To meet third-party funding requirements, NSF emphasizes "money in the bank." Vaguely worded letters of commitment are not adequate. Beyond actually having the third-party funding in hand, the only commitments that are acceptable are those that are "date certain," that is, commitments that define a series of payments that are scheduled to be made by the third-party investor on specified dates, with evidence provided that the grant recipient actually receives payment as promised.

8.5.3 Application and Selection Procedures

About 30 percent of Phase II grantees have applied for Phase IIB grants. Approximately 80 percent of these applicants have been successful in receiving a Phase IIB grant. As indicated earlier, the 20 percent of applicants that do not receive grants are not officially recorded as failed applications. Rather, applicants who cannot successfully find third-party financing withdraw their application, and those who do find the financing receive the Phase IIB grant. It does not appear that any applicants who met the third-party investment requirement had been turned down.

Applicants must submit Phase IIB proposals using the NSF FastLane system, and they must follow other proposal preparation and submittal directives posted at the NSF SBIR website. The additional work proposed must expand on the technical work being performed in the present Phase II project and must fall within the scope of that project. (See Section 8.11.1 for more on FastLane.)

All Phase IIB proposals are reviewed in-house by NSF; there is no review for Phase IIB proposals external to the NSF's SBIR program office. Each proposal is reviewed based on the criteria given in Section 8.5.1. They are identical to the criteria applied to Phase I and Phase II proposals except that they are fleshed out to relate specifically to Phase IIB.

Additionally, if the requested amount exceeds $250,000, a representative from the SBIR grantee company and one from the private-sector third-party contributor are expected to make a presentation to a panel made up of NSF SBIR program officers.[60] Further, if the requested amount exceeds $250,000, the final

---

[60]One of the case study companies (Language Weaver) made the point that there might be issues a grantee would not wish to discuss in front of an investor (third-party funding source) and that the presentation process should take this into account. The ATP handled a similar issue by allowing potential partners and investors to give portions of their presentations separately.

grant recommendation may be subject to SBA approval. (Supplements equal to or less than $250,000 require no presentation and no SBA approval.)

According to both NSF program officers and Phase IIB applicants who were interviewed in the course of the case studies, proposal selection for the Phase IIB option heavily emphasizes commercial potential as signaled by third-party funding. Essentially, rather than allocate all of its Phase II funding based on the reviews conducted just after Phase I the NSF's SBIR program imposes a second, market-based test. As was shown in Chapter 4, the effect is to distribute the money differently than if all of it had been awarded based on the initial Phase II selection process.

Within 60 days of the submission deadline for Phase IIB proposals, the applicant is notified of the results. If a proposal is recommended for grant, the company must submit proof of the bank transaction showing that the third-party has exercised its commitment or the date-certain agreement for transfers of third-party funding.

8.5.4 Role of Program Manager in Phase IIB

NSF SBIR program managers play a key role in the selection of Phase IIB grants. A minimum of two program managers review each Phase IIB proposal and make a recommendation either to fund or not to fund to the SBIR office director. A program manager oversees each Phase IIB grant throughout its life and reviews all documentation. The assigned program manager reviews all financial documentation for compliance with program requirements. If funding were made available for site visits, this would provide an additional role for the program manager.[61]

8.6 THE GRANT CYCLE AND FUNDING GAPS

8.6.1 Phase I to Phase II Gap

Table 8.6-1 shows the NSF's SBIR program grant cycle over a ten-year period—from 2002 through 2012—for four topic areas and for Phase I, Phase II, and Phase IIB grants. According to program staff, the program intends to follow this same basic time line for the foreseeable future.[62]

The ten-year cycle covers solicitations, proposal submittal, review, award of Phase I, Phase II, and Phase IIB grants, and post-grant commercialization reports. The illustrative cycle provided by the NSF's SBIR program covers separate solicitations for two topic sets, each set consisting of two major technology

---

[61]According to discussions with program staff, program managers sometimes visit grantee companies now when they are in the area for other reasons that have financial support and have the opportunity to "piggy-back" a company visit. However, such visits are not systematic.

[62]Discussions in 2004 with Ritchie Coryell, NSF SBIR Program Staff. (Note that Mr. Coryell has since retired.)

TABLE 8.6-1 NSF SBIR Significant Dates for a Recent Grant Cycle

SBIR Phase	Event	Relevant Dates for Topics 1 & 2 Advanced Materials and Information Technology	Relevant Dates for Topics 3 & 4 Electronics and Biotechnology
Phases I, II & IIB	Topic formulation	January 2002	
	Solicitation topics posted on Web	March 1, 2002	
Phase I	Solicitation opens	May 1, 2002	
	Proposals due	June 12, 2002	January 22, 2003
	Review panels	August–September 2002	March–April 2003
	Grants & declines	December 2002	June 2003
	Grant start dates	January 1, 2003	July 10, 2003
	Grantees workshops	April 7–9, 2003	~October 2003
	Final reports	July 15, 2003	January 15, 2004
Phase II	Proposal due dates	July 29, 2003	~January 2004
	Review panels	August–October 2003	March–May 2004
	Grants & declines	December 2003	June–July 2004
	Grant start dates	January–March 2004	June–November 2004
	Grantees workshops	~January 2004, 2005 & 2006	
Phase IIB[a]	Proposal due date	March 1, 2005	November 1, 2005
Phases II & IIB	Final reports	~2007	~2007
	Postgrant annual commercialization reports	2008–2012	

[a]Does not reflect the schedule for the "supersized" Phase IIB grants.

SOURCE: Ritchie Coryell, NSF.

topics. Each of the Phase I solicitations extends over a cycle of about 19 months, covering the period from the opening of the solicitation to the final Phase I reports for the second topic group.

For each topics set, approximately a year elapsed from the Phase I proposal due date to the submission of the project's final report. About half the elapsed time occurred between the proposal due date and the project start date, and about half occurred between the project start date and the due date for the final reports. Thus, processing the proposal took approximately the same amount of time as carrying out the project. For example, proposals for the Advanced Materials and

Information Technology topics set were due on June 12, 2002; the projects started the first of the year in 2003; and the final reports were due in July 2003.

Interwoven with the Advanced Materials and Information Technology set of topics was the time line for the other topic set, Electronics and Biotechnology. For this second set, the proposals were due in late January 2003, about seven months after the proposals for the first set. The second set of projects started in July 2003, close to six months after the proposals were due, and their final reports were due in mid-January 2004. Again, it takes about the same amount of time to process the Phase I proposals and identify the grants as it takes the companies to complete the Phase I research and prepare a final project report—approximately six months for each activity.

The third section of the table pertains to Phase II grants. The proposals for the follow-on Phase II projects are due about the time the Phase I projects end. The Phase II projects start five to ten months after the Phase II proposal due dates and should be completed within two years—unless a follow-on Phase IIB is received.

Phase IIB proposals are due approximately a year after Phase II projects start. For Phase IIB grants not in excess of $250,000, the projects are due to be completed and the final reports filed about a year after the Phase II proposals would have been due. For grants in excess of $250,000, another year is added to the completion time, so the projects should be completed within four years.

From an analysis of the schedule depicted in Table 8.6-1, the main persisting funding gap occurs between completion of the Phase I grant and the start of the Phase II follow-on grant. According to the grant schedule, this period is approximately six to ten months. Any unscheduled delays will make the gap even longer for individual companies. A delay may be critical for very small companies. Yet, without provision for bridge funding of some sort, a gap may be unavoidable because time is required for the program to review Phase II proposals and decide which Phase I projects are worthy of continuation. With a growing number of proposals and administrative resource constraints, reducing the gap by tightening the schedule appears problematic. At the same time, as explained below, firms can hold the gap to a minimum by adhering to the program's reporting and application schedule.

Eighty-one percent of NRC Phase II Survey respondents reported they had experienced a funding gap between the end of Phase I and the start of Phase II; 19 percent said they had not. For those reporting that they experienced a gap between the end of Phase I and the start of Phase II, the average gap was 8 months. For 5 percent of the respondents, the gap was 2 years or more.

The funding gap had troubling implications for many respondents. As a result of the funding gap between Phase I and Phase II, 28 percent of NRC Phase II Survey respondents stopped work on the project. Thirty-seven percent continued work but at a reduced pace during the gap, and 1 percent ceased all operations until funding resumed. Five percent of the respondents said they experienced a funding gap but were unaffected by it. A small percentage reported receiving bridge funding. As the case studies revealed, companies that had built up a revenue stream in advance were able to continue work during the funding gap by self-funding research.

According to program officials, companies have an opportunity to cut the funding gap themselves by taking the following steps: (a) promptly completing and filing their reports for Phase I; (b) promptly applying for Phase II at the first of two opportunities; and (c) taking steps in advance to ensure that there will not be problems with their financial audits that would delay the Phase II award.[63]

The program's flexible approach to preaward work and its payment plan may also serve to abate funding gaps. Phase I grant recipients are allowed to begin preaward spending up to ninety days before the effective date of the grant. On the effective date, the grant recipient receives two-thirds of the money, i.e., $66,666 (assuming the current maximum Phase I grant), and the remaining third, i.e., $33,334 at the project's end upon submission of a final report. Recently, the program has begun providing what it calls a Phase IB option to bridge the gap in funding between Phase I and Phase II. The Phase IB option provides additional funds to Phase I grantees that obtain third-party investment to support their projects, thereby extending the R&D beyond the initial Phase I grant for six months.[64]

A Phase II grant recipient is also allowed to begin preaward spending. This is at the company's own risk, however, because any grant is conditioned on the successful completion of a financial audit. To speed financial audits, the program has hired CPA firms. Phase II recipients receive a lump sum up front, with the remainder spread out over the work period. However, the program holds final payment until the final report is submitted.

Given the time required for proposal solicitation and project selection, the program's limited administration resources, and its lack of procurement possibilities, it appears difficult—if not impossible—for this program to avoid altogether a funding gap between Phase I and Phase II. Hence, it seems particularly important that the program staff ensure that grantees understand the various steps they themselves can take to minimize the length of the funding gap.

8.6.2 Other Funding Sources

The NRC Phase II Survey asked respondents whether matching funds or other types of cost sharing were received for the Phase II proposals. Approximately one-third of the survey projects received such funding and two-thirds did not. Table 8.6-2 shows the sources of the third-party funding for the 48 projects that did receive it. The major source was "another company," followed by the

---

[63]Telephone interview with Joseph Hennessey, NSF, October 18, 2005.
[64]See NSF SBIR program information at <*http://www.nsf.gov/eng/iip/sbir/phaseibinfo.pdf*>.

TABLE 8.6-2 Third-Party Sources of Matching and Co-investment Funding in Phase II

Source of Matching Funds	Percentage of Projects with Matching Funds Obtained from Each Source (%)
Another company provided funding	58
Our own company provided funding (includes borrowed funds)	38
An angel or other private investment source provided funding	21
A federal agency provided non-SBIR funds	4
Venture capital provided funding	2

NOTE: The percentages are computed for the approximately one-third of NRC Phase II Survey projects that reported third-party sources of matching and co-investment funding.

SOURCE: NRC Phase II Survey.

company itself, and then by "an angel or other private investor." Note that few of the projects received funding from venture capitalists. Clearly, respondents interpreted this question as extending beyond third-party financing for Phase IIB because self-funding is not eligible for this purpose. The other sources included are eligible for meeting third-party financing requirements.

NRC Phase II Survey respondents were also asked whether any additional developmental funding had been received or invested in the referenced project. Nearly two-thirds said they did receive such funding and slightly more than one-third said they did not. As shown in Table 8.6-3, the major source of the additional funding was "non-SBIR federal funds," followed by "other private equity." Venture capitalists provided approximately 6 percent of the reported additional funds. On average, each referenced project in the survey received over half a million dollars in additional funding for further development.

The case studies (see Appendix D) provide examples of companies that avoided the funding gap after Phase I by attracting funding from other sources. Some leveraged SBIR grants to obtain the larger ATP awards then available. Some had developed products and services for sale. Some obtained SBIR grants from other agencies. Some had formed partnerships that provided various forms of financial support, including licensing revenue.

While research results provide insights about funding in the Phase II period, neither the NRC Phase II Survey results nor the case studies show quantitatively how adequate the additional funding is in meeting financial requirements for development and commercialization. Frequent comments by case study interviewees about the large requirements for capital and the difficulty in obtaining it suggest that they experienced a financing shortfall in most cases.

Because the NSF generally does not procure the technologies it funds to serve its mission, there is essentially no support from the NSF to grantees in the

TABLE 8.6-3 Funding Sources and Average Amounts of Additional
Developmental Funding for the Referenced Projects

Funding Source	Average Amount of Developmental Funding per Survey Project ($)
Non-SBIR federal funds	248,077
Private investment:	
(1) U.S. venture capital	39,450
(2) Foreign investment	19,290
(3) Other private equity	196,141
(4) Other domestic private company	57,925
Other sources:	
(1) State or local governments	19,938
(2) College or universities	617
Not previously reported:	
(1) Your own company (including borrowed money)	54,617
(2) Personal funds	22,154
Total average additional funding for development, all sources, per referenced project	658,214

NOTE: The average amount of developmental funding was computed by dividing the total dollar
amounts reported for each funding source by the 162 respondents in the survey, not just the number
of respondents reporting additional development funding.

SOURCE: NRC Phase II Survey.

post–Phase II period from agency purchases of grantee goods and services. As
was shown earlier, however, there is evidence that other federal agencies procure
the results of some NSF-funded innovations, and these procurements may help
reduce a financing shortfall.

8.6.3 Bridge Funding Programs (After Phase II)

Other than the Phase IIB funding program described in detail in Section 8.5,
the NSF's SBIR program does not operate bridge funding programs.

Program officials created the Matchmaker program in 2002 to help grantees
deal with the sometimes lengthy process of securing third-party funding and
commercial partners after Phase II. Aimed at helping grantees find funding from
the private sector, Matchmaker is an online database where prospective private-
sector investors and partners, as well as SBIR firms looking for investors and
partners, can register and connect with one another. SBIR program managers
provide a brokerage function by helping to match registered qualified prospective
investors/partners with Phase II grantees that fit the profile of interest. Although
the Matchmaker service reportedly has not yet been as effective in speeding the
process of finding third-party funding as had been hoped, it offers potential as a

mechanism for reducing the funding gap at both the Phase II-Phase IIB stages and the post Phase IIB stage.

8.7 REPORTING REQUIREMENTS

8.7.1 Reports Submitted to the Agency by SBIR Winners

For Phase I grants, the deliverable at the end of the grant period is a technical report that summarizes the experimental and theoretical accomplishments of the project. Phase I grant recipients must submit a Phase I Final Report within 15 days after the expiration of the grant. This report must be submitted using the report template provided by the FastLane system prior to submitting a Phase II proposal. According to instructions provided to applicants, failure to provide the final technical report will delay NSF review and processing of pending proposals submitted by the principal investigator. Principal investigators are reminded in NSF instructions to examine the formats of the required reports in advance to assure the availability of required data.

Phase II grantees are required to provide three kinds of reports: (1) interim reports of progress, (2) a final project report, and (3) annual commercialization reports for 5 years following completion of the Phase II grant. The interim reports are due at the end of 6, 12, and 18 months, unless the grantee receives a supplementary Phase IIB grant, in which case additional interim reports at the end of 24 and 30 months are added (and more interim reports are added at 6-month intervals, as needed, to cover the duration of a supersized Phase IIB grant). These reports also are submitted using the FastLane system.

If a Phase IIB supplement is added, a final report is due at the end of Phase IIB. The final report is to include an account of cumulative project milestones, documentation of technical accomplishments, and a commercialization section. A final report following completion of a Phase IIB grant must include a combined Phase II and IIB final report and a commercialization report inclusive of an updated commercialization plan. When the grant closes, regardless of whether it is a Phase II or a Phase IIB grant, the grantee company is responsible for continuing to report annually on commercialization for five years, as noted earlier. A warning is issued to companies stating that failure to submit required reports may deter their selection by the program for future grants.[65] (See Appendix F for the format of the annual report.)

A new reporting requirement in the postcompletion period has recently been added. NSF SBIR program staff reported that there are to be commercialization reports on the third, fifth, and eighth anniversaries after the start of a Phase II grant. A program manager is charged with conducting telephone interviews with eligible companies who have completed Phase II projects. The information is

---

[65]Yet, as noted earlier, there has been a substantial lack of compliance by grantees.

being collected monthly using a survey template. The companies are provided the survey questions in advance. This new reporting requirement was launched in July 2005.

The first new monthly report ("SBIR Metrics"), submitted by the contractor for July 2005, provided results for 24 of 30 companies who had third-, fifth-, and eighth-year anniversaries in that month. Simple metrics on "successes" are embedded in a one-page textual report.[66]

8.7.2 Report Utilization and Utility to the Agency

Reportedly, compliance by grantees in filing postcompletion reports has been sketchy. Furthermore, no evidence was found that program management has used data from the annual postcompletion reports to provide performance metrics or otherwise assist in program evaluation.[67]

As described above, the NSF's SBIR program office has taken a step to increase company compliance for reporting postproject data by implementing a new postcompletion telephone interview. The program office reportedly plans to use the results of the telephone survey to gain insight into commercialization progress. However, without a more standardized reporting format which systematically organizes and reports the data in ways that allow trends to be established and easily detected, this new effort would seem to yield limited results.

8.8 EVALUATION AND ASSESSMENT

8.8.1 Annual and Intermittent Agency Evaluation of Its SBIR Program

NSF SBIR program managers routinely monitor the technical outcomes of Phase I projects using the Phase I final reports. If a Phase II proposal is submitted, other reviewers will also assess the Phase I technical outcomes.[68] Routine monitoring of commercial data contained in proposals and final project reports is also done at the Phase II and Phase IIB stages by program managers on an individual basis.

Although a variety of data that might comprise outcome metrics is contained in individual reports and in the occasional internal studies, the program does not systematically compile this data and does not use aggregate results as outcome indicator metrics.[69]

---

[66]Dr. George Vermont, an NSF SBIR program manager, conducted the telephone survey of postcompletion companies and provided this study a copy of the first monthly report on "SBIR Metrics."

[67]Conclusions based on discussions with SBIR program officials at the NSF, August 17, 2005.

[68]NRC Program Manager Survey, op cit.

[69]For example, the 1996 Tibbetts Study included a table that included outcome data for 50 companies. Response to the NRC Program Manager Survey was that the NSF routinely collects outcome indicator data, listed by type, but follow-up revealed that the data in question remain disaggregated,

The NSF's SBIR program office informally collects anecdotal information from grantees and unsuccessful applicants to assess their satisfaction with the program. The results provide guidance to the program office in considering how to improve the program.[70]

No existing evaluation studies of the NSF's SBIR program, either conducted or commissioned by the agency, were found in the open published literature.[71] However, as discussed from the standpoint of commercial results in Chapter 5, several ad hoc unpublished studies of the program were made available by OII for this NRC study. Two of these were carried out internally by SBIR program managers, and the other (in draft) was commissioned as a contractor study. Of the two internal studies, one was conducted in 2004 and the other in 1996. The contractor report, incomplete and undated, was reportedly intended as a continuation of the 1996 internally conducted study. Additional evaluation activities include the production of "success stories," or "nuggets," which are regularly used. A Committee of Visitors (COV) provides an expert review of NSF's SBIR/STTR program every three years. Each of these efforts is discussed in more detail below.

Once every three years since 1998, the Committee of Visitors has been convened to review NSF's SBIR program. The COV uses a peer review, or expert review, methodology. The COV is asked to assess all program elements and to recommend ways to improve the program. It is a significant ongoing assessment activity that prompts documented program changes.

In its most recent review (2004), the COV reviewed the NSF's SBIR/STTR program forthe period from 2001 through 2003. It focused primarily on (a) processes and management, and (b) outputs and outcomes but also examined several other topics, including the positioning of the SBIR program within the NSF, the value placed by the agency on the program, and the development and maintenance of supporting databases.[72]

As a resource for assessing the program's processes, the COV reviewed 123 proposal jackets from among the 5,814 proposals processed by the NSF over the three years. The nonrandom selection, which aimed at including proposals across the three years, geographic regions, and underrepresented groups, yielded 78 Phase I proposals, 36 Phase II proposals, and 9 Phase IIB proposals.

are found in different forms in individual proposals and project reports, and are not systematically collected and compiled by the program office for use as indicator data. Telephone interview with Joseph Hennessey, NSF, October 18, 2005.

[70]Ibid.

[71]The NSF at large has commissioned evaluation studies carried out by external evaluators that are available in the open literature, but these studies were not of the SBIR program.

[72]Descriptive information about the COV is available at the NSF's online account of COV reports and annual updates at <http:// www.nsf.gov/eng/general/cov>.

TABLE 8.8-1 Criteria for Judging Success of the NSF's SBIR Program

Criteria for Judging Success	Used by the NSF's Program	Not Used by the NSF's Program
Efficient program management (i.e., grants made on time)	X	
Commercial outcomes	X	
Outcomes that support specific agency missions		X
Customer (grantee) satisfaction	X	
Technical contributions to the state of the art	X	

SOURCE: NRC Program Manager Survey.

8.8.2 Operational Benchmarks for the NSF's SBIR Program

Table 8.8-1 shows how OII rated its use of four of the listed criteria to judge the success of its program. The one criterion listed in the table that is not used as a success measure is "Outcomes that support specific agency missions."

Operational benchmarks used by OII that were identified by the study pertain to the length of time it takes to process proposals. One stated benchmark is the goal of holding the elapsed time from the Phase I proposal due date to notification of Phase I winners to six months. Another stated benchmark is to complete all Phase I panels within three and a half months. Regarding Phase II proposal processing, a goal is to complete proposal review in no more than four months and to notify winners within six months. The latter goal, however, is conditional on company financial audits being completed without problems. At last report from the NSF, about half the Phase II proposals are processed within six months and half are taking longer than six months due to audit problems.

8.8.3 Evaluators (Internal and External)

NSF SBIR staff members have been the main performers of program-sponsored grantee surveys. By discipline, training, and experience, these staff members are not primarily evaluators; rather, they have assumed evaluative activities in addition to their usual program manager duties.

Only one instance was found of the program's use of an outside contractor to conduct an evaluation. The firm hired, Dawnbreaker, Inc., is largely known for its expertise in providing business assistance, not evaluation. However, Dawnbreaker does regularly conduct follow-up assessments of the success of the firms it coaches in attracting investment funding.

The writers of nuggets are principally professional writers, not evaluators. They can skillfully present an accomplishment by an innovative company to a varied audience that includes nontechnical people, but their focus is not on presenting a thorough assessment of benefits and costs over time.

The COV members are external experts selected with an eye to diversity. In 2004 the eight members included four from SBIR firms, three from academia, one from the investment sector, and one from state government. Six were male and two were female; one was African American, one was Hispanic, and one was Native American. Two of the COV members were also members of the SBIR/STTR Advisory Committee.[73]

8.8.4 Annual Evaluation and Assessment Budget

According to the OII at the NSF, a limited amount of funds are regularly made available for program evaluation. For example, it was reported that during the period FY2000–2002, on average less than 1 percent of the NSF SBIR budget was spent each year on evaluation and assessment.[74]

8.9 FLEXIBILITY

8.9.1 Program Manager Discretion

NSF SBIR program managers appear to have a large amount of discretion and flexibility, as well as a large amount of responsibility. With the program director's concurrence, they appear to have substantial discretion in establishing topic and subtopic areas, in arranging for proposal reviews within their areas of responsibility, and in making recommendations as to what will be funded at each stage. The program managers also appear to have considerable autonomy in running meetings within their topic areas and in holding meetings with grantees.

The program managers have primary responsibility, without the advice of peer reviewers, in deciding when supplemental funding is warranted under the provisions of Phase IIB grants. In practice, however, the requirement for third-party funding appears to be the governing factor.

Interviewees for the case studies frequently spoke positively about the fact that NSF program managers are empowered and flexible. Many specifically commented that the NSF SBIR program manager system is a program strength that should be maintained.

8.9.2 Program Manager Funding Discretion

Because of the centralized nature of the NSF's SBIR program and the lack of subagency divisions often found in other agency SBIR programs, there is no

---

[73]This profile of COV members was excerpted from a memo—dated June 23, 2005—reporting on the diversity, independence, balance, and resolution of conflicts among the SBIR/STTR program COV members. Available online at <*http://www.nsf.gov/od/oia/activities/cov/eng/2004/SBIRcov.pdf*>.

[74]NRC Program Manager Survey, op. cit.

issue of shifting funding from one subagency to another. There is, however, the issue of shifting funding across topic areas, particularly given the fact that two topic areas have solicitations for proposals within the same time frame. Program guidelines point to flexibility in funding decisions but do not directly address the issue of whether funding is shifted across topic areas to achieve "balancing," and if so, how.

8.9.3 Program Manager Perceptions of Constraints

Informal discussions with program managers revealed no complaints regarding a lack of discretion and flexibility, except in one area: the constraint on their ability to conduct site visits of their grant recipients. It should be noted that this constraint has resulted from a lack of designated travel funds for this purpose rather than the ability of program managers to conduct site visits should funding be provided.

8.10 SIZE—FUNDING AMOUNTS AND SOURCES

This section explores the size of funding amounts available to small businesses through the NSF's SBIR program and other sources. It first revisits the formal and effective limits on the size and duration of Phase I, Phase II, and Phase IIB grants available from the NSF's SBIR program. Then it examines for a sample of Phase II projects the additions to funding that grantees obtained prior to and after the survey's referenced grant. It also investigates the sources of these funds.

8.10.1 Formal and Effective Limits on Size and Duration of Grants

The program imposes formal and effective limits on the size and duration of individual Phase I, Phase II, and Phase IIB grants, characterized as follows:

- Phase I: Funding for feasibility research
  — Six-month research period
  — A no-cost time extension—typically three months—can be granted[75]
  — Up to $100,000 of SBIR funding

- Phase II: Initial Phase II funding for research toward developing a prototype
  — Twenty-four-month research period
  — Up to $500,000 of SBIR funding

---

[75]Additional no-cost time extensions may be granted, but if they go past the second open date for a company to submit its Phase II proposal, the company will miss out on the Phase II opportunity. Interview with Joseph Hennessey, NSF, October 18, 2005.

— A no-cost time extension—in six-month increments for up to two years—can be granted[76]

- Phase IIB: Supplemental Phase II funding in match against outside investment funding
    - A Phase IIB grant has a minimum size of $50,000 and a maximum size of $500,000.
    - Twelve-month research period added to the Phase II period if the supplement does not exceed $250,000
    - Twenty-four-month research period added to the Phase II period if the supplement exceeds $250,000
    - From roughly 1998 to November 1, 2003, up to $250,000 of Phase IIB funding per grant was available.
    - As of November 1, 2003, funding over $250,000 and up to $500,000 is available per "supersized" Phase IIB grant.
    - NSF will match up to 50 percent of third-party investment funding received.
    - To receive the minimum amount of $50,000 from the NSF, the applicant must show a minimum of $100,000 in third-party funding.
    - To receive the maximum amount of $500,000 from the NSF requires a minimum of $1,000,000 in third-party funding.
    - For grant amounts in excess of $250,000, the third-party funding must come from nongovernmental private-sector sources.
    - Phase IIB helps bridge the gap between research and commercialization
    - A one-time opportunity

- Phase III: A non-SBIR funding phase during which companies are encouraged to develop commercial products/processes/service and take them to market. This phase has:
    - No time limits
    - No SBIR funding provided

The total amount of funding set aside for the SBIR is a function of the SBIR legislated rate, currently set at 2.5 percent of an agency's R&D budget, and the amount of the agency's R&D budget to which the rate is applied. Thus, the level of SBIR funding may rise or fall over time. It appears that the NSF uses all of its available SBIR funding each year.

---

[76]Ibid.

TABLE 8.10-1 Phase I Grants, 1992–2005

Fiscal Year	Number of Grants	Average ($)	Total ($)
1992	208	49,755	10,349,115
1993	256	49,727	12,730,184
1994	309	64,103	19,807,945
1995	301	64,571	19,436,011
1996	252	74,283	18,719,287
1997	261	74,262	19,382,371
1998	215	98,371	21,149,814
1999	236	98,749	23,304,757
2000	233	99,405	23,161,372
2001	219	99,353	21,758,218
2002	286	99,162	28,360,340
2003	437	99,275	43,383,103
2004	244	100,000	25,117,992
2005	149	98,575	14,687,703

SOURCE: NSF SBIR program data.

8.10.2 Distribution of Funding to
Phase I and Phase II Grants within the Specified Limits

The distribution of NSF SBIR funds to Phase I and Phase II grants reflects the amount of funding available and a time-lag effect of the previous year's allocation decisions. For example, a year of a large rise in the allocation to Phase I grants tends to be followed the next year by a rise in the allocation to Phase II grants. Discussions with NSF SBIR program officials did not reveal a formulaic approach for allocating available funding among the different grant categories. However, the solicitations mention an approximate amount of funding available, indicating advanced planning for the funding allocation.

Actual Size and Duration of Phase I Grants. The average size of NSF Phase I grants, stated in current dollars, increased in a series of steps from nearly $50,000 in 1992 and 1993 to an average of approximately $64,000 in 1994 and 1995, to approximately $74,000 in 1996 and 1997; and thereafter to an average of nearly $100,000, the current maximum Phase I grant amount.

Total annual Phase I funding rose stepwise over much of the fourteen-year period. Over the years 1992 to 1993, annual Phase I funding ranged from $10 million to $12 million. Over the years 1994 to1997, Phase I funding totaled roughly $19 million annually. Over the years 1998 to 2001, Phase I funding ranged from $21 million to $23 million annually. Then, in 2002, the total jumped to $28 million, followed by an even more dramatic jump to $43 million in 2003. In 2004 there was a drop in total Phase I funding back to $25 million, and a further drop

TABLE 8.10-2 Phase II Grants, 1992–2005

Fiscal Year	Number of Grants	Average ($)	Total ($)
1992	57	246,373	14,043,289
1993	52	269,668	14,022,757
1994	21	279,983	5,879,644
1995	48	277,136	13,302,542
1996	90	297,584	26,782,533
1997	122	289,390	35,305,525
1998	117	344,958	40,360,087
1999	89	394,854	35,141,968
2000	95	400,827	38,078,524
2001	91	475,018	43,226,632
2002	67	495,645	33,208,238
2003	77	498,505	38,384,851
2004	131	498,254	65,278,995
2005	132	499,715	65,962,445

SOURCE: NSF SBIR program data.

in 2005 to $14.7 million. Table 8.10-1 shows the annual number, average size, and total amount of Phase I grants from 1992 to 2005.

The mean annual duration of Phase I grants was essentially half a year throughout the ten-year period, with only minor fluctuations. A mean duration of half a year would be expected since this is a standard feature of Phase I grants for the NSF's SBIR program, as established by the SBA, which has oversight responsibility for the program.

Actual Size and Duration of Phase II Grants. The average size of a Phase II grant grew from $246,000 in 1992 to over $300,000 by 1998, to over $400,000 by 2000, and to nearly $500,000 from 2001 to 2005. Effective November 1, 2003, the SBA increased the limit of Phase II (including the combination of Phase II/IIB grants) from $750,000 to $1,000,000. The NSF, however, has chosen to limit the funding of its Phase II grants to $500,000 and to make the remaining allowable $500,000 available as a supplement to the initial amount through follow-on Phase IIB grants.

Like its total Phase I funding, NSF total annual Phase II funding rose over the thirteen-year period, but this rise was more irregular. Total Phase II funding was at the $14-million level in 1992 and 1993. It dropped in 1994, rose back to the $13-million level in 1995, and jumped to nearly $27 million in 1996. Between 1997 and 2003, total annual Phase II funding fluctuated between $33 million to $43 million. In 2004, Phase II funding surged to $65 million as Phase I funding fell back to its 2002 level. Table 8.10-2 shows the annual number, average size, and total amount of Phase II grants from 1992 to 2005.

TABLE 8.10-3 Phase IIB Grants, 1998–2005

Fiscal Year	Number of Grants	Average ($)	Total ($)
1998	4	99,986	399,944
1999	21	95,170	1,998,574
2000	9	184,466	1,660,191
2001	14	307,334	4,302,682
2002	39	245,861	9,588,580
2003	24	237,096	5,690,294
2004	22	273,883	6,025,436
2005	28	330,731	9,260,464

NOTE: Effective November 1, 2003, the limit of Phase IIB grants was increased from $250,000 to $500,000. Yet the average is shown to exceed $250,000 in 2001. This apparent dichotomy is explained by the fact that if the Phase II grant were below the allowable size, the program would sometimes increase the amount of the Phase IIB grant to bring the combined Phase II/Phase IIB amount up to the allowable limit—in this case, $750,000. The record keeping for this practice would cause the Phase IIB award amount to appear to exceed the limit. Telephone interview with Joseph Hennessey, NSF, October 18, 2005. Dollar amounts are in current dollars.

SOURCE: NSF SBIR program data.

According to the SBA, Phase II grants may last "as many as two years." The annual mean duration for the initial Phase II grants over the ten-year period ranged from two to two and a half years. Reasons for the average exceeding the official limit likely reflect accounting closeout issues or requests for extensions.

Actual Size and Duration of Phase IIB Grants. In 1998 and 1999 the average Phase IIB grant ranged between $92,000 and $100,000. The average jumped to $184,000 in 2000. Thereafter, the average Phase IIB grant fluctuated between $234,000 and $294,000 each year.

Total Phase IIB grants rose dramatically from $399,900 in 1998, to nearly $2 million in 1999, which was followed by a drop to $1.7 million in 2000. The total jumped to over $4.3 million the next year, and in 2002 it more than doubled to $9.6 million, the highest level of the entire period. In 2003 and 2004 the total was close to $6 million, and it rose to $9.3 million in 2005. Table 8.10-3 gives the number, average size, and total amount of Phase IIB grants.

According to NSF program officials, they made the decision to increase the amount of funding to Phase IIB as opposed to increasing the amount for Phase I or Phase II grants in an attempt to leverage the government's investment more effectively. Program officials thought offering additional funding through Phase IIB grants would accelerate development and commercialization of technology by encouraging Phase II grantees to seek third-party funding.

Program officials pointed out that changes in the average size of Phase IIB grants in response to the move to provide supersized grants depend on how suc-

TABLE 8.10-4 Total Funding Provided to Grantees by the NSF's SBIR Program, 1992–2005

Fiscal Year	Total Grants of All Types	Total Grant Outlays ($)
1992	265	24,392,404
1993	308	26,752,941
1994	330	25,687,589
1995	349	32,738,553
1996	342	45,501,820
1997	383	54,687,896
1998	336	61,909,845
1999	346	60,445,299
2000	337	62,900,087
2001	324	69,287,532
2002	392	71,157,158
2003	538	87,458,248
2004	397	96,422,423
2005	309	89,910,612

SOURCE: NSF SBIR program data.

cessful grantees are in securing third-party investments. "We do not anticipate a significant increase in the amount of total funding, because the grantee will be required to secure $1 million in outside investment in order for NSF to match with an additional $500,000. For instance, we received 15 Phase IIB proposals in November [2004], and only three were able to secure $1 million in outside commitments."[77]

Total Amount of Funding Provided by the NSF. Table 8.10-4 shows the total funding the NSF's SBIR program provided to small businesses each year from 1992 to 2005. It began at a level of $24.4 million, rose fairly steadily to a level of $96.4 million in 2004, and fell back to $89.9 million in 2005.

8.11 ONLINE CAPABILITIES AND PLANS

8.11.1 FastLane System

Grant application and processing at the NSF is now entirely electronic via the NSF's online FastLane system. FastLane is used for proposal submittal, panel review, proposal management, tracking payments, checking proposal status, making travel arrangements, submitting reports, and other purposes. According

---

[77]NSF SBIR Response to NRC Questions, op. cit.

to program administrators, as a result of FastLane, "Doing business with NSF is simpler, faster, more accurate, and less expensive."[78]

Proposers are required to prepare and submit all proposals through the FastLane system. To facilitate proposal preparation, FastLane has smart forms capability that pulls in all individual and organizational information available in the NSF database to minimize the amount of information that must be typed in by the user. Detailed instructions for proposal preparation and submission via FastLane are available at <*http://www.fastlane.nsf.gov/a1/newstan.htm*>.

Principal investigators are also required to use the NSF electronic project reporting system, available through FastLane, for preparation and submission of project reports. This system facilitates electronic submission and updating of project reports, including information on project participants (individual and organizational), activities and findings, publications, and other specific products and contributions.

According to program administrators, adoption of electronic business Web-based approaches is consistent with the broader commitment to e-business by the NSF. At the time of the study, the NSF was the only agency to receive the highest status rating (green) in two of the government-wide President's Management Agenda initiatives. During 2002 the NSF became the first federal agency to receive the top rating for the e-government initiative.[79]

8.11.2 Barriers to Online Capabilities and Plans

No barriers obstructing implementation of electronic filing and processing of proposals were found. The provision of online capabilities appears to have been successfully achieved by the NSF's SBIR program for all aspects of grant application, processing, and reporting.

8.12 ADMINISTRATIVE RESOURCES

8.12.1 Funding of Program Administration

According to NSF SBIR program officials, the program's dedicated annual administrative budget was $2 million in a recent year, not including program officer salaries and other administrative support that does not come out of the $2 million. Items included in the $2 million are contracts for support to supplement NSF personnel, costs of running review panels, contractor costs to organize and run national conferences or operate outreach Web sites, and other items.[80]

---

[78]NSF SBIR Response to NRC Questions, op. cit.

[79]Ibid.

[80]Note that the NSF SBIR staff originally reported its 2004 administrative budget at $3 million. Ibid. This figure was later changed to $2 million. Telephone interview with Joseph Hennessey, NSF, October 18, 2005.

The COV in 2004 commented in a broad way on the inadequacy of the resources that are available for managing and administering the program. It noted that as the workload of the program has increased, pressures on resources have intensified.

8.12.2 Administration Budget as a Percentage of Agency SBIR Funding

The administrative budget was 2 percent of agency SBIR funding in 2004 (\$2 million out of a total SBIR funding of \$96.4 million). Because the administrative budget has been relatively stable over several years, while the SBIR funding has increased, the administrative budget as a percentage of the agency's SBIR funding has dropped. For example, the administrative budget was 2.8 percent of total funding in 2002 and 2.3 percent in 2003. Because the administrative budget is not free to vary with changes in the SBIR funding, program managers have to do more with less as the program grows. If NSF budgets increase in coming years, the pinch in administrative funding can be expected to tighten further unless a change is made.

8.12.3 Evaluation and Assessment Funding

As noted in Section 8.8.4, program officials have estimated that the program's annual expenditures on evaluation and assessment in recent years total less than 1 percent of the NSF SBIR program budget.[81] However, no actual dollar value was placed on the amount spent, so it might be anywhere between \$0 and nearly \$100,000 per year. Considering the fact that externally conducted evaluation would have to come out of the small administration budget, it is not surprising that the program generally has not commissioned such evaluations. Given that in-house staff do not come out of this budget, it is likewise not surprising that staff members have generally carried out the evaluations that have been done.

8.13 BEST PRACTICES AND PROGRAM EVOLUTION

8.13.1 Adoption of Best Practices from Other Agencies

There is evidence that the NSF's SBIR program has adopted at least three to four best practices from other agencies. The following are illustrative rather than exhaustive.

A notable example of how the NSF's SBIR program adopted a "best-practice" from another agency was the introduction of the NSF Phase IIB program in 1998. According to an NSF staff member, in the mid-1990s the Department of Defense (DoD) introduced "Fast Track" as an incentive for partnering between

---

[81]NRC Program Manager Survey op. cit.

the small business and investment communities. The DoD's Fast Track required third-party funding at the front end as a prerequisite for a faster transition from Phase I to Phase II. The DoD program manager would automatically raise the ranking of Fast Track proposals to the top of the funding category. Recognizing the value of attracting investors early on, the OII decided to adapt the third-party financing requirements of Fast Track to the NSF's SBIR/STTR program. Though derived from the DoD's Fast Track, the NSF put its own stamp on its Phase IIB program, placing the third-party financing initiative after Phase II and using it to move grantees closer to commercialization.

A second practice borrowed from another agency is NSF's Commercialization Assistance Program (CP^2). The DoE's SBIR program initiated a commercialization assistance program to assist its grantees early on. The NSF program later recognized the value of such assistance, and adapted the concept for its own grantees, recognizing that many of the grantees tend to be strong in science but less so in business.

A third practice, very recently borrowed by the NSF from the DoE and Navy SBIR programs, is to provide grantees the opportunity to network with and present to potential investors in organized events. In 2005 the NSF began to sponsor some of its own grantees in participating in a DoE-sponsored "Opportunity Forum."

A fourth practice, the NSF's Matchmaker service, may have been modeled on ATP's R&D Alliance Network, which was established in the mid-1990s. It is not clear that NSF actually did model its service on ATP's, however, the services are similar and ATP's service predated NSF's.

8.13.2 Evolution of the NSF's Program During the NRC Study

It has often been noted that the effects of program evaluation and assessment typically are felt before a study is completed and recommendations are made. As an outside body examines a program, the subject program almost immediately intensifies its self-examination, which leads to internally generated changes. This phenomenon may be at work in the present situation, as the rate of evolution in the NSF's SBIR program appears to have intensified during the NRC study. Of course, during this same period, the program has also been responding to recommendations of its COV.

An example of a program improvement that has occurred during the course of the program is improved data compilation and analysis. At the beginning of this study, requests for program data were met with delays, delivery of partial and incorrect data, and a lack of evidence that the program administration routinely and systematically used program data effectively to manage the program. During the course of the study, the program appeared to have made progress in its data management and analysis and became more responsive to the study's data requests. Furthermore, as evidenced by the program's recent Strategic Report, the program is making use of data to profile and enhance understanding of program developments and trends.

OII Operational Plans for Improvements. OII included plans for additional program improvements in its recent Strategic Plan. In part, these plans anticipate and potentially address some of the issues identified in this study.[82] Here we provide a brief overview of these plans, particularly as they relate to the issues identified in this study.

OII's Strategic Plan addressed four major goals: (1) to identify, nurture, and lead the small business community to technological innovation arising from the frontiers of academic research; (2) to improve the commercial success of small high technology businesses; (3) to grow the small business community as a major employer of U.S. scientists and engineers; and (4) to deliver the highest value to the nation's small technology business community. For each of these major goals, a set of objectives is given and an operation plan is presented listing specific tasks for accomplishing the objectives. Here we briefly identify a selection of those objectives and a selection of operational plans.

A. OII Objective: Enhance training and assistance to small business grantees for the commercialization of SBIR/STTR grants

Operational Plan:
1. Offer additional contract assistance to Phase II grantees
   - Develop plans to work with incubators, business schools, and other resources
   - Concentrate on specific technology incubators, for example, Biotech incubators in Maryland
   - Provide innovation management courses to grantees
2. Revise Phase I requirements (to include more "meat" up front)
   - Bring business reviewers into the Phase I process
3. Provide training for international patent strategies

B. OII Objective: Create small business partnerships with investors and corporate partners and provide incentives to accelerate commercialization

Operational Plan:
1. Bring investors and corporate partners to grantees conferences and workshops
2. Use business school MBAs for business assistance
3. Conduct workshops on "narrow" topics for partnering

---

[82]According to the preface in the Strategic Plan, the motivating force driving the OII Strategic Plan is the Engineering Directorate's strategic directions to "Strengthen Technological Innovation" in response to the National Innovation Initiative Report "Innovate America."

4. Proactively channel NSF grantees to other agencies as potential subcontractors to primes

C. OII Objective: Judiciously select SBIR/STTR solicitation topics

Operational Plan:
1. Continuously review, refine, rejuvenate, and revise the investment business focused topics of Electronics Technology, Biotechnology, and Information-Based Technology
2. Continuously review, refine, rejuvenate, and revise the industrial market-driven topics of Advanced Materials and Manufacturing and Chemical-Based Technology
3. Be flexible and nimble to have solicitations at short notice on technologies that respond to national needs

D. OII Objective: Encourage entrepreneurship by underrepresented groups

Operational Plan:
1. Expand SBIR/STTR program beyond the newly initiated partnering with CREST (predominantly minority academic research institutions)
2. Increase subcontractor efforts to expose underrepresented small businesses to all business resources including National Outreach Conferences
3. Seek ways to increase underrepresented participation in SBIR/STTR
4. Target underrepresented community pockets
5. Target both the physically disabled community as well as the technologies for that community

E. OII Objective: Grant all Phase I and Phase II grants within six (6) months of the solicitation deadline

Operational Plan:
The plan includes a group of tasks that center on organizing [in order] to handle a high volume of proposals with limited staff and the need to employ contractors and convene panels of reviewers.

F. OII Objective: Implement a robust grants management program

G. OII Objective: Redefine outreach

H. OII Objective: Conduct an in-depth outcomes assessment

Operational Plan:
Develop a plan to define, measure, and test SBIR/STTR program metrics

The objectives in OII's Strategic Plan point to the program's efforts to improve the program on a continuing basis. Moreover, the history of the program is replete with changes to make the program better.

8.14 CONCLUSION

This report is one element of the study that Congress requested of the SBIR program as a part of the 2000 program reauthorization.[83] It focuses on the smallest, and the oldest, of the "big-five" federal SBIR programs addressed by the NRC study, namely, the NSF's SBIR program. Drawing on results of survey, case study, data and document analyses, and interviews of program staff and officials, the report has examined how the NSF's SBIR program is meeting its four mandated purposes. In the aggregate, the report provides an account of a program that is well managed and is delivering results. At the same time, it identifies challenges and recommends operational improvements to strengthen the SBIR program at the National Science Foundation.

---

[83]SBIR Reauthorization Act of 2000 (H.R. 5667, Section 108).

Appendixes

Appendix A

NSF SBIR Program Data[1]

TABLE App-A-1 Phase I Grants by Year, 1992–2005

Fiscal Year	Number of Grants	Average Grant ($)	Total Grant Amount ($)
1992	208	49,755	10,349,115
1993	256	49,727	12,730,184
1994	309	64,103	19,807,945
1995	301	64,571	19,436,011
1996	252	74,283	18,719,287
1997	261	74,262	19,382,371
1998	215	98,371	21,149,814
1999	236	98,749	23,304,757
2000	233	99,405	23,161,372
2001	219	99,353	21,758,218
2002	286	99,162	28,360,340
2003	437	99,275	43,383,103
2004	244	100,000	25,117,992
2005	149	98,575	14,687,703

---

[1]All data in Appendix A were provided by or derived from data provided by the National Science Foundation.

191

TABLE App-A-2 Phase I Applications and Success Rates, 1994–2005

Fiscal Year	Phase I Proposals Not Funded	Phase I Proposals Funded	Total Phase I Proposals	Success Rate: Percentage of Proposals Granted
1994	1,587	309	1,896	16.3
1995	1,552	301	1,853	16.2
1996	1,530	252	1,782	14.1
1997	1,390	261	1,651	15.8
1998	1,367	215	1,582	13.6
1999	1,286	236	1,522	15.5
2000	1,110	233	1,343	17.3
2001	942	219	1,161	18.9
2002	1,125	286	1,411	20.3
2003	1,653	437	2,090	20.9
2004	1,186	244	1,430	17.1
2005	923	149	1,072	13.9

TABLE App-A-3 Phase I Grants to Woman-owned Businesses, 1994–2005

Fiscal Year	Phase I Grants to Woman-owned Businesses	All Phase I Grants to Businesses	Woman-owned Share of Phase I Grants (%)	Average Size of Phase I Grants to Woman-owned Businesses ($)	Average Size of All Phase I Grants ($)	Average Size of Woman-owned Phase I Grants as a Percentage of Average Size of Overall Average Grants
1994	25	309	8.1	63,496	64,103	99.1
1995	33	301	11.0	64,680	64,571	100.2
1996	27	252	10.7	74,051	74,275	99.7
1997	24	261	9.2	74,252	74,262	100.0
1998	22	215	10.2	99,341	98,371	101.0
1999	16	236	6.8	96,863	98,096	98.7
2000	21	233	9.0	99,378	97,690	101.7
2001	17	219	7.8	99,299	99,353	99.9
2002	38	286	13.3	99,345	100,559	98.8
2003	47	437	10.8	99,789	99,272	100.5
2004	23	244	9.4	97,678	99,254	98.4
2005	20	149	13.4	99,136	98,575	100.6

TABLE App-A-4 Phase I Applications and Success Rates for Woman-owned Companies, 1994–2005

Fiscal Year	Proposals Declined	Proposals Granted	Total Applications	Woman-owned Success Rate (%)	Overall Success Rate (%)	Woman-owned as Percentage of Overall Success Rate
1994	249	25	274	9.1	16.3	56.0
1995	212	33	245	13.5	16.2	82.9
1996	197	27	224	12.1	14.1	85.2
1997	204	24	228	10.5	15.8	66.6
1998	177	22	199	11.1	13.6	81.3
1999	179	16	195	8.2	15.5	52.9
2000	157	21	178	11.8	17.3	68.0
2001	133	17	150	11.3	18.9	60.1
2002	165	38	203	18.7	20.3	92.4
2003	209	47	256	18.4	20.9	87.8
2004	168	23	191	12.0	17.1	70.4
2005	135	20	154	13.0	13.9	93.4

TABLE App-A-5 Phase I Grants to Disadvantaged Businesses, 1994–2005

Fiscal Year	Grants to Disadvantaged Businesses	Grants to All Businesses	Grants to Disadvantaged Businesses as a Percentage of Grants to All Businesses	Average Grant Size to Disadvantaged Businesses ($)	Average Grant Size to All Businesses ($)	Average Grant Size to Disadvantaged Businesses as Percentage of Average Grant Size to All Businesses
1994	35	309	11.3	63,597	64,103	99.2
1995	40	301	13.3	64,849	64,571	100.4
1996	30	252	11.9	74,949	74,275	100.9
1997	35	261	13.4	74,746	74,262	100.7
1998	27	215	12.6	99,589	98,371	101.2
1999	36	236	15.3	98,431	98,096	100.3
2000	32	233	13.7	99,773	97,690	102.1
2001	35	219	16.0	99,571	99,353	100.2
2002	50	286	17.5	99,720	100,559	99.2
2003	54	437	12.4	99,515	99,272	100.2
2004	28	244	11.5	98,348	99,254	99.1
2005	16	149	10.7	98,601	98,575	100.0

NOTE: The NSF's term "disadvantaged" is used here, whereas the main body of the report has used the SBA counterpart, "minority." As concerns this NRC study, the two terms are considered equivalent, despite differences in federal contract usage.

TABLE App-A-6 Phase I Applications and Success Rates of Disadvantaged Businesses, 1994–2005

Fiscal Year	Disadvantaged Business Phase I Applications Declined	Disadvantaged Business Phase I Applications Granted	Disadvantaged Business Total Phase I Applications	Success Rate of Disadvantaged Businesses (%)	Success Rate of All Businesses (%)	Disadvantaged Business Success Rate as Percentage of Overall Success Rate
1994	306	35	341	10.3	16.3	63.0
1995	302	40	342	11.7	16.2	72.0
1996	283	30	313	9.6	14.1	67.8
1997	263	35	298	11.7	15.8	74.3
1998	220	27	247	10.9	13.6	80.4
1999	211	36	247	14.6	15.5	94.0
2000	159	32	191	16.8	17.3	96.6
2001	167	35	202	17.3	18.9	91.9
2002	242	50	292	17.1	20.3	84.5
2003	359	54	413	13.1	20.9	62.5
2004	207	28	235	11.9	17.1	69.7
2005	154	16	170	9.4	13.9	67.7

NOTE: The NSF's term "disadvantaged" is used here, whereas the main body of the report has used the SBA counterpart, "minority." As concerns this NRC study, the two terms are considered equivalent, despite differences in federal contract usage.

TABLE App-A-7 Distribution of Phase I Grants among Firms, 1993–2002

Number of Grants	Number of Companies Receiving This Number of Grants	Total Grants Received by These Companies
40	1	40
36	1	36
33	1	33
32	1	32
31	1	31
24	2	48
22	2	44
21	1	21
18	2	36
16	4	64
15	1	15
14	2	28
13	5	65
12	7	84
11	6	66
10	1	10
9	4	36
8	11	88
7	11	77
6	27	162
5	24	120
4	47	188
3	92	276
2	249	498
1	1,115	1,115

TABLE App-A-8 Phase I Previous Winners and New Winners, 1996–2003

Fiscal Year	Grants	Unique Winners	Percentage of Grants to Multiple Winners Within Each Year	Previous Winners	New Winners	Percentage New Winners
1996	252	207	17.9	98	109	52.7
1997	261	215	17.6	99	116	54.0
1998	215	183	14.9	83	100	54.6
1999	236	194	17.8	113	81	41.8
2000	233	205	12.0	97	108	52.7
2001	219	187	14.6	92	95	50.8
2002	286	232	18.9	99	133	57.3
2003	427	374	12.4	138	236	63.1

NOTE: The excess of grants over unique winners indicates the number of grants that went to firms receiving more than one grant within the designated year.

TABLE App-A-9 Phase I Grants by State, 1992–2003

State	Grants	Percentage of All	State	Grants	Percentage of All
CA	656.0	19.0	MO	28	0.8
MA	476.0	13.8	DE	22	0.6
CO	192.0	5.6	IN	25	0.7
NY	183.0	5.3	WI	26	0.8
TX	140.0	4.0	OK	20	0.6
NJ	129.0	3.7	HI	21	0.6
CT	119.0	3.4	AR	24	0.7
VA	123.0	3.6	VT	18	0.5
MD	115.0	3.3	ME	20	0.6
OH	105.0	3.0	SC	15	0.4
PA	88.0	2.5	WY	13	0.4
AZ	81.0	2.3	SD	12	0.3
NM	78.0	2.3	KS	16	0.5
MN	63.0	1.8	AK	12	0.3
MI	69.0	2.0	WV	11	0.3
WA	63.0	1.8	ND	12	0.3
FL	60.0	1.7	LA	11	0.3
IL	53.0	1.5	ID	9	0.3
OR	46.0	1.3	IA	8	0.2
NC	53.0	1.5	NV	6	0.2
UT	40.0	1.2	KY	7	0.2
GA	44.0	1.3	NE	6	0.2
AL	34.0	1.0	MS	7	0.2
TN	30.0	0.9	RI	2	0.1
NH	29.0	0.8	PR	3	0.1
MT	32.0	0.9	DC	2	0.1

TABLE App-A-10 Phase I Applications and Success Rates by State, 1992–2003

State	Declines	Grants	Total Applications	Success Rate (%)
AK	29	12	41	29.3
AL	173	25	198	12.6
AR	34	18	52	34.6
AZ	442	72	514	14.0
CA	2420	602	3022	19.9
CO	743	184	927	19.8
CT	457	118	575	20.5
DC	25	2	27	7.4
DE	109	22	131	16.8
FL	332	53	385	13.8

continued

TABLE App-A-10 Continued

State	Declines	Grants	Total Applications	Success Rate (%)
GA	152	38	190	20.0
GU	1		1	0.0
HI	56	17	73	23.3
IA	39	8	47	17.0
ID	49	7	56	12.5
IL	267	52	319	16.3
IN	116	19	135	14.1
KS	43	12	55	21.8
KY	17	6	23	26.1
LA	121	10	131	7.6
MA	1512	472	1984	23.8
MD	611	106	717	14.8
ME	35	14	49	28.6
MI	338	60	398	15.1
MN	259	63	322	19.6
MO	87	26	113	23.0
MS	28	4	32	12.5
MT	78	26	104	25.0
NC	154	41	195	21.0
ND	24	11	35	31.4
NE	6	4	10	40.0
NH	89	31	120	25.8
NJ	581	117	698	16.8
NM	295	79	374	21.1
NV	32	5	37	13.5
NY	638	171	809	21.1
OH	515	88	603	14.6
OK	60	19	79	24.1
OR	141	47	188	25.0
PA	338	84	422	19.9
PR		2		
RI	27	3	30	10.0
SC	29	14	43	32.6
SD	42	10	52	19.2
TN	126	30	156	19.2
TX	759	140	899	15.6
UT	157	38	195	19.5
VA	773	113	886	12.8
VI	1	0	1	0.0
VT	35	15	50	30.0
WA	273	58	331	17.5
WI	109	22	131	16.8
WV	48	11	59	18.6
WY	27	12	39	30.8

NOTE: Through 2003.

TABLE App-A-11 Phase II Grants, 1992–2005

Fiscal Year	Grants	Average ($)	Maximum ($)	Minimum ($)	Total ($)
1992	57	246,373	288,934	203,984	14,043,289
1993	52	269,668	321,875	224,503	14,022,757
1994	21	279,983	300,000	242,194	5,879,644
1995	48	277,136	300,000	6,000	13,302,542
1996	90	297,584	399,998	213,271	26,782,533
1997	122	289,390	325,817	66,327	35,305,525
1998	117	344,958	400,000	262,889	40,360,087
1999	89	394,854	400,000	200,000	35,141,968
2000	95	400,827	453,585	300,000	38,078,524
2001	91	475,018	500,000	200,000	43,226,632
2002	67	495,645	500,000	329,984	33,208,238
2003	77	498,505	500,000	417,779	38,384,851
2004	131	498,254	500,000	398,768	65,278,995
2005	132	499,715	749,947	320,102	65,962,445

TABLE App-A-12 Phase II Applications and Success Rates, 1993–2005

Fiscal Year	Phase II Proposals Not Funded	Phase II Proposals Funded	Total Phase II Proposals	Success Rate: Percentage of Proposals Granted
1993	41	52	93	55.9
1994	106	21	127	16.5
1995	159	48	207	23.2
1996	133	90	223	40.4
1997	115	122	237	51.5
1998	93	117	210	55.7
1999	91	89	180	49.4
2000	92	95	187	50.8
2001	83	91	174	52.3
2002	91	67	158	42.4
2003	110	77	187	41.2
2004	190	131	321	40.8
2005	83	132	215	61.4

TABLE App-A-13 Phase II Grants to Woman-owned Businesses, 1995–2005

Fiscal Year	Phase II Grants to Woman-owned Businesses	Phase II Grants to All Businesses	Woman-owned Share of All Phase II Grants (%)	Average Size of Phase II Grants to Woman-owned Businesses ($)	Average Size of All Phase II Grants ($)	Average Size of Woman-owned Phase II as a Percentage of Average Size of All Phase II Grants
1995	1	48	2.1	299,994	277,136	1.1
1996	5	90	5.6	289,799	297,584	1.0
1997	7	122	5.7	292,329	289,390	1.0
1998	16	117	13.7	327,549	344,958	0.9
1999	8	89	9.0	399,801	394,854	1.0
2000	9	95	9.5	404,585	400,827	1.0
2001	6	91	6.6	466,323	475,018	1.0
2002	4	67	6.0	499,407	495,645	1.0
2003	4	77	5.2	499,903	498,505	1.0
2004	5	131	3.8	499,903	497,997	1.0
2005	14	132	10.6			

TABLE App-A-14 Phase II Applications and Success Rates for Woman-owned Businesses, 1995–2005

Fiscal Year	Woman-owned Business Phase II Applications Declined	Woman-owned Business Phase II Applications Granted	Woman-owned Business Total Phase II Applications	Success Rate of Woman-owned Businesses (%)	Overall Success Rate (%)	Success Rate of Woman-owned Businesses as Percentage of Overall Success Rate
1995	1	1	2	50.0	23.2	215.6
1996	15	5	20	25.0	40.4	61.9
1997	14	7	21	33.3	51.5	64.8
1998	8	16	24	66.7	55.7	119.7
1999	14	8	22	36.4	49.4	73.5
2000	5	9	14	64.3	50.8	126.5
2001	9	6	15	40.0	52.3	76.5
2002	11	4	15	26.7	42.4	62.9
2003	22	4	26	15.4	41.2	37.4
2004	14	5	19	26.3	40.8	64.5
2005	11	14	25	56.0	61.4	91.2

TABLE App-A-15 Phase II Grants to Disadvantaged Businesses, 1995–2005

Fiscal Year	Phase II Grants to Disadvantaged Businesses	Phase II Grants to All Businesses	Phase II Grants to Disadvantaged Businesses as Percentage of Phase II Grants to All Businesses
1995	1	48	2.1
1996	9	90	10.0
1997	30	122	24.6
1998	14	117	12.0
1999	20	89	22.5
2000	16	95	16.8
2001	7	91	7.7
2002	8	67	11.9
2003	8	77	10.4
2004	12	131	9.2
2005	10	132	7.6

TABLE App-A-16 Phase II Applications and Success Rates for Disadvantaged Businesses, 1995–2005

Fiscal Year	Disadvantaged Business Phase II Applications Declined	Disadvantaged Business Phase II Applications Granted	Disadvantaged Business Total Phase II Applications	Success Rate of Disadvantaged Businesses (%)	Success Rate of All Businesses (%)	Success Rate of Disadvantaged Businesses as Percentage of Overall Success Rate
1995	8	1	9	11.1	23.2	47.9
1996	20	9	29	31.0	40.4	76.9
1997	17	30	47	63.8	51.5	124.0
1998	13	14	27	51.9	55.7	93.1
1999	9	20	29	69.0	49.4	139.5
2000	12	16	28	57.1	50.8	112.5
2001	11	7	18	38.9	52.3	74.4
2002	22	8	30	26.7	42.4	62.9
2003	24	8	32	25.0	41.2	60.7
2004	35	12	47	25.5	40.8	62.6
2005	10	10	20	50.0	45.2	110.7

TABLE App-A-17 Phase II Grants by State, 1992–2004

State	Grants	Percentage of All	State	Grants	Percentage of All
CA	228	27.8	NH	10	0.0
MA	139	16.9	MT	12	0.0
CO	57	6.9	VT	8	0.0
NY	56	6.8	DE	8	0.0
CT	38	4.6	WI	8	0.0
VA	39	4.8	HI	6	0.0
MD	36	4.4	ND	7	0.0
TX	35	4.3	OK	6	0.0
OH	32	3.9	SD	5	0.0
NJ	29	3.5	WY	6	0.0
MI	24	2.9	AK	4	0.0
NM	24	2.9	IN	4	0.0
MN	21	2.6	KS	5	0.0
WA	20	2.4	ME	4	0.0
PA	21	2.6	SC	6	0.0
AZ	17	2.1	AR	6	0.0
IL	20	2.4	KY	3	0.0
TN	15	1.8	MO	5	0.0
FL	17	2.1	ID	2	0.0
GA	15	1.8	LA	3	0.0
NC	12	1.5	MS	1	0.0
OR	14	1.7	NE	1	0.0
UT	13	1.6	NV	1	0.0
AL	11	1.3	RI	1	0.0
			DC	1	0.0
			IA	1	0.0

TABLE App-A-18 Phase II Applications and Success Rates by State, 1993–2003

State	Applications Declined	Applications Granted	Total Applications	Success Rate (%)
CA	194	196	390	50.3
MA	174	125	299	41.8
CO	61	51	112	45.5
NY	66	49	115	42.6
CT	45	36	81	44.4
VA	43	35	78	44.9
MD	41	32	73	43.8
TX	44	31	75	41.3
OH	38	29	67	43.3
NJ	45	25	70	35.7
MI	17	24	41	58.5
NM	36	22	58	37.9
MN	22	18	40	45.0
WA	19	18	37	48.6
PA	26	16	42	38.1
AZ	35	15	50	30.0
IL	11	15	26	57.7
TN	7	14	21	66.7
FL	20	12	32	37.5
GA	13	12	25	48.0
NC	8	12	20	60.0
OR	12	12	24	50.0
UT	15	12	27	44.4
AL	15	10	25	40.0
NH	8	10	18	55.6
MT	5	9	14	64.3
VT	7	8	15	53.3
DE	10	7	17	41.2
WI	4	7	11	63.6
HI	5	6	11	54.5
ND	1	6	7	85.7
OK	6	5	11	45.5
SD	5	5	10	50.0
WY	3	5	8	62.5
AK	1	4	5	80.0
IN	6	4	10	40.0
KS	3	4	7	57.1
ME	3	4	7	57.1
SC	5	4	9	44.4
AR	5	3	8	37.5
KY	2	3	5	60.0
MO	6	3	9	33.3
ID	2	2	4	50.0
LA	5	2	7	28.6

TABLE App-A-18 Continued

State	Applications Declined	Applications Granted	Total Applications	Success Rate (%)
MS	3	1	4	25.0
NE	0	1	1	100.0
NV	4	1	5	20.0
RI	1	1	2	50.0
IA	2	0	2	0.0
PR	1	0	1	0.0
WV	4	0	4	0.0

TABLE App-A-19 Sample of Multiple Phase I Grant Recipients and Their Success in Getting Follow-on Phase II Grants

Company	Phase II Grants	Phase I Grants	Conversion Rate (%)
Optivision Incorporated	6	6	100.0
T/J Technologies, Inc.	7	9	77.8
Displaytech Incorporated	6	8	75.0
Advanced Fuel Research, Inc.	10	18	55.6
INRAD, Inc.	6	11	54.5
Reveo Incorporated	7	13	53.8
GINER, INC.	7	13	53.8
Quantum Magnetics, Inc.	6	12	50.0
HYPRES, Inc.	6	12	50.0
Science Research Laboratory, Inc.	10	22	45.5
Envirogen, Inc.	6	14	42.9
TPL, Inc.	6	15	40.0
CFD Research Corporation	7	18	38.9
Physical Optics Corporation	12	33	36.4
TDA Research, Inc.	12	40	30.0
Materials Modification, Inc.	6	24	25.0
Eltron Research, Inc.	6	24	25.0
Materials & Electrochemical Research Corp. (MER)	7	31	22.6
Lynntech, Inc.	8	36	22.2
Spire Corporation	7	32	21.9

TABLE App-A-20 Phase IIB Grants, 1992–1995

Fiscal Year	Phase IIB Grants	Total Outlay Phase IIB ($)
1992	0	0
1993	0	0
1994	0	0
1995	0	0
1996	0	0
1997	0	0
1998	4	399,944
1999	21	1,998,574
2000	9	1,660,191
2001	14	4,302,682
2002	39	9,588,580
2003	24	5,690,294
2004	22	6,025,436
2005	28	9,260,464

TABLE App-A-21 Distribution of NSF SBIR Phase IIB Grants by State in Descending Order of Grants Received, 1998–2005

State	Grants	Percent of All	State	Grants	Percent of All
CA	36	22.4	DE	2	1.2
MA	21	13.0	KS	2	1.2
NY	11	6.8	NH	2	1.2
NC	7	4.3	NM	2	1.2
OH	7	4.3	TN	2	1.2
MN	6	3.7	UT	2	1.2
FL	5	3.1	VT	2	1.2
MD	5	3.1	WI	2	1.2
VA	5	3.1	AK	1	0.6
AL	4	2.5	AZ	1	0.6
IL	4	2.5	HI	1	0.6
MI	4	2.5	MT	1	0.6
PA	4	2.5	ND	1	0.6
GA	3	1.9	OK	1	0.6
TX	3	1.9	OR	1	0.6
WA	3	1.9	RI	1	0.6
WY	3	1.9	SC	1	0.6
CO	2	1.2	SD	1	0.6
CT	2	1.2			

Appendix B

NRC Phase II Survey and
NRC Firm Survey

The first section of this appendix describes the methodology used to survey Phase II SBIR awards (also referred to as projects). The second part presents the results—first of the awards, or project, survey (NRC Phase II Survey), and then of the firm survey (NRC Firm Survey). (Appendix C presents the NRC Phase I survey.)

ABOUT THE SURVEYS

Starting Date and Coverage

The survey of SBIR Phase II awards was administered in 2005 and included awards made through 2001. This allowed most of the Phase II awarded projects (nominally two years) to be completed and provided some time for commercialization. The selection of the end date of 2001 was consistent with a GAO study, which in 1991, surveyed awards made through 1987.

A start date of 1992 was selected. The year 1992 for the earliest Phase II project was considered a realistic starting date for the coverage, allowing inclusion of the same (1992) projects as the Department of Defense (DoD) 1996 survey, and of the 1992 and 1993 projects surveyed in 1998 for the Small Business Administration (SBA). This adds to the longitudinal capacities of the study. The 10 years of Phase II coverage spanned the period of increased funding set-asides and the impact of the 1992 reauthorization. This time frame allowed for extended periods of commercialization and for a robust spectrum of economic conditions. Establishing 1992 as the cutoff date for starting the survey helped to avoid the problems from which older awards suffer, including meager early data collection as well as potentially irredeemable data loss; the fact that some firms and principal investigators (PIs) are no longer in place; and fading memories.

Award Numbers

While adding the annual awards numbers of the five agencies would seem to define the larger sample, the process was more complicated. Agency reports usually involve some estimating and anticipation of successful negotiation of selected proposals. Agencies rarely correct reports after the fact. Setting limitations on the number of projects to be surveyed from each firm required knowing how many awards each firm had received from all five agencies. Thus, the first step was to obtain all of the award databases from each agency and combine them into a single database. Defining the database was further complicated by variations in firm identification, location, phone numbers, and points of contact within individual agency databases. Ultimately, we determined that 4,085 firms had been awarded 11,214 Phase II awards (an average of 2.7 Phase II awards per firm) by the five agencies during the 1992–2001 time frame. Using the most recent awards, the firm information was updated to the most current contact information for each firm.

Sampling Approaches and Issues

The Phase II Survey used an array of sampling techniques to ensure adequate coverage of projects, to address a wide range both of outcomes and potential explanatory variables, and also to address the problem of skew. That is, a relatively small percentage of funded projects typically account for a large percentage of commercial impact in the field of advanced, high-risk technologies.

- **Random Samples.** After integrating the 11,214 awards into a single database, a random sample of approximately 20 percent was sampled. Then a random sample of 20 percent was ensured for each year; e.g., 20 percent of the 1992 awards, of the 1993 awards, etc. Verifying the total sample one year at a time allowed improved ability to adapt to changes in the program over time, as otherwise the increased number of awards made in recent years might dominate the sample.

- **Random Sample by Agency.** Surveyed awards were grouped by agency; additional respondents were randomly selected as required to ensure that at least 20 percent of each agency's awards were included in the sample.

- **Firm Surveys.** After the random selection, 100 percent of the Phase IIs that went to firms with only one or two awards were polled. These are the hardest firms to find for older awards. Address information is highly perishable, particularly for earlier award years. For firms that had more than two awards, 20 percent were selected, but no less than two.

- **Top Performers.** The problem of skew was dealt with by ensuring that all Phase IIs known to meet a specific commercialization threshold (total of $10 million in the sum of sales plus additional investment) were surveyed (derived from the DoD commercialization database). Since 56 percent of all awards were in the random and firm samples described above, only 95 Phase IIs were added in this fashion.

- **Coding.** The project database tracks the survey sample, which corresponds with each response. For example, it is possible for a randomly sampled project from a firm that had only two awards to be a top performer. Thus, the response could be analyzed as a random sample for the program, a random sample for the awarding agency, a top performer, and as part of the sample of single or double winners. In addition, the database allows examination of the responses for the array of potential explanatory or demographic variables.

- **Total Number of Surveys.** The approach described above generated a sample of 6,410 projects and 4,085 firm surveys—an average of 1.6 award surveys per firm. Each firm receiving at least one project survey also received a firm survey. Although this approach sampled more than 57 percent of the awards, multiple award winners, on average, were asked to respond to surveys covering about 20 percent of their projects.

Administration of the Survey

The questionnaire drew extensively from the one used in the 1999 National Research Council assessment of SBIR at the Department of Defense, *The Small Business Innovation Research Program: An Assessment of the Department of Defense Fast Track Initiative.*[1] That questionnaire in turn built upon the questionnaire for the 1991 GAO SBIR study. Twenty-four of the twenty-nine questions on the earlier NRC study were incorporated. The researchers added twenty-four new questions to attempt to understand both commercial and noncommercial aspects, including knowledge base impacts, of SBIR, and to gain insight into impacts of program management. Potential questions were discussed with each agency, and their input was considered. In determining questions that should be in the survey, the research team also considered which issues and questions were best examined

---

[1]National Research Council, *The Small Business Innovation Research Program: An Assessment of the Department of Defense Fast Track Initiative*, Charles W. Wessner, ed., Washington, DC: National Academy Press, 2000.

in the case studies and other research methodologies. Many of the resultant 33 Phase II Survey questions and 15 Firm Survey questions had multiple parts.

The surveys were administered online, using a Web server. The formatting, encoding and administration of the survey was subcontracted to BRTRC, Inc., of Fairfax, Virginia.

There are many advantages to online surveys (including cost, speed, and possibly response rates). Response rates become clear fairly quickly, and can rapidly indicate needed follow up for nonrespondents. Hyperlinks provide amplifying information, and built-in quality checks control the internal consistency of the responses. Finally, online surveys allow dynamic branching of question sets, with some respondents answering selected subsets of questions but not others, depending on prior responses.

Prior to the survey, we recognized two significant advantages of a paper survey over an online one. For every firm (and thus every award), the agencies had provided a mailing address. Thus surveys could be addressed to the firm president or CEO at that address. That senior official could then forward the survey to the correct official within the firm for completion. For an online survey we needed to know the email address of the correct official. Also each firm needed a password to protect its answers. We had an SBIR point of contact (POC) and an email address and a password for every firm, which had submitted for a DoD SBIR 1999 survey. However, we had only limited email addresses and no passwords for the remainder of the firms. For many, the email addresses that we did have were those of principal investigators rather than an official of the firm. The decision to use an online survey meant that the first step of survey distribution was an outreach effort to establish contact with the firms.

Outreach by Mail

This outreach phase began with establishing a National Academy of Sciences (NAS) registration Web site which allowed each firm to establish a POC, an email address, and a password. Next, the Study Director, Dr. Charles Wessner, sent a letter to those firms for which email contacts were not available. Ultimately, only 150 of the 2,080[2] firms provided POC/email after receipt of this letter. The U.S. Postal Service returned 650 of those letters as invalid addresses. Each returned letter required thorough research by calling the agency-provided phone number for the firm, then using the Central Contractor Registration database, Business. com (powered by Google), and Switchboard.com to try to find correct address information. When an apparent match was found, the firm was called to verify that it was, in fact, the firm, which had completed the SBIR. Two hundred thirty-seven of the 650 missing firms were so located. Another ten firms that had gone out of business and had no POC were located.

---

[2]The letter was also erroneously sent to an additional 43 firms that had received only STTR awards.

Two months after the first mailing, a second letter from the study director was mailed to firms whose first letter had not been returned, but which had not yet registered a POC. This letter also went to 176 firms, which had a POC email but no password, and to the 237 newly corrected addresses. The large number of letters (277) from this second mailing that were returned by the U.S. Postal Service, indicated that there were more bad addresses in the first mailing than were indicated by its returned mail. (If the initial letter was inadvertently delivered, it may have been thrown away.) Of the 277 returned second letters, 58 firms were located using the search methodology described above. These firms were asked on the phone to go to the registration Web site to enter POC/email/password. A total of 93 firms provided POC/email/password on the registration site subsequent to the second mailing. Three additional firms were identified as out of business.

The final mailing, a week before survey, was sent to those firms that had not received either of the first two letters. It announced the study/survey and requested support of the 1,888 CEOs for which we had assumed good POC/email information from the DoD SBIR submission site. That letter asked the recipients to provide new contact information at the DoD submission site if the firm information had changed since their last submission. One hundred seventy-three of these letters were returned. We were able to find new addresses for 53 of these, and to ask those firms to update their information. One hundred fifteen firms could not be found, and 5 more were identified as out of business.

The three mailings had demonstrated that at least 1,100 (27 percent) of the mailing addresses were in error, 734 of which firms could not be found, and 18 were reported to be out of business.

Outreach by Email

We began Internet contact by emailing the 1,888 DoD POCs to verify their email and to give them an opportunity to identify a new POC. Four hundred ninety-four of those emails bounced. The next email went to 788 email addresses that we had received from agencies as PI emails. We asked that the PI have the correct company POC identify themselves at the NAS Update registration site. One hundred eighty-eight of these emails bounced. After a more detailed search of the list used by the National Institutes of Health (NIH) to send out their survey, we identified 83 additional PIs and sent them the PI email discussed above. Email to the POCs not on the DoD Submission site resulted in 110 more POCs/emails/passwords being registered on the NAS registration site.

We began the survey at the end of February with an email to 100 POCs as a beta test and followed that with another email to 2,041 POCs (total of 2,141) a week later.

Survey Responses

By August 5, 2005, five months after release of the survey, 1,239 firms had begun and 1,149 firms had completed at least 14 of 15 questions on the Firm Survey. Project surveys were begun on 1,916 Phase II awards. Of the 4,085 firms that received Phase II SBIR awards from DoD, NIH, NASA, NSF, or DoE from 1992 to 2001, an additional 7 firms were identified as out of business (total of 25), and no email addresses could be found for 893. For an additional 500 firms, the best email addresses that were found were also undeliverable. These 1,418 firms could not be contacted and thus had no opportunity to complete the surveys. Of these firms, 585 had mailing addresses known to be bad. The 1,418 firms that could not be contacted were responsible for 1,885 of the individual awards in the sample.

Using the same methodology as the GAO had used in the 1992 report of their 1991 survey of SBIR, undeliverables and out of business firms were eliminated prior to determining the response rate. Although 4,085 firms were surveyed, 1,418 firms were eliminated as described. This left 2,667 firms, of which 1,239 responded, representing a 46 percent response rate by firms,[3] which could respond. Similarly, when the awards, which were won by firms in the undeliverable category, were eliminated (6408 minus 1885), this left 4,523 projects, of which 1,916 responded, representing a 42 percent response rate. Figure App-B-1 displays by agency the number of Phase II awards in the sample, the number of those awards, which by having good email addresses had the opportunity to respond, and the number that responded.[4] Percentages displayed are the percentage of awards with good addresses, the percentage of the sample that responded, and the responses as a percentage of awards with the opportunity to respond.

The NRC Methodology report had assumed a response rate of about 20 percent. Considering the length of the survey and its voluntary nature, the rate achieved was relatively high and reflects both the interest of the participants in the SBIR program and the extensive follow-up efforts. At the same time, the possibility of response biases that could significantly affect the survey results must be recognized. For example, it may be possible that some of the firms that could not be found have been unsuccessful and folded. It may also be possible that unsuccessful firms were less likely to respond to the survey.

---

[3]Firm information and response percentages are not displayed in Figure 1, which displays by agency, since many firms received awards from multiple agencies.

[4]The average firm size for awards, which responded, was 37 employees. Nonresponding awards came from firms that averaged 38 employees. Since responding Phase II were more generally more recent than nonresponding, and awards have gradually grown in size, the difference in average award size ($655,525 for responding and $649,715 for nonresponding) seems minor.

TABLE App-B-1 NRC Phase II Survey Responses by Agency as of August 4, 2005

Agency	Phase II Sample Size	Awards with Good Email Addresses	Percentage of Sample Awards with Good Email Addresses	Answered Survey as of 8/4/2005	Surveys as a Percentage of Sample	Surveys as a Percentage of Awards Contacted
DoD	3,055	2,191	72	920	30	42
NIH	1,680	1,127	67	496	30	44
NASA	779	534	69	181	23	34
NSF	457	336	74	162	35	48
DoE	439	335	76	157	36	47
Total	6,408	4,523	70	1,916	30	42

NRC Phase II Survey Results

NOTE: RESULTS ARE FOR NSF. SURVEY RESPONSES APPEAR IN BOLD, AND EXPLANATORY NOTES ARE IN A TYPEWRITER FONT.

Project Information 162 respondents answered the first question. Since respondents are directed to skip certain questions based on prior answers, the number that responded varies by question. Also, some respondents did not complete their surveys. 146 completed all applicable questions. Unless otherwise indicated, 162 is used as the denominator in computing averages. Where appropriate, the basis for calculations is provided after the question.

Part I. Current status of the Project.

1. What is the current status of the project funded by the referenced SBIR grant? *Select the one best answer.* Percentages are based on the 162 respondents who answered this question
    a. **1%** Project has not yet completed Phase II. *Go to question 21.*
    b. **20%** Efforts at this company have been discontinued. No sales or additional funding resulted from this project. *Go to question 2.*
    c. **10%** Efforts at this company have been discontinued. The project did result in sales, licensing of technology, or additional funding. *Go to question 2.*
    d. **28%** Project is continuing postPhase II technology development. *Go to question 3.*
    e. **19%** Commercialization is underway. *Go to question 3.*
    f. **22%** Products/Processes/ Services are in use by target population/customer/consumers. *Go to question 3.*

2. Did the reasons for discontinuing this project include any of the following? ***PLEASE SELECT YES OR NO FOR EACH REASON AND NOTE THE ONE PRIMARY REASON.***
48 projects were discontinued. The % below is the percent of the discontinued projects that responded with the indicated response.

	Yes	No	Primary Reason
a. Technical failure or difficulties	33%	67%	19%
b. Market demand too small	42%	58%	17%
c. Level of technical risk too high	21%	79%	0%
d. Not enough funding	48%	52%	13%
e. Company shifted priorities	40%	60%	8%
f. Principal investigator left	17%	83%	6%
g. Project goal was achieved (e.g., prototype delivered for federal agency use)	31%	69%	6%
h. Licensed to another company	21%	79%	10%
i. Product, process, or service not competitive	29%	71%	6%
j. Inadequate sales capability	15%	85%	0%
k. Other (please specify): _____	19%	81%	15%

The next question to be answered depends on the answer to Question 1. If c., go to Question 3. If b, skip to Question 16.

Part II. Commercialization activities and planning.

Questions 3-7 concern actual sales to date resulting from the technology developed during this project. **Sales** includes all sales of a product, process, or service, to federal or private sector customers resulting from the technology developed during this Phase II project. A sale also includes licensing, the sale of technology or rights, etc.

3. Has your company and/or licensee had any actual sales of products, processes, services or other sales incorporating the technology developed during this project? *Select all that apply.* This question was not answered for those projects still in Phase II (1%) or for projects, which were discontinued without sales or additional funding (20%). The denominator for the percentages below is all projects that answered the survey. Only 79% of all projects, which answered the survey, could respond to this question.

   a. **25%** No sales to date, but sales are expected. *Skip to Question 8*
   b. **6%** No sales to date nor are sales expected. *Skip to Question 11*
   c. **38%** Sales of product(s)
   d. **3%** Sales of process(es)
   e. **22%** Sales of services(s)
   f. **10%** Other sales (e.g. rights to technology, licensing, etc.)

   From the combination of responses 1b, 3a, and 3b, we can conclude that 26% had no sales and expect none, and that 25% had no sales but expect sales.

4. For your company and/or your licensee(s), when did the first sale occur, and what is the approximate amount of total sales resulting from the technology developed during this project? If multiple SBIR grants contributed to the ultimate commercial outcome, report only the share of total sales appropriate to this SBIR project. *Enter the requested information for your company in the first column and, if applicable and if known, for your licensee(s) in the second column. Enter approximate dollars. If none, enter 0 (zero)*

	Your Company	**Licensee(s)**

a. Year when first sale occurred

46% reported a year of first sale. 65% of these first sales occurred in 2000 or later. 21% reported a licensee year of first sale. 56% of these

b. Total Sales Dollars of Product (s) Process(es) **$ 332,364 $1,413,708** or Service(s) to date. (average of 162 survey respondents)

Although 74 reported a year of first sale, only 30 reported sales >0. Their average sales were $ 769,186. Over half of the total sales dollars were due to 8 projects, each of which had $2,300,000 or more in sales. The highest reporting project had $4,757,000 in sales. Similarly only of the 34 projects that reported a year of first licensee sale, only 12 reported actual licensee sales >0. Their average sales were $19,085,065. Over half of the total sales dollars was due to 1 project, which had $200,000,000 or more licensee sales.

c. Other Total Sales Dollars (e.g., Rights to   **$ 65,876 $ 4,148** technology, Sale of spin-off company, etc.) to date (average of 162 survey respondents)

Combining the responses for b and c, the average for each of the 162 projects that responded to the survey is thus sales of over $398,000 by the SBIR company and over one million four hundred thousand dollars in sales by licensees.

Display this box for Q 4 & 5 if project commercialization is known.

Your company reported sales information to DoD as a part of an SBIR proposal or to NRC as a result of an earlier NRC request. This information may be useful in answering the prior question or the next question. You reported as of *(date)*: DoD sales *($ amount),* Other federal sales *($ amount),* Export sales *($ amount),* Private-Sector sales *($ amount)*, and other sales *($ amount).*

---

5. To date, approximately what percent of total sales from the technology developed during this project have gone to the following customers? *If none enter 0 (zero). Round percentages. Answers should add to about 100%.*[5] 162 firms responded to this question as to what percent of their sales went to each agency or sector

Domestic private sector	**57%**
Department of Defense (DoD)	**11%**
Prime contractors for *DoD or NASA*	**5%**
NASA	**1%**
Agency that granted the Phase II	**1%**
Other federal agencies *(Pull down)*	**2%**
State or local governments	**4%**
Export markets	**11%**
Other *(Specify)*_____	**9%**

The following questions identify the product, process, or service resulting from the project supported by the referenced SBIR grant, including its use in a fielded federal system or a federal acquisition program.

---

6. Is a federal system or acquisition program using the technology from this Phase II?
    If yes, please provide the name of the federal system or acquisition program that is using the technology. **4% reported use in a federal system or acquisition program.**

7. Did a commercial product result from this Phase II project? **13% reported a commercial product.**

8. If you have had no sales to date resulting from the technology developed during this project, what year do you expect the first sales for your company or its licensee? Only firms that had no sales but answered that they expected sales got this question.

    15% expected sales. The year of expected first sale is ⬚⬚⬚⬚
    85% of those expecting sales expected sales to occur before 2008.

---

[5]Please note: If a NASA SBIR grant, the prime contractor's line will state "Prime contractors for NASA." The "Agency that granted the Phase II" will only appear if it is not DoD or NASA. The name of the actual granting agency will appear.

9. For your company and/or your licensee, what is the approximate amount of total sales expected between now and the end of 2006 resulting from the technology developed during this project? *If none, enter 0 (zero).* This question was seen by those who already had sales and those w/o sales who reported expecting sales; however, averages are computed for all who took the survey since all could have expected sales.

   a. Total sales dollars of product(s), process(es) or services(s) expected between now and the end of 2006. **$ 657,849** (average of 162 projects)

   b. Other Total Sales Dollars (e.g., rights to technology, sale of spin-off company, etc.) expected between now and the end of 2006. **$ 185,833** (average of 162 projects)

   c. Basis of expected sales estimate. *(Select all that apply.)*
      a. **26%** Market research
      b. **38%** Ongoing negotiations
      c. **34%** Projection from current sales
      d. **5%** Consultant estimate
      e. **32%** Past experience
      f. **40%** Educated guess

10. How did you (or do you expect to) commercialize your SBIR grant?
    a. **2%** No commercial product, process, or service was/is planned.
    b. **32%** As software
    c. **54%** As hardware (final product, component, or intermediate hardware product)
    d. **32%** As process technology
    e. **24%** As new or improved service capability
    f. **0%** As a drug
    g. **3%** As a biologic
    h. **21%** As a research tool
    i. **12%** As educational materials
    j. **8%** Other, please explain _____

11. Which of the following, if any, describes the type and status of marketing activities by your company and/or your licensee for this project? *Select one for each marketing activity.* This question answered by 122 firms, which completed Phase II and have not discontinued the project, w/o sales or additional funding.

Marketing activity	Planned	Need Assistance	Underway	Completed	Not Needed
a. Preparation of marketing plan	13%	4%	26%	34%	23%
b. Hiring of marketing staff	11%	6%	12%	24%	47%
c. Publicity/advertising	16%	7%	30%	19%	28%
d. Test marketing	11%	5%	19%	22%	43%
e. Market research	6%	7%	32%	28%	28%
f. Other (Specify)	2%	0%	3%	4%	22%

Part III. Other outcomes.

12 As a result of the technology developed during this project, which of the following describes your company's activities with other companies and investors? *Select all that apply.* Percentage of the 121 who answered this question.

Activities	U.S. Companies/Investors		Foreign Companies/Investors	
	Finalized Agreements	Ongoing Negotiations	Finalized Agreements	Ongoing Negotiations
Licensing Agreement(s)	20%	21%	10%	7%
Sale of company	2%	3%	0%	1%
Partial sale of company	2%	6%	2%	2%
Sale of technology rights	5%	16%	4%	3%
Company merger	0%	4%	2%	1%
Joint Venture agreement	3%	10%	1%	2%
Marketing/distribution agreement(s)	16%	12%	8%	2%
Manufacturing agreement(s)	8%	10%	3%	2%
R&D agreement(s)	17%	17%	5%	7%
Customer alliance(s)	12%	18%	3%	4%
Other (Specify) _____	2%	2%	1%	0%

13. In you opinion, in the absence of this SBIR grant, would your company have undertaken this project?
    Select one. Percentage of the 121 who answered this question.
    a. **4%** Definitely yes
    b. **10%** Probably yes *(If selected a. or b. , Go to question 14.)*
    c. **19%** Uncertain
    d. **43%** Probably not
    e. **24%** Definitely not *(If c., d. or e., skip to quest. 16.)*

14. If you had undertaken this project in the absence of SBIR, this project would have been Questions 14 and 15 were answered only by the 15% who responded that they definitely or probably would have undertaken this project in the absence of SBIR.
    a. **11%**   Broader in scope
    b. **5%**   Similar in scope
    c. **84%**   Narrower in scope

15. In the absence of SBIR funding, (Please provide your best estimate of the impact.)
    a. The start of this project would have been delayed about **an average of 13** months.
       58% of the 11 firms expected the project would have been delayed. 60% (9 firms) expected the delay would be at least 12 months. 26% anticipated a delay of at least 24 months.
    b. The expected duration/time to completion would have been
       1) **63%**   Longer
       2) **21%**   The same
       3) **0%**   Shorter
          16%   No response
    c. In achieving <u>similar</u> goals and milestones, the project would be
       1) **0%**   Ahead
       2) **5%**   The same place
       3) **68%**   Behind
          26%   No response

16. Employee information. *Enter number of employees. You may enter fractions of full time effort (e.g., 1.2 employees). Please include both part-time and full-time employees, and consultants, in your calculation.*

Number of employees (if known) when Phase II proposal was submitted.	**Ave = 21** **5% report 0** **48% report 1 - 5** **23% report 6 - 20** **10% report 21 - 50** **8% report >100**
Current number of employees.	**Ave = 38** **3% report 0** **30% report 1 - 5** **35% report 6 - 20** **11% report 21 - 50** **10% report >100**
Number of current employees <u>who were hired</u> as a result of the technology developed during this Phase II project.	**Ave = 1.5** **48% report 0** **48% report 1 - 5** **5% report 6 – 20** **0% report >20**
Number of current employees <u>who were retained</u> as a result of the technology developed during this Phase II project.	**Ave = 2** **42% report 0** **51% report 1 - 5** **1% report 6 – 20** **0% report >20**

17. The Principal Investigator for this Phase II Grant was a (check all that apply)
    a. **12%** Woman
    b. **13%** Minority
    c. **79%** Neither a woman or minority

18. Please give the number of patents, copyrights, trademarks, and/or scientific publications for the technology developed as a result of this project. *Enter numbers. If none, enter 0 (zero).* Results are for 151 respondents to this question.

Number Applied For/Submitted		Number Received/Published
159	Patents	101
49	Copyrights	42
42	Trademarks	33
266	Scientific Publications	250

Part IV. Other SBIR funding.

19. How many SBIR grants did your company receive prior to the Phase I that led to this Phase II?
    a. Number of previous Phase I grants. **Average of 11. 38% had no prior Phase I, and another 38% had 5 or less prior Phase I.**
    b. Number of previous Phase II grants. **Average of 4. 55% had no prior Phase II, and another 30% had 5 or less prior Phase II.**

20. How many SBIR grants has your company received that are related to the project/technology supported by this Phase II grant ?
    a. Number of related Phase I grants   **Average of 1; 49% had no prior related Phase I and another 47 % had 5 or less prior related Phase I**
    b. Number of related Phase II grants   **Average of 1; 61% had no prior related Phase II and another 37 % had 5 or less prior related Phase II**

Part V. Funding and other assistance.

21. Prior to this SBIR Phase II grant, did your company receive funds for research or development of the technology in this project from any of the following sources? Of 151 respondents.
    a. **20%** Prior SBIR *(Excluding the Phase I, which proceeded this Phase II.)*
    b. **13%** Prior non-SBIR federal R&D
    c. **2%** Venture Capital
    d. **15%** Other private company
    e. **14%** Private investor
    f. **35%** Internal company investment (including borrowed money)
    g. **5%** State or local government
    h. **4%** College or University
    i. **9%** Other *Specify* _____

Commercialization of the results of an SBIR project normally requires additional developmental funding. Questions 22 and 23 address additional funding. Additional Developmental Funds include non-SBIR funds from federal or private sector sources, or from your own company, used for further development and/or commercialization of the technology developed during this Phase II project.

---

22. Have you received or invested any additional developmental funding in this project?

    a. **63%** Yes *Continue.*
    b. **37%** No *Skip to Question 24.*

23. To date, what has been the total additional developmental funding for the technology developed during this project? Any entries in the **Reported** column are based on information previously reported by your firm to DoD or NAS. They are provided to assist you in completing the **Developmental funding** column. Previously reported information did not include investment by your company or personal investment. *Please update this information to include breaking out Private investment and Other investment by subcategory. Enter dollars provided by each of the listed sources. If none, enter 0 (zero).* The dollars shown are determined by dividing the total funding in that category by the 162 respondents who started the survey to determine an average funding. Ninety-three of these respondents reported any additional funding.

Source	Reported	Developmental Funding
a. Non-SBIR federal funds	$_ _, _ _ _, _ _ _	$ 248,077
b. Private investment	$_ _, _ _ _, _ _ _	
(1) U.S. venture capital		$ 39,450
(2) Foreign investment		$ 19,290
(3) Other Private equity		$ 196,141
(4) Other domestic private		
company		$ 57,925
c. Other sources	$_ _, _ _ _, _ _ _	
(1) State or local		
governments		$ 19,938
(2) College or universities		$ 617
d. Not previously reported		
(1) Your own company		
(Including money		
you have borrowed)		$ 54,617
(2) Personal funds		$ 22,154
Total average additional developmental funding, all sources, per grant		$ 658,214

24. Did this grant identify matching funds or other types of cost sharing in the Phase II Proposal?[6]
   a. **68%** No matching funds/co-investment/cost sharing were identified in the proposal.
      If a, skip to question 26.
   b. **32%** <u>Although not a DoD Fast Track</u>, matching funds/co-investment/ cost sharing were identified in the proposal.
   c. **0%** <u>Yes. This was a DoD Fast Track proposal.</u>

25. Regarding sources of matching or co-investment funding that were proposed for Phase II, check all that apply. The percentages below are computed for those 48 projects, which reported matching funds.
   a. **38%** Our own company provided funding (includes borrowed funds).
   b. **4%** A federal agency provided non-SBIR funds.
   c. **58%** Another company provided funding.
   d. **21%** An angel or other private investment source provided funding.
   e. **2%** Venture Capital provided funding.

26. Did you experience a gap between the end of Phase I and the start of Phase II?
   a. **81%** Yes *Continue.*
   b. **19%** No *Skip to question 29.*
   The average gap reported by 121 respondents was 8 months. 5% of the respondents reported a gap of two or more years.

27. Project history. Please fill in for all dates that have occurred. This information is meaningless in aggregate. It has to be examined project by project in conjunction with the date of the Phase I end and the date of the Phase II grant to calculate the gaps.

   Date Phase I ended *Month/ year*

   Date Phase II proposal submitted *Month /year*

28. If you experienced funding gap between Phase I and Phase II for this grant, *select all answers that apply.*
   a. **28%** Stopped work on this project during funding gap.
   b. **37%** Continued work at reduced pace during funding gap.
   c. **5%** Continued work at pace equal to or greater than Phase I pace during funding gap.
   d. **2%** Received bridge funding between Phase I and II.
   e. **1%** Company ceased all operations during funding gap.

---

[6]The words <u>underlined</u> appear only for DoD grants.

29. Did you receive assistance in Phase I or Phase II proposal preparation for this grant? Of 136 respondents.
    a. **3%** State agency provided assistance.
    b. **1%** Mentor company provided assistance.
    c. **0%** Regional association provided assistance.
    d. **5%** University provided assistance.
    e. **91%** We received no assistance in proposal preparation.

    Was this assistance useful?
    a. **62%** Very useful
    b. **38%** Somewhat useful
    c. **0%** Not useful

30. In executing this grant, was there any involvement by universities faculty, graduate students, and/or university developed technologies? Of 146 respondents.
    47% Yes
    53% No

31. This question addresses any relationships between your firm's efforts on this Phase II project and any University (ies) or College (s). The percentages are computed against the 146 who answered question 30, not just those who answered yes to question 30. *Select all that apply.*
    a. **1%** The Principal Investigator (PI) for this Phase II project was at the time of the project a faculty member.
    b. **5%** The Principal Investigator (PI) for this Phase II project was at the time of the project an adjunct faculty member.
    c. **37%** Faculty member(s) or adjunct faculty member (s) work on this Phase II project in a role other than PI, e.g., consultant.
    d. **27%** Graduate students worked on this Phase II project.
    e. **25%** University/College facilities and/or equipment were used on this Phase II project.
    f. **5%** The technology for this project was licensed from a University or College.
    g. **14%** The technology for this project was originally developed at a University or College by one of the percipients in this phase II project.
    h. **17%** A University or College was a subcontractor on this Phase II project.

    In remarks enter the name of the University or College that is referred to in any blocks that are checked above. If more than one institution is referred to, briefly indicate the name and role of each.

32. Did commercialization of the results of your SBIR grant require FDA approval? Yes **4%**

    In what stage of the approval process are you for commercializing this SBIR grant?
    a. **0%** Applied for approval
    b. **1%** Review ongoing
    c. **0%** Approved
    d. **3%** Not Approved
    e. **0%** IND: Clinical trials
    f. **1%** Other

NRC Firm Survey Results

NOTE: ALL RESULTS APPEAR IN BOLD. TWO SETS OF RESULTS ARE REPORTED—THOSE FOR ALL 5 AGENCIES (DOD, NIH, NSF, DOE, AND NASA) AND THOSE FOR ONLY NSF. NSF RESULTS APPEAR IN PARENTHESES.
(137) 1,239 firms began the survey. 1,149 (137) completed through question 14. 1,108 (130) completed all questions.

If your firm is registered in the DoD SBIR/STTR Submission Web site, the information filled in below is based on your latest update as of September 2004 on that site. Since you may have entered this information many months ago, you may edit this information to make it correct. In conjunction with that information, the following additional information will help us understand how the SBIR program is contributing to the formation of new small businesses active in federal R&D and how they impact the economy. Questions A-G are autofilled from Firm database, when available.

A. Company Name: _____

B. Street Address: _____

C. City: _____ State: _____ Zip: _____

D. Company Point of Contact: _____

E. Company Point of Contact Email: _____

F. Company Point of Contact Phone: (___) ___ - ____ Ext: _____

G. The year your company was founded: _____

1. Was your company founded because of the SBIR Program?
    a. 79% (80%)   No
    b. 8% (7%)   Yes
    c. 13% (13%)   Yes, In part

2. Information on company founders. *Please enter zeros or the correct number in each pair of blocks.*
    a. Number of founders.
       5% (1%)   unknown
       40% (47%)   1
       30% (25%)   2
       13% (16%)   3
       8% (7%)   4
       2% (1%)   5
       2% (3%)   >5
       Average = 2 (2) founders/firm

b. Number of other companies started by one or more
    of the founders.   ⬚⬚
    5% (0%)   unknown
    46% (44%)   started no other firms
    23% (26%)   started 1 other firm
    13% (14%)   started 2 other firms
    7% (4%)   started 3 other firms
    3% (6%)   started 4 other firms
    3% (7%)   started 5 or more other firms
    Average number of other firms founded is one. (1%)

c. Number of founders who have a business background.   ⬚⬚
    5% (0%)   Unknown
    50% (47%)   No founder known to have business background
    30% (37%)   One founder with business background
    14% (15%)   More than one founder with business background

d. Number of founders who have an academic background   ⬚⬚
    5% (0%)   Unknown
    29% (32%)   No founder known to have academic background
    38% (38%)   One founder with academic background
    28% (30%)   More than one founder with academic background

**(Note: Business only=20%; Academic only=36%; Both business and
academic=31%; Neither=12%).**

3. What was the most recent employment of the company founders prior to
    founding this company? *Select all that apply.* **Total >100% since many
    companies had more than one founder.**
    a. **65% (68%)**   Other private company
    b. **36% (40%)**   College or University
    c. **9% (4%)**   Government
    d. **10% (8%)**   Other

4. How many SBIR and/or STTR awards has your firm received from the fed-
    eral government?
    a. Phase I: _____ **Average number of Phase I reported was 14.**
        13% (14%)   **1 Phase I**
        34% (32%)   **2 to 5 Phase I**
        24% (15%)   **6 to 10 Phase I**
        14% (14%)   **11 to 20 Phase I**
        11% (10%)   **21 to 50 Phase I**
        3% (7%)   **51 to 100 Phase I**
        2% (8%)   **>100 Phase I   Five firms reported >300 Phase I**
        (Note: total=4,387; av=31; range 1 to 462.)

What year did you receive your first Phase I Award? _____
3% (7%)	**reported 1983 or sooner.**
33% (49%)	**reported 1984 to 1992.**
40% (45%)	**reported 1993 to 1997.**
24% (36%)	**reported 1998 or later.**

b. Phase II: _____ **Average number of Phase II reported was 7**
27% (23%)	**1 Phase II**
44% (40%)	**2 to 5 Phase II**
15% (12%)	**6 to 10 Phase II**
8% (7%)	**11 to 20 Phase II**
5% (10%)	**21 to 50 Phase II**
1% (7%)	**>50 Phase II Four firms reported >100 Phase II**

(Note: total=1,873; average=14; range=1 to 182.)

What year did you receive your first Phase II Award? _____
3% (2%)	**reported 1983 or sooner.**
22% (28%)	**reported 1984 to 1992.**
35% (30%)	**reported 1993 to 1997.**
41% (56%)	**reported 1998 or later.**

5. What percentage of your company's growth would you attribute to the SBIR program after receiving its first SBIR award?
   a. **31% (27%)**   Less than 25%
   b. **25% (28%)**   25% to 50%
   c. **20% (24%)**   51% to 75%
   d. **24% (22%)**   More than 75%

6. Number of company employees (including all affiliates):
   a. At the time of your company's first Phase II Award: ____
56% (60%)	**5 or less**
28% (28%)	**6 to 20**
9% (7%)	**21 to 50**
8% (6%)	**> 50 Fourteen firms 1.3% had greater than 200 employees at time of first Phase.**

   (Note: total of 1,728, average of 13; range of 1-175; 12% had only 1 employee.)

b. Currently: _____
   29% (31%)   5 or less
   37% (35%)   6 to 20
   17% (15%)   21 to 50
   13% (15%)   51 to 200
   5% (3%)   > 200 Eleven firms report over 500 current
   employees.
   (Note: total =4,951; av of 36; range 1-750; 7% had only 1 employee.)

7. What Percentage of your Total R&D Effort (Man-hours of Scientists and
   Engineers) was devoted to SBIR activities during the most recent fiscal
   year?___%
   22% (24%)   0% of R&D was SBIR during most recent fiscal
   year.
   16% (10%)   1% to 10% of R&D was SBIR during most
   recent fiscal year.
   11% (10%)   11% to 25% of R&D was SBIR during most
   recent fiscal year.
   18% (20%)   26% to 50% of R&D was SBIR during most
   recent fiscal year.
   14% (20%)   51% to 75% of R&D was SBIR during most
   recent fiscal year.
   19% (16%)   >75% of R&D was SBIR during most recent
   fiscal year.

8. What was your company's total revenue for the last fiscal year?
   a. **10% (13%)**   <$100,000
   b. **18% (18%)** $100,000 - $499,999
   c. **16% (16%)** $500,000 - $999,999
   d. **33% (34%)** $1,000,000 - $4,999,999
   e. **14% (18%)** $5,000,000 - $19,999,999
   f. **6% (10%)** $20,000,000 - $99,999,999
   g. **1% (4%)** $100,000,000 +
   h. **0.4% (0%)**   Proprietary information

9. What percentage of your company's revenues during its last fiscal year is federal
   SBIR and/or STTR funding (Phase I and/or Phase II)? _____
   30% (28%)   0% of revenue was SBIR (Phase I or II) during
   most recent fiscal year.
   17% (14%)   1% to 10% of revenue was SBIR (Phase I or II)
   during most recent fiscal year.
   11% (7%)   11% to 25% of revenue was SBIR (Phase I or II)
   during most recent fiscal year.

13% (14%)	**26% to 50% of revenue was SBIR (Phase I or II) during most recent fiscal year.**
13% (21%)	**51% to 75% of revenue was SBIR (Phase I or II) during most recent fiscal year.**
13% (11%)	**76% to 99% of revenue was SBIR (Phase I or II) during most recent fiscal year.**
4% (4%)	**100% of revenue was SBIR (Phase I or II) during most recent fiscal year.**

10. **This question eliminated from the survey as redundant.**

11. Which, if any, of the following has your company experienced as a result of the SBIR Program? *Select all that apply.*

    a. Fifteen (4) firms made an initial public stock offering in calendar year
    Seven reported prior to 2000; two in 2000; four in 2004; and one in both 2006 and 2007?
    (1 in 2004; 1 in 2000; 1 in 1994; 1 in 1983)

    b. **Six (2)** planned an initial public stock offering for 2005/2006.

    c. **14% (18%)** Established one or more spin-off companies.

    How many spin-off companies?
    242 (49) Spin-off companies were formed.

    d. **84% (77%)** reported None of the above.

12. How many patents have resulted, at least in part, from your company's SBIR and/or STTR awards?
    43% (26%) reported no patents resulting from SBIR/STTR.
    16% (16%) reported one patent resulting from SBIR/STTR.
    27% (29%) reported 2 to 5 patents resulting from SBIR/STTR.
    13% (20%) reported 6 to 25 patents resulting from SBIR/STTR.
    1 % (8%) reported >25 patents resulting from SBIR/STTR.

    A total of over 3,350 (842) patents were reported; an average of (6) almost 3 per firm; (range=0 to 66)

The remaining questions address how market analysis and sales of the commercial results of SBIR are accomplished at your company.

13. This company normally first determines the potential commercial market for an SBIR product, process or service
   a. **66% (69%)**   Prior to submitting the Phase I proposal
   b. **21% (22%)**   Prior to submitting the Phase II proposal
   c. **9% (9%)**   During Phase II
   d. **3% (1%)**   After Phase II

14. Market research/analysis at this company is accomplished by: (*Select all that apply.*)
   a. **28% (28%)**   The Director of Marketing or similar corporate position
   b. **7% (5%)**   One or more employees as their primary job
   c. **41% (40%)**   One or more employees as an additional duty
   d. **23% (26%)**   Consultants
   e. **53% (55%)**   The Principal Investigator
   f. **67% (71%)**   The company President or CEO
   g. **1% (0%)**   None of the above

15. Sales of the product(s), process(es) or service(s) that result from commercialising an SBIR award at this company are accomplished by: *Select all that apply.*
   a. **35% (39%)**   An in-house sales force
   b. **52% (50%)**   Corporate officers
   c. **30% (28%)**   Other employees
   d. **30% (36%)**   Independent distributors or other company (ies) with which we have marketing alliances
   e. **26% (34%)**   Other company (ies), which incorporate our product into their own.
   f. **9% (12%)**   Spin-off company (ies)
   g. **26% (39%)**   Licensing to another company
   h. **11% (8%)**   None of the above

Appendix C

NRC Phase I Survey

This section describes a survey of Phase I SBIR awards over the period 1992–2001. The intent of the survey was to obtain information on those which did not proceed to Phase II, although most that did receive a Phase II were also surveyed.

Over that period the five agencies (DoD, DoE, NIH, NASA, and NSF) made 27,978 Phase I awards. Of the total number for the five agencies, 7,940 Phase I awards could be linked to one of the 11,214 Phase II awards made from 1992–2001. To avoid putting an unreasonable burden on the firms which had many awards, we identified all firms which had over ten Phase I awards that apparently had not received a Phase II. For those firms, we did not survey any Phase I awards that also received a Phase II. This amounted to 1,679 Phase Is that were not surveyed.

We chose to survey the principal investigator (PI) rather than the firm both to reduce the number of surveys that any person would have to complete, and because if the Phase I had not gone on to a Phase II, the PI was more likely to have any memory of it than would the firm officials. There were no PI email addresses for 5,030 Phase I, a fact that reduced the number of surveys sent since the survey was conduced by email.

Thus there were 21,269 surveys (27,978 minus 1,679 minus 5,030 = 21,269) emailed to 9,184 PIs. Many PIs had received multiple Phase Is. Of these surveys, 6,770 were bounced (undeliverable) email. This left possible responses of 14,499. Of these, there were 2,746 responses received. The responses received represented 9.8 percent of all Phase I awards for the five agencies, or 12.9 percent of all surveys emailed, and 18.9 percent of all possible responses.

The agency breakdown, including Phase I Survey results, is given in Table App-C-1.

TABLE App-C-1 Agency Breakdown for Phase I Survey

Phase I Surveys by Agency	Phase I awards, 1992–2001	Answered Survey (Number)	Answered Survey (%)
DoD	13,103	1,198	9
DoE	2,005	281	14
NASA	3,363	303	9
NIH	7,049	716	10
NSF	2,458	248	10
Total	27,978	2,746	10

NRC Phase I Survey Results

NOTE: ALL RESULTS APPEAR IN BOLD. TWO SETS OF RESULTS ARE REPORTED—THOSE FOR RESPONDENTS ACROSS 5 AGENCIES (DOD, NIH, NSF, DOE, AND NASA) AND THOSE FOR ONLY NSF. NSF RESULTS APPEAR IN PARENTHESES. EXPLANATORY NOTES ARE IN A TYPEWRITER FONT.

2,746 (248) responded to the survey. Of these 1,380 (113) received the follow on phase II. 1,366 (135) received only a Phase I.

1. Did you receive assistance in preparation for this Phase I proposal?

Phase I only	Received Phase II
95% (97%) No *Skip to Question 3.*	**93% (93%)** No
5% (3%) Yes *Go to Question 2.*	**7% (8%)** Yes

2. If you received assistance in preparation for this Phase I proposal, put an X in the first column for any sources that assisted and in the second column for the most useful source of assistance. Check all that apply. Answered by 74 (4) Phase I only and 91 (9) Phase II who received assistance.

Phase I only		Received Phase II	
Assisted/Most Useful		Assisted/Most Useful	
10/3 (0/0)	State agency provided assistance	**11/10 (2/2)**	
15/9 (0/0)	Mentor company provided assistance	**21/15 (1/0)**	
31/17 (1/1)	University provided assistance	**34/22 (5/4)**	
16/8 (2/2)	Federal agency SBIR program Managers or technical representatives provided assistance	**25/19 (2/2)**	

3. Did you receive a Phase II award as a sequential direct follow-on to this Phase I award? *If yes, please check yes. Your survey would have been automatically submitted with the HTML format. Using this Word format, you are done after answering this question. Please email this as an attachment to jcahill@brtrc.com or fax to Joe Cahill 703 204 9447. Thank you for you participation.* 2,746 (248) responses.

   50% (54%) No. We did not receive a follow-on Phase II after this Phase I.
   50% (46%) Yes. We did receive the follow-on Phase II after this Phase I.

4. Which statement correctly describes why you did not receive the Phase II award after completion of your Phase I effort. *Select best answer.* All questions which follow were answered by those 1,366 (135) who did not receive the follow on Phase II. % based on 1,366 (135) responses.

    33% (32%) The company did not apply for a Phase II. Go to question 5.
    63% (58%) The company applied, but was not selected for a Phase II. Skip to question 6.
    1% (1%) The company was selected for a Phase II, but negotiations with the government failed to result in a grant or contract. Skip to question 6.
    3% (8%) Did not respond to question 4.

5. The company did not apply for a Phase II because: *Select all that apply.* % based on 446 (43) who answered "The company did not apply for a Phase II" in question 4.

    38% (53%) Phase I did not demonstrate sufficient technical promise.
    11% (37%) Phase II was not expected to have sufficient commercial promise.
    6% (5%) The research goals were met by Phase I. No Phase II was required.
    34% (14%) The agency did not invite a Phase II proposal.
    3% (7%) Preparation of a Phase II proposal was considered too difficult to be cost effective.
    1% (0%) The company did not want to undergo the audit process.
    8% (7%) The company shifted priorities.
    5% (5%) The PI was no longer available.
    6% (2%) The government indicated it was not interested in a Phase II.
    13% (5%) Other—explain: *Commercial partner reluctance; death of key personnel*

6. Did this Phase I produce a noncommercial benefit? Check all responses that apply. % based on 1,366 (135).

    59% (41%) The awarding agency obtained useful information.
    83% (79%) The firm improved its knowledge of this technology.
    27% (27%) The firm hired or retained one or more valuable employees.
    17% (18%) The public directly benefited or will benefit from the results of this Phase I. *Briefly explain benefit.*

25 responses explaining benefits.

Published and presented scientific results, led to advances in field (5)
Work referenced by many others (1)
New modeling capability of 2-phase flows commercially available (1)
Led to subsequent Phase I and Phase II awards (1)
Improved software for analyzing unstructured data (1)
Demonstrated upgrade to electric transmission distribution grid (1)
Environmentally safe alternative (1)
Led eventually to product produced by Adelphi spinoff (1)
Assessment materials were produced that were incorporated into the company's textbooks. An assessment framework was developed that guided commercial textbook development (1)
Production of less expensive natural gas (1)
Will be helpful to develop an efficient NOx emission control technology
Project clearly established student ability to use assistive technology to perform classroom activities (1)
Led to another application (1)
Recycling post consumer bumpers (1)
Publicized hot (hard??) to make dense nanostructured nickel oxide battery electrodes (1)
Sensors used to protect the Liberty Bell from damage (1)
Provided training for interns and student workers (1)
We became more proficient at carrying out SBIR research, and went on to other projects were we succeeded in getting to Phase 2. We are now a thriving commercially viable business (1)
Products were commercialized (1)
Technology was used to develop current products (1)
Knowledge of nanoparticle dispersion (1)

13% **(16%)** This Phase I was essential to founding the firm or to keeping the firm in business.
8% **(8%)** No

7. Although no Phase II was awarded, did your company continue to pursue the technology examined in this Phase I? *Select all that apply.* % based on 1,366 (135).

46% **(45%)** The company did not pursue this effort further.
22% **(22%)** The company received at least one subsequent Phase I SBIR award in this technology.
14% **(20%)** Although the company did not receive the direct follow-on Phase II to this Phase I, the company did receive at least one other subsequent Phase II SBIR award in this technology.

12% **(10%)** The company received subsequent federal non-SBIR contracts or grants in this technology.

9% **(12%)** The company commercialized the technology from this Phase I.

2% **(4%)** The company licensed or sold their rights in the technology developed in this Phase I.

16% **(13%)** The company pursued the technology after Phase I, but it did not result in subsequent grants, contracts, licensing or sales.

Part II. Commercialization

8. How did you, or do you, expect to commercialize your SBIR award? *Select all that apply.* % based on 1,366 (135).

   33% (26%) No commercial product, process, or service was/is planned.
   16% (14%) As software
   32% (32%) As hardware (final product component or intermediate hardware product)
   20% (36%) As process technology
   11% (13%) As new or improved service capability
   15% (12%) As a research tool
   4% (2%) As a drug or biologic
   3% (6%) As educational materials

9. Has your company had any actual sales of products, processes, services, or other sales incorporating the technology developed during this Phase I? *Select all that apply.* % based on 1,366 (135).

   5% **(4%)** Although there are no sales to date, the outcome of this Phase I is in use by the intended target population.
   65% (59%) No sales to date, nor are sales expected. Go to question 11.
   15% (10%) No sales to date, but sales are expected. Go to question 11.
   9% **(15%)** Sales of product(s)
   1% **(3%)** Sales of process(es)
   6% **(10%)** Sales of services(s)
   2% **(4%)** Other sales (e.g., rights to technology, sale of spin-off company, etc.)
   2% **(3%)** Licensing fees

10. For your company and/or your licensee(s), when did the first sale occur, and what is the approximate amount of total sales resulting from the technology developed during this phase I? If other SBIR awards contributed to the ultimate commercial outcome, estimate only the share of total sales appropriate to this

Phase I project. (Enter the requested information for your company in the first column and, if applicable and if known, for your licensee(s) in the second column. Enter dollars. If none, enter 0 [zero]; leave blank if unknown.)

	Your Company	Licensee(s)
a. Year when first sale occurred	**89 of 147 after 1999**	**11 of 13 after 1999**
b. Total Sales Dollars of Product(s) Process(es) or Service(s) to date	**(12 of 22 after 1999)**	**(1 of 1 after 1999)**
Sale Averages	$**84,735** ($93,167)	$**3,947** ($741)

Top 5 Sales	1. $**20,000,000** ($2,033,589)
Accounts for **43%** (**62%**)	2. $**15,000,000** ($2,000,000)
of all sales	3. $**5,600,000** ($1,893,000)
	4. $**5,000,000** ($1,200,000)
	5. $**4,200,000** ($700,000)

c. Other Total Sales Dollars
    (e.g., Rights to technology,
       Sale of spin-off company,
       etc.) to date

Sale Averages	$**1,878** ($2,222)	$**0** (0)

Sale averages determined by dividing totals by 1,366 (135) responders

11. If applicable, please give the number of patents, copyrights, trademarks and/or scientific publications for the technology developed as a result of Phase I. (Enter numbers. If none, enter 0 [zero]; leave blank if unknown.)

#Applied For or Submitted / # Received/Published

319 (48) / 251 (35)   Patent(s)
50 (16) / 42 (16)   Copyright(s)
52 (15) / 47 (14)   Trademark(s)
521 (52) / 472 (49)   Scientific Publication(s)

12. In your opinion, in the absence of this Phase I award, would your company have undertaken this Phase I research? *Select only one lettered response. If you select c., and the research, absent the SBIR award, would have been different in scope or duration, check all appropriate boxes.* Unless otherwise stated, % are based on 1,366 (135).

a. **5%** **(8%)** Definitely yes
b. **7%** **(5%)** Probably yes, similar scope and duration
c. **16%** **(21%)** Probably yes, but the research would have been different
in the following way
% based on 218 (29) who responded probably
yes, but research would have . . .
1) **75%** **(72%)** Reduced scope
2) **4%** **(0%)** Increased scope
   21% **(28%)** No response to scope
3) **5%** **(3%)** Faster completion
4) **51%** **(59%)** Slower completion
   44%   No response to completion rate
d. **14%** **(15%)** Uncertain
e. **40%** **(34%)** Probably not
f. **16%** **(11%)** Definitely not
g. **4%** **(0%)** No response to question 12

Part III. Funding and other assistance

Commercialization of the results of an SBIR project normally requires additional
developmental funding. Questions 13 and 14 address additional funding. Additional Developmental Funds include non-SBIR funds from federal or private
sector sources, or from your own company, used for further development and/or
commercialization of the technology developed during this Phase I project.

13. Have you received or invested any additional developmental funding in this
Phase I? % based on 1,366 (135).

   25% **(27%)** Yes. Go to question 14.
   72% **(68%)** No. Skip question 14 and submit the survey.
   3% **(5%)** No response to question 13.

14. To date, what has been the approximate total additional developmental funding for the technology developed during this Phase I? (Enter numbers. If
none, enter 0 [zero]; leave blank if unknown.)

Source	# Reporting that source	Developmental Funding (Average Funding)
a. Non-SBIR federal funds	**79 (17)**	**$72,697 ($12,306)**
b. Private Investment		
(1) U.S. Venture Capital	**13 (12)**	**$4,114 ($1,037)**
(2) Foreign investment	**8 (19)**	**$4,288 ($185)**
(3) Other private equity	**20 (13)**	**$7,605 ($2,963)**
(4) Other domestic private company	**39 (20)**	**$8,522 ($6,667)**
c. Other sources		
(1) State or local governments	**20 (23)**	**$1,672 ($1,110)**
(2) College or Universities	**6 (18)**	$293 **(0)**
d. Your own company		
(Including money you have borrowed)	**149 (30)**	**$21,548 ($16,733)**
e. Personal funds of company owners	**54 (21)**	**$4,955 ($963)**

Average Funding determined by dividing totals by 1,366 (135) responders.

Appendix D

Selected Case Studies

Rosalie Ruegg
TIA Consulting

CASE STUDY COMPANIES AND CONTACT INFORMATION

Case studies were performed for the 10 companies listed below. Each listing provides the company name, location, telephone number, principal interviewee and his or her title, and email contact address.

Faraday Technology, Inc.
Clayton, OH
937-836-7749
Dr. Jennings Taylor, CEO and IP Director
jenningstaylor@faradaytechnology.com

Immersion Corporation
801 Fox Lane
San Jose, CA
408-350-8835
Dr. Chris Ullrich, Director of Applied Research
cullrich@immersion.com

ISCA Technology, Inc.
2060 Chicago Ave #C2
Riverside, CA
951-686-5008
Dr. Agenor Mafra-Neto
President@iscatech.com

Language Weaver
4640 Admiralty Way, Suite 1210
Marina del Rey, CA 90292
310-437-7300
Mr. William Wong, Director of Technology Transfer
jwolfe@languageweaver.com (office manager)

MER Corporation
Tucson, AZ
520-574-1980
Dr. Roger Storm, CEO
rstorm@mercorp.com

MicroStrain, Inc.
Williston, VT
1-802-862-6629
Dr. Steven Arms, President
swarms@microstrain.com

National Recovery Technologies, Inc. (NRT)
Nashville, TN
615-734-6400
Dr. Ed Sommer, President and CEO
ejsommer@nrt-inc.com

NVE Corporation
Eden Prairie, MN
952-996-1603
Mr. Robert Schneider, Director of Marketing
bobsch@nve.com

T/J Technologies, Inc.
Ann Arbor, MI
734-213-1637, ext. 11
Ms. Maria Thompson, President and CEO
mthompson@tjtechnologies.com

WaveBand Corporation
Irvine, CA
949-253-4019, ext. 123
Ms. Toni Quintana, Director of Business Development
tquintana@waveband.com

CASE SELECTION PROCESS

The companies listed in Section D.2 were 10 of 12 companies who were contacted in the effort to obtain a targeted set of 10 cases. Of the 12, one company, Alderon Biosciences, Inc., of Durham, NC, declined the request for an interview without saying why. A second company, Triangle Research and Development Corporation of Research Triangle Park, NC, was contacted by phone and a review requested. The company principal was in the process of moving, making a site visit impractical. Although he was willing to discuss his company's SBIR experience by phone, insufficient information was obtained to develop a full case study. The 10 companies listed all agreed to participate in the study and provided extended in-person interviews (generally from 1.5 to 2 hours in length) and usually also provided lab tours and company reports. All but three of the interviews were conducted at company headquarters. Three were conducted in Reston, VA, at a Navy Opportunity Forum.

The selection of the 12 companies contacted was not random. The companies were selected to provide companies of different age and size, pursuing different technologies, located in different parts of the country, with differing forms of ownership, and with some, although varying degrees of, commercial success. Some of the companies are university spin-offs; some are company spin-offs; some are neither. Some received many SBIR grants; some relatively few. Some continue to obtain a high percentage of their funding from government sources; others have reduced the percentage to low numbers.

The 12 companies who were asked for an interview were drawn sequentially from the following four lists:

1. A list of 12 companies designated "stars" by the NSF SBIR Office was compiled at the request of the interviewer. The list showed companies sorted on the basis of whether they had received no Phase IIB grants, only one Phase IIB grant, or multiple Phase IIB grants. At the request of the interviewer, the companies were also selected to provide variation in state location, company age, years to first SBIR, sales volume (with categories ranging from $1 million-or-less to more than $10 million), and to provide at least one minority or woman-owned company. The "star" designation was said by NSF SBIR administrators to mean that the NSF Program Managers expected the companies eventually to achieve "better than average success." The following six companies were selected from this NSF list of 12: Faraday Technologies, Immersion Corporation, ISCA Technologies, National Recovery Technologies, NVE Corporation, and T/J Technologies.

2. A list of 47 companies that had received NSF SBIR grants and were showing associated commercialization results was compiled at the request of the interviewer by the NRC research team member with responsibility for

existing survey databases. From this list of 47, one company—Language Weaver, Inc.—was added to the existing case study set to provide a company that was very recently founded. Two already selected companies—Immersion Corporation and NVE Corporation—were noted to be also on this second list.

3. A list of four companies located near the interviewer was compiled by the same NRC research team member with responsibility for survey databases, at the request of the interviewer. The intention was to reduce travel costs. Of the four companies, the two showing the most SBIR activity—Alderon Biosciences, Inc. and Triangle Research and Development Corporation—were selected as potential cases, but, as noted previously, neither led to actual case studies.

4. A list of NSF SBIR recipients that would be presenting at a Navy Opportunity Forum in Reston, VA, in May 2005 was provided to the interviewers. Three companies were identified as having received NSF grants and had not already been selected for case study by other NRC research team members who were developing DoD-focused cases. These were MicroStrain, Inc., WaveBand, and MER Corp. The latter company was found also to be on the second list above.

Thus the 10 companies selected do not represent a random sample. Yet drawing them from different lists reduces bias present in any single list. A bias that remains—particularly due to the fact that the third list did not yield cases—is that they all may be regarded as providing examples of revenue-earning SBIR-funded companies. They nevertheless provide considerable diversity.

The 10 selected companies are located in seven states: California, Ohio, Tennessee, Minnesota, Arizona, Vermont, and Michigan. They are developing 10 different technologies, including technologies in the areas of software, electrochemical processes, information technology for pest monitoring and control, electronics, manufacturing processes, and nanomaterials. They range from a more than 20-year-old company founded in 1983, and using SBIR to make a new technology start, to a very "new" company founded in 2002 and already realizing significant revenue. They include very small companies with only about a dozen employees, as well as several companies with 70 or more employees, and one with nearly 150 employees. They include companies that were able to commercialize product very early and those whose technologies will take considerable time. Annual revenue among the companies ranges from $2 million to about $24 million. Among the companies are a university spin-off, a large company's spin-off, a small company's spin-off, a company started by a graduate student, one started by a retiring large-company executive, several started by university researchers, several started by scientists/entrepreneurs, and one started by a professor-husband and entrepreneur-wife team. They include two woman-owned companies, one of which is actively operated by a woman who is also a member of a minority group. They include companies whose share of annual revenue contributed by SBIR and other government grants ranges from a low of 4 percent to a high of 70 percent. Commercialization strategies include licensing agreements, contract research, sale of product produced in-house, commercialization partnerships with larger companies, as well as sale of the technology or of the company to other companies.

While they differ in these many respects, the companies in the case study set are similar in at least three respects: (1) All of the 10 companies have positive annual revenue, reflecting the selection bias that favored relatively successful grant recipients, though not the most successful grant recipients. (2) They share the expressed view that SBIR grants were critical to their ability either to get started at all or to develop capabilities critical to their businesses. (3) Without exception, they sought and received grants not only from NSF but also from other agency SBIR programs, and, in multiple cases, from other government funding programs as well—principally the Advanced Technology Program (ATP) and the Defense Advanced Research Projects Agency (DARPA).

Given the limited number of case studies performed, and the diversity that characterizes them, the cases are illustrative only and cannot be taken as necessarily representative of any of the particular features they exhibit. Yet, a number of common themes run through them, as discussed in the body of the report. Interviewees were asked their views on how the SBIR program might be improved. These views are brought out in Section 4. First, brief synopses of the cases are presented and then the cases are presented in full.

CASE SYNOPSES

Faraday Technology. This case study shows how SBIR grants enabled a scientist-founded company in Ohio to develop an underlying electrochemical technology platform, and, through continuing innovation, to leverage it into multiple lines of business. The technology provides cleaner, faster, more precise, and cost-effective processes to add or remove materials from many different kinds of media, ranging from metal coatings to fabricated parts, to electronic components, to contaminants in soil. The challenges of laying down uniform coatings in tiny holes of many layers of stacked circuits, for example, differ sufficiently from those of producing super smooth surface finishing for titanium jet engine components. The range of these challenges justifies application-specific research to develop the necessary processes. SBIR grants enabled the company to develop the technical capability needed to pursue these many application areas. The company uses an aggressive patenting strategy and licensing to generate business revenue. The relatively modest licensing fees rest on a much larger revenue stream realized by Faraday's customers (their licensees). At the time of the interview, nearly half the company's revenue came from government sources.

Immersion Corporation. This case study illustrates how government funding was used by a university spin-off to leverage private funding to develop technology inspired by NASA technology. The technology adds the sense of touch to diverse computer applications—enhancing entertainment experiences, increasing the productivity of computer use, training doctors, and more. With SBIR assistance, the company has developed a large intellectual property portfolio, which it licenses to other companies, increasing the value of clients' hardware and software. Over its first decade, the company has grown the business to approximately 141 employees and $24 million in annual revenue. Immersion is the largest of the companies included in the case study set. Government R&D support at the time of the survey comprised only about four percent of current revenues.

ISCA Technology, Inc. This case study illustrates how SBIR grants helped a young company that was started with export sales survive a collapse of those sales. Using SBIR, the company, founded by a university researcher, was able to innovate, bringing new technology to the important but then largely static field of pest monitoring and control. The company developed better lures and smarter traps, integrating them with advanced communication tools. ISCA developed new markets in the United States and reestablished export markets. The effect of the company's technologies has been to reduce grower need for insecticides, cutting costs to growers, reducing unwanted effects on insects, lowering pollution, and improving the quality of produce. Another effect is to provide early warning of mosquito outbreaks, providing potential health benefits. Government funding sources at the time of the case study comprised about 40 percent of company revenues.

Language Weaver. This case study shows how an NSF SBIR grant was critical to bootstrapping a technology with national security and economic potential out of a university into use on a fast-track basis. The would-be company, unable to obtain private funding was about to shelve the idea, when the idea arose to seek a grant from NSF. The SBIR grant afforded the technology the credibility required to obtain the management, additional funding, and strategic partners it needed to make a viable business. The technology is statistical machine translation that Language Weaver has applied to translating Arabic, Farsi, Chinese, and other languages. It is being used to create translations of Arabic broadcasts and for other military-related purposes. In addition to licensing its technology to the military, the company also has civilian customers for its software licenses. Without the NSF SBIR grant, a technology that turned out to be extremely timely would not have been developed in the same time frame. From its founding in 2002, Language Weaver has moved from being almost entirely dependent on government grants to cutting the share of government grants to less than half of company revenue by the time of the case study. In 2005, the majority of revenue came from licensing.

MER (Materials and Electrochemical Research) Corporation. This case study illustrates the continuing role played by SBIR grants in the research of a grant-winning company started 20 years ago. The SBIR program was said to be particularly important to the owners as a means for not losing control of the company. It has allowed the company steadily to improve and advance its R&D capabilities in advanced composites, powders, coatings, reinforcements, nanotubes, manufacturing processes to produce near net shape metals and alloys, and energy conversion systems. In parallel with its R&D activities, MER is commercializing its technologies, primarily through military channels. One current focus of the company is on commercializing its rapid manufacturing near net shape processing technology. The process allows a variety of very complex shapes to be produced without tooling, without waste of materials, with desirable joining features, and at a cost advantage to machining techniques. At the time of the case study, roughly 60% of the company's funding comes from government sources, and the remaining from engineering services and product sales.

MicroStrain, Inc. This case study shows a still-small company that leveraged EPSCoR grants to obtain SBIR grants. With SBIR support, it developed an innovative line of microminiature, digital, wireless sensors which it manufacturers. These sensors can autonomously and automatically collect and report data in a variety of applications. They have been used to protect the Liberty Bell during a move and to determine the need for major retrofit of a bridge linking Philadelphia and Camden. Current development projects include power harvesting wireless sensors for use aboard Navy ships, and damage tracking wireless sensors for use on Navy aircraft. Unlike most research companies, MicroStrain, started by a graduate student, has emphasized product sales since its inception in 1985. For a relatively small cost for installing a wireless sensor network, the company has demonstrated that millions of dollars can be saved. At the time of the case study, little more than a quarter of the company's revenue came from government sources.

National Recovery Technologies, Inc. (NRT). This case study shows how a company founded more than 20 years ago used the SBIR to rejuvenate its technology platform in order to enter new growth markets. NRT used SBIR grants early in its history to support R&D underlying its first line of business—mixed municipal solid waste recycling and plastics recycling—lines which did not achieve the original projected growth. In its second decade, NRT used SBIR grants to leverage its existing technological base in a directional change that would offer the potential of increased growth. One of these areas of research was to develop an optoelectronic process for sorting metals at ultra-high speeds into pure metals and alloys. Another area of research was to combine fast-throughput materials detection technology with data compilation, retrieval, analysis, and reporting to provide an airport security system that represents an improvement over the current nonautomated, manual inspection system. At the time of the study, the Transportation Security Administration (TSA) was evaluating NRT's system, a necessary step in qualifying it for use in airport security. While it develops the metals reprocessing and security product lines, NRT has maintained a steady revenue stream of several million dollars annually, primarily from sales of plastics analysis and sorting equipment.

NVE Corporation. This case study shows how a company, which traces its origins to a large company, used SBIR and other federal grants to help launch the company, to keep it from failing, and to improve its ability to attract capital from other sources. Since its founding, the company has pursued development of MRAM technology that uses electron spin to store data and that promises non-volatile, low-power, high-speed, small-size, extended-life, and low-cost computer memory. NVE has developed substantial intellectual property in MRAM. As NVE pursued MRAM development, it saw related potential applications, such as magnetic field sensors. The company has licensing arrangements with a number of other companies. Approximately half of the company's funding comes from government funding, and the remainder from commercial sales, up-front license fees, and royalties. The company is now traded on the NASDAQ Small Cap Market.

T/J Technologies, Inc. This case study features an innovative materials research company, facing a relatively long time to commercialization, which has used a "building block" strategy, leveraging off SBIR and other federal grants to get started and build needed capacity. With this increased capacity, the company was able to go after government research contracts. The next step towards commercialization, underway at the time the case study was conducted, was to form partnerships with global companies for testing and demonstrating its advanced materials for electrochemical energy storage and conversion, and eventually to reach civilian markets. The case shows a company struggling to move up the value chain in order to receive more value for its technology in an environment where funding is scarce and negotiations are difficult. At the helm of the company is a minority woman, during a time that woman-owned businesses received less than 10 percent of SBIR Phase II grants, and minority woman-owned businesses received an even smaller percentage. The company, at the time of the case study, was receiving approximately 15–20 percent of its revenue from the SBIR/STTR program and the remainder primarily from contract research.

WaveBand Corporation. This case study illustrates the role played by SBIR grants in the creation of a company as a spin-off of another small company. It also shows how the company used SBIR and other research funding sources to develop a portfolio of technologies attractive to a larger company that recently acquired it. The case illustrates the dual, unique roles played by highly targeted SBIR grants from defense agencies and by less targeted grants from NSF. The company specializes in antennas that rely on an electron-hole plasma grating to provide rapid beam steering and beam forming without the use of bulky mechanically moved reflectors, which are slow, and electronically steered phase shifters, which are fast but expensive. WaveBand's antennas reportedly offer a price advantage 100 times more favorable to buyers than traditional systems. At the time of the study, approximately half of WaveBand's revenue comes from SBIR and other government grants.

Faraday Technology, Inc.[1]

THE COMPANY

After a stint in a large company research lab where few of the research ideas actually became products, Dr. E. Jennings Taylor was eager to test the waters in a small company environment. He subsequently worked at first one, then another small research company in the Boston area. During this period the entrepreneurial bug bit, and he added an M.S. in technology strategy and policy at Boston University to his Ph.D. in material science from the University of Virginia. Shortly afterwards, he left Boston for Ohio where he launched his own company, Faraday Technology.

Dr. Taylor chose Ohio for two reasons: It was his home state, and, while at Boston University, he had heard about the Ohio Thomas Edison Program, which offered an incubator system for business start-ups. The incubator turned out to be an old school building in Springfield. Basement space was provided at the rate of about $2.00 per ft^2, plus telephone answering and part-time use of a conference facility. It was modest assistance, but it gave the company inexpensive space to get started. Two years later, the company was able to move into a research park near Dayton, and, two years after this, into a custom-built facility, which has since been expanded. The custom facility provides space for the development of pilot-scale prototypes of electrochemical-based processes.

The staff of approximately 10 full-time and 9 part-time employees includes researchers and experienced manufacturing engineers. Dr. Taylor, who is a registered patent agent, serves not only as CTO, but also as IP Director. The company has developed core business competencies in patent analysis. The staff also includes a full-time marketing director who oversees implementation of the company's strategic marketing plan for developing new implementation areas and customers.

The company has collaborative arrangements with a number of universities, including Columbia University, Case Western Reserve University, University of South Carolina, University of Dayton Research Institute, University of Cincinnati, Ohio State University, Wright State University, University of Nebraska, University of California-San Diego, United States Naval Academy, University of Virginia, and others. It often employs students, professors, and postdocs in a research

---

[1]The following informational sources informed the case study: interview at the company with company founder, CTO, and IP Director, Dr. E. Jennings Taylor; telephone discussion with company marketing director, Mr. Phillip Miller; company Web site: *<http://www.faradaytechnology.com>*; company brochures and other company documents; news articles; Dun & Bradstreet Company Profile Report; and earlier interview results compiled by Ritchie Coryell, NSF (retired).

FARADAY TECHNOLOGY, INC.: COMPANY FACTS AT A GLANCE

- **Address:** 315 Huls, Clayton, OH 45315
- **Telephone:** 937-836-7749
- **Year Started:** 1991 (incorporated in 1992)
- **Ownership:** private; majority woman-owned
- **Revenue:** Approx. $2 million annually
  Approx. $6.6 in direct cumulative commercial sales
  Approx. $22.9 in cumulative licensee sales
  — Revenue share from SBIR/STTR grants & contracts: 48 percent
  — Revenue share from sales, licensing, & retained earnings: 52 percent
- **Number of Employees:** 10 full-time, 9 part-time
- **Issued Patent Portfolio:** 23 U.S., 3 foreign
- **Issued Patents per Employee:** 1.4
- **3 Year Issued Patent Growth:** 130 percent
- **SIC:** Primary SIC: 8731, Commercial Physical Research
  87310300, Natural Resource Research
  Secondary SIC: 8732, Commercial Nonphysical Research
  87320108, Research Services, Except Laboratory
- **Technology Focus:** Electrochemical technologies
- **Application Areas:** Electronics, edge and surface finishing, industrial coatings, corrosion countermeasures, environmental systems, and emerging areas, e.g., fuel cell catalysis and MEMS manufacturing.
- **Funding Sources:** State and federal government grants and contracts, government sales, commercial sales, licensing fees, reinvestment of retained earnings, and private investment.
- **Number of SBIR grants:** 47
  — From NSF: 10
  — From other agencies: 37

capacity. The company also has collaborated with national laboratories, including Los Alamos National Laboratory.

Asked what drives the company, Dr. Taylor responded, "What drives us is we are technologists and we want to see our stuff implemented. . . . A company like Faraday is an innovation house for a number of companies that are not well positioned to innovate themselves."

THE TECHNOLOGY AND ITS USE

The company's mission, which has not changed over time, is to develop and commercialize novel electrochemical technology. Called the Faradayic™ Process, the company's platform technology is an electrically mediated manufacturing process that offers advantages of robust control, enhanced performance, cost effectiveness, and reduced hazards to the environment and to workers as compared with using chemical controls. Electrical mediation entails the sophisticated manipulation of non-steady-state electric fields as a process control method for inventing and innovating electrochemical processes, which add to or remove material from targeted devices and other media.

The technology's advantages of cleaner, faster, more precise, and cost-effective results save money for the company's customers and support value-added manufacturing for them. It will allow, for example, electronics manufacturers to make smaller circuit boards with 20 or more layers stacked on a single board, with each layer connected by tiny holes uniformly plated with copper.

Faraday is applying its technology platform in multiple applications. Developing each new application area entails a new set of technical problems and is research intensive. Over the company's first decade, it has developed more than six application areas. About 25 percent of the business is currently in the electronics sectors, and about 28 percent is in edge and surface finishing. Environmental applications account for 15 percent of the business and include effluent recycling and monitoring. Industrial coatings account for about 6 percent of the company's business. Counter measures to corrosion account for another 20 percent. Emerging technologies, including nanocoatings, 3-D MEMs manufacturing, and fuel cell catalysis make up the remaining 6 percent of the business. The company is always looking for the next manufacturing process for which it can solve a problem using its Faradayic™ process, and attract new customers.

THE ROLE OF SBIR IN COMPANY FUNDING

Dr. Taylor became aware of the SBIR program from working in two SBIR-funded companies during the early part of his career. Were it not for this experience, he is doubtful that he would have become aware of the SBIR. With his knowledge of the program, he applied for an SBIR grant early in 1993, soon after the company was incorporated. The first SBIR grant was from DoE to harness electrical mediation for monitoring contaminants in soils and groundwaters. A follow-on Phase II application was not successful. Next the company received an SBIR grant from the Navy for a sensor application, followed by non-SBIR funding of more than $1 million from DARPA for developing process technology to clean up circuit board waste. At that point the company received several EPA SBIR grants to address additional environmental problems. A Phase I SBIR grant from the Air Force followed, and still later the company received SBIR grants from other agencies including NSF. In total, the company has received 47 SBIR grants, 28 of them

TABLE App-D-1 Faraday Technology, Inc.: SBIR/STTR Grants from NSF and Other Agencies

NSF Awards	Number	Amount ($)	Other Agency Awards	Number	Total Amount ($)	Number	Total Amount ($)
SBIR Phase I	5	470,000	SBIR Phase I	23	1,875,000	28	2,345,000
SBIR Phase II[a]	4	1,986,000	SBIR Phase II[a]	12	6,624,315	16	8,610,315
SBIR Phase IIB/ Enhancements	1	350,000	SBIR Phase IIB/ Enhancements	2	749,900	3	1,099,900
STTR Phase I	2	200,000	STTR Phase I	0	0	2	200,000
STTR Phase II	1	532,000	STTR Phase II	0	0	1	532,000
STTR Phase IIB/ Enhancements	0	0	STTR Phase IIB/ Enhancements	0	0	0	
Totals	13	3,538,000		37	9,249,215	50	12,787,215

[a]Exludes Phase IIB/Enhancement awards which are listed separately.

SOURCE: Faraday Technology, Inc.

Phase I, 16 Phase II, and 3 Phase IIB or Phase II enhancements. From NSF, it has received 5 Phase I grants, 4 Phase II grants, and 1 Phase IIB grant. Table App-D-1 summarizes the company's SBIR/STTR grants in number and amount.

SBIR funding has been an essential component of the company's funding, particularly in the early years when nearly all the funding came from SBIR grants. In fact, according to Dr. Taylor, SBIR grants are involved in all areas of application pursued by the company. There are concentrations of SBIR funding in certain areas. Some things started out under SBIR, but later other sources of funding supported further research. Some things started out under other sources of funding, but later entailed an SBIR funding component for further development.

Taking into account all funding sources, the company obtained financial support from state and federal government grants and contracts, government sales, commercial sales, licensing, retained earnings, and private investment. Historically, SBIR/STTR grants and federal research contracts have comprised approximately 48 percent of total revenue. The next largest share at 28 percent has come from commercial sales. Sales to the government have comprised about 15 percent. Licensing has provided approximately 3 percent. Reinvestment of retained earnings and facilities reinvestment has comprised another 9 percent and 2 percent, respectively.

According to company sources, the SBIR program has enabled the company in a variety of ways to do what it otherwise would not have done. Reportedly, it has allowed the company to undertake research that otherwise would not have been undertaken. It has sped the development of proof of concepts and pilot-scale prototypes, opened new market opportunities for new applications, and led to the formation of new business units in the company. It has enabled the company to increase licensing agreements for intellectual property. It has led to key strategic alliances with other firms. It has also enabled the hiring of key professional and technical staff.

The SBIR conveys more than dollars to the grantee, according to Dr. Taylor. "It is well structured to allow taking on higher risk, and it is highly competitive. The larger government programs tend to have specific deliverables instead of looking at the feasibility of high-risk activities. So to me, it [the SBIR program] is very unique. It is understood that the program is highly competitive, therefore, there is prestige associated with gaining an SBIR grant."

BUSINESS STRATEGY, COMMERCIALIZATION, AND BENEFITS

Historically the company's business strategy has been to determine a market need that can potentially be met by an adaptation of its Faradayic™ Process. Then it has pursued an SBIR grant or other sources of research funding to support the necessary research and to develop a pilot-scale prototype of the process. The company actively files patents to protect intellectual property as it develops new technical capabilities. It also investigates who is citing Faraday's patents in different application areas, obtaining a patent file wrapper to see the documentations that occur in the prosecution of each citing patent. This allows Faraday to see how other companies have claimed around Faraday's patents, and gives Faraday background knowledge about potential customers in different areas of interest. Patents and the fees they generate are the central focus of Faraday's business strategy. Thus far, the company has 23 U.S issued patents and three foreign issued patents, which, historically, amounts to 1.4 issued patents per employee.

A major route to commercialization has been to license "fields of use" to interested customers. Company staff members regularly participate in conferences and trade shows to help inform potential customers of Faraday's existing and newly emerging capabilities. The company's strategic marketing plan identifies potential customers for further contact. Once engaged, potential customers issue a purchase order to Faraday to adapt its technology for the customers' needs. In addition to the purchase order, the customer typically pays additional consideration to Faraday contractually to encumber the technology into a "no-shop/stand-still" position, effectively taking the product off the market during the period of adaptation and evaluation. The potential customer has an option to acquire exclusive rights to the technology by paying a negotiated up-front fee and a license fee in the range of 3 percent to 5 percent of the user's generated revenue for the life of the patent. The company also performs contract research for hire, and engages in product design and vending for equipment manufacturers. In the future, Faraday hopes more frequently to form strategic partnerships at the outset of a research program both for the purpose of securing research funding, but also to have a better defined path to market.

Although the company's annual revenue of roughly $2 million is relatively modest, Dr. Taylor makes the point that its license fees signal on the order of 20 to

30 times as much revenue generated by customers who are using Faraday's manufacturing processes. In Dr. Taylor's words, "Based on calculations we have done using our royalty and licensee revenue and associated multipliers, we believe that we have created on the order of $30 million in market value. Of course, Faraday only reaped a small part of this." Over its history, Faraday has generated direct commercial sales of approximately $6.6 million. Customers realize value from Faraday's processes in several main ways: lower cost manufacturing processes, higher quality output, and a combination of the two.

Beyond developing technical capabilities that lead to revenue for Faraday and value-added for its customers, Dr. Taylor pointed to a whole "undercurrent" of effects that the SBIR is having that nobody is really able to capture. "For example, one of our customers likely was going to move off-shore if it could not find a cost-reducing solution to a manufacturing problem it had. Faraday was able to meet the need through innovative research. Of course, I can't prove it, but I think the technology solution figured importantly in their decision not to move. So what's the value of having a company remain in the United States? That's an example of the benefit of innovation funded by SBIR that is not usually factored into the value of the SBIR program. I can't quantify the value; yet I feel strongly that it is true based on what I know."

As an example of another difficult-to-capture type of benefit, the technology also offers the potential of environmental effects in several ways. For one thing, by using electricity to achieve results, the Faradayic™ Process reduces the need for polluting chemical catalysts. For another, the process enables the capture of materials from industrial process waste streams. Yet another emerging application is to control the flow of contaminants through soil for more cost-effective capture and cleanup.

Educational benefits also result from the company's activities. Largely as a result of participating in NSF's annual conference, the company became active in encouraging young people to pursue careers in science. It has provided internships to three junior high students; it has employed several high-school teachers during the summer; and it annually hosts a high-school science day. Moreover, as a result of the company's many collaborative relationships with universities, it has employed about 20 undergraduate and graduate students, one of whom did a Ph.D. dissertation and another, a master's thesis under the Dr. Taylor's supervision.

The innovation process and the multi-faceted roles played by the SBIR are complex and nonlinear, noted Dr. Taylor. He recalled several Phase I grants that did not go on to Phase II. By one standard, these would be considered "failed grants." Yet, he explained, after some twists and turns, the concepts explored in these earlier Phase I grants eventually came to fruition and became important application areas. For example, a "failed" Phase I grant provided the seed for later electronics work that now provides 35 percent of the company's business and accounts for 8 of its patents. "It is not a tidy path; it is a cumulative process."

VIEWS ON THE SBIR PROGRAM AND PROCESSES

Dr. Taylor made a number of observations about the SBIR program and its processes that may serve to improve the program. These are summarized as follows:

Need to Recognize Multiple Paths to Commercialization

Dr. Taylor expressed the hope that it will be recognized that there are multiple paths to commercialization that have merit. He pointed out that it is particularly important for agencies "to understand the various ways to get to the commercial end game—which could involve venture capitalists, could involve strategic partners, could involve an ongoing company trying to augment its business. Agencies need to be flexible. It would be a myopic view if we were to conclude that SBIR funding should only go to companies that are going to do no more than, say, four years of SBIR work and then go public. That is a model, but another model is that innovation is an ongoing thing. . . . The idea of some limit to the number of grants a company can receive cannot be addressed well in absolute terms. Rather, it is important to look at a company's history and see if it is accomplishing something in the longer run—helping to meet R&D needs of an agency or seeding work that eventually turns into something useful."

He went on to raise the issue of a company that receives many DoD and NASA SBIR grants, posing the question: "Would that make it a mill? Well, I would expect that they are providing a research service that DoE and NASA want. . . . If the grants process is modeled correctly, with effective criteria and review, and if it is functioning well, there should be no mills without value added, because nobody would keep funding them unless they do have value." In short, the existence of a mill implies a program breakdown, where SBIR is not taken seriously, where inadequate attention is given to proposal review and project selection.

Mr. Phillip Miller, company marketing director, noted that most SBIR grantees are not OEM suppliers of product; that most grantees develop technology and intellectual property and in turn sell the innovations to customers through a variety of means—not just through products shipped. Yet, the agencies who collect information about the SBIRs impact, typically ask only about products.

Recommendation for Simpler Accounting

Dr. Taylor's opinion is that there are a lot of misconceptions among prospective applicants about accounting requirements, particularly in terms of the indirect rate and what is allowable and what is not. A company's overhead rate may look higher than others because it puts items in it that others put in the direct rate. It is important to look at the overall rate in comparing costs across companies. Furthermore, Dr. Taylor mused, "If the program is geared toward commercialization, and patenting is an important component of this, why wouldn't they allow you to charge patent costs? After all, the government has so called 'march-in' rights,' Furthermore, what is a patent cost? Clearly filing fees and maintenance fees are patent costs. But, are patent attorney's fees associated with evaluating and assessing the technical and patent literature also patent costs? We often use consultants and professors to do the same work and allocate these costs to professional services." He also questioned why the DoD SBIR forms allow you to charge fees and to pay royalties, but the other programs he is familiar with do not have these features. Noting that it is the financial side that is the most daunting to technical people, he urged the SBIR program to give more attention to education on financial issues. He noted that some small companies have very poor accounting systems, and they could benefit from learning how to set up an appropriate system.

SBIR Application Process

According to Dr. Taylor, other agencies' online submission application processes are easier to manage than the system implemented at NSF. His opinion is that NSF's application seems a bit strange from a business standpoint because the form is geared to universities, which comprise the majority of NSF's customers. It is not a dedicated, customized form for SBIR.

Commercialization Issues

Dr. Taylor had heard that a commercialization index is being used to rate SBIR companies, but he did not know how it is computed. He expressed the view that license revenue should be treated differently in computing the index than product sales and other revenue, because a license fee represents on the order of 3 percent to 5 percent of the revenue generated by the licensee. This means that a multiplier of 20 to 30 would need to be applied to license fees to put them on a comparable basis with product sales.

He also emphasized the importance that should be placed on matching funds as a way to indicate commercial potential. "Bringing in cash matching funds is a more powerful signal of commercialization than any review panel's opinion."

Misconceptions About the SBIR and Other Government R&D Partnerships

Dr. Taylor noted that many of the other companies he has worked with have had no awareness of the SBIR program. He recalled a strategic partner who had misunderstandings and misconceptions about the SBIR that interfered with negotiations on a strategic alliance. Another partner, he noted, was afraid of march-in rights, and this was a major encumbrance to making a deal. "More public education might help, or even the elimination of the march-in rights clause."

Turning to the NSF's SBIR program, Dr. Taylor commented on some of its features as follows:

Support of Manufacturing Innovations

According to Dr. Taylor, the NSF SBIR program is unique in its strong support of manufacturing innovations. In his words: "NSF seems more supportive of manufacturing-type innovations. . . . NSF seems to actively appreciate the importance of innovating in manufacturing." He contrasted this interest with a lack of interest in manufacturing on the part of most other SBIR programs.

Specification of Topics

Dr. Taylor saw NSF's SBIR as having more topic flexibility than the other programs. He indicated that this flexibility is helpful for a business like his that wishes to pursue various application areas. Further, he noted that the NSF SBIR is very responsive to national priorities and needs in crafting its topics.

Portfolio or Program Managers

"NSF's special strength is in what they call their portfolio managers. They have people who manage the different technology sectors. NSF is more proactive in helping grantees through the commercial stuff. . . . To me, it is a very, very good, solid interactive group. ...They hold you to task, but I like that. . . . I am impressed with the NSF group because they themselves are an innovative entity—they are looking for ways to continually improve the program; improve their grantee's conference. . . . For example, NSF is trying to expand its matchmaker program—which was geared towards the venture capital community—to include strategic industrial partners . . . and to me it just makes sense. They had about 12 potential strategic industrial partners in Phoenix" [location of the last NSF annual grantee conference].

Commercialization Assistance

A three-day patenting workshop offered at the NSF annual conferences won special praise from Dr. Taylor. At the same time, he found it interesting that many grantees said they couldn't afford the time to attend. To him, this indicated not only an excellent opportunity missed, but a lack of seriousness about patenting, and, therefore, a possible lack of effective commercialization plans.

Lack of Travel Funds

Noting that "NSF hasn't had one person out here," Dr. Taylor expressed his view that it is unfortunate that NSF staff does not have the funding to travel. At the same time, he acknowledged that NSF requires grantees to attend the annual conference, and the conference provides a forum for interacting with NSF program managers and other people "who are at a similar stage of business as you." He added that NSF does not appear to be unique in a lack of travel funding.

Proposal Review Process

Commenting on NSF's review process, Dr. Taylor noted that he has served on panels, and "it is an extensive process." He explained that the technical and business reviewers sat together on the panels in which he participated, and commented that this combining of the reviewers, which may be unique to NSF, is a valuable approach.

SUMMARY

This case study shows how SBIR grants enabled a start-up company in Ohio, Faraday Technology, Inc., to develop an underlying electrochemical technology platform and, through continuing innovation, to leverage it into multiple lines of business. The company's main focus is on developing cleaner, faster, more precise, and more cost-effective processes to add to or remove target materials from many different kinds of media, ranging from metal coatings to fabricated parts, to electronic components, to contaminants in soil. By offering innovative processes that reduce costs for customers and support value-added manufacturing, the company serves as an innovation house for a number of manufacturing companies that are not well positioned to innovate themselves.

The case illustrates that a basic technology platform can be leveraged through additional research into many novel applications to solve specific problems. Because the application areas are so different, they each have required advances in scientific and technical knowledge for success. The challenges of devising a process to uniformly coat tiny holes through 20 or more layers of circuits stacked on a printed wiring board, for example, are quite different from the challenges of developing a process to produce super smooth surface finishing for titanium jet engine components.

The case illustrates a business model that relies primarily on an aggressive patenting strategy and licensing in multiple fields of use to generate business revenue. Essential to leveraging the technology platform is the alignment of intellectual property and marketing strategies. The company continually assesses market drivers to identify needs that may be addressed by Faraday's platform technology. The company works closely with patent firms, has a patent professional on staff and has its engineers and scientists trained in patent drafting. Furthermore, the company has long employed a full-time marketing director.

The case also demonstrates how relatively modest licensing fees rest on a much larger revenue stream realized by the innovating firm's customers. Further, by leveraging advances from one application area into the next, customers in different industry sectors benefit from the company's past advances in other industry sectors. Economists studying the rationale for government support of scientific and technical research have identified the licensing of technology as one of the factors conducive to generating higher than average spillover benefits.

Finally, this case compares and contrasts aspects of different agency SBIR programs. It suggests ways for improving the SIBR program.

Immersion Corporation[2]

THE COMPANY

As a Stanford graduate student in mechanical engineering, Louis Rosenberg, investigated computer-based and physical simulations of remote space environments to provide a bridge across the sensory time gap created when an action is performed remotely and the resulting effect is known only after a time delay. For example, a satellite robot tightens a screw and scientists on the ground find out with a delay if the screw was stripped. As earlier described by Dr. Rosenberg, "I was trying to understand conceptually how people decompose tactile feeling. How do they sense a hard surface? Crispness? Sponginess? . . . Vision and sound alone do not convey all the information a person needs to understand his environment. Feel is an important information channel."[3]

From aerospace researcher, Dr. Rosenberg turned entrepreneur with a focus on the less-studied sensory problem of feel, which was also closely attuned to his specialization in mechanical engineering. He took as his first business challenge to convert a $100,000, dishwasher-sized NASA test flight simulator into a $99 gaming joystick. To take advantage of his breakthroughs, he founded Immersion Corporation in 1993 in San Jose, initially drawing heavily on other Stanford graduates to staff the company. Reflecting the early NASA-inspired challenge, Immersion's first products were computer games with joysticks and steering wheels that move in synch with video displays. Other application areas followed.

The company has now grown to 141 employees. Growth over the first seven years reflected internal gains mainly in the entertainment area. Then, in 2000, Immersion grew mainly by acquiring two companies: Haptic Technologies, located in Montreal, Canada, and Virtual Technologies, Inc.[4], located in Palo Alto, California, both acquisitions now an integral part of Immersion Corporation. And, in 2001, Immersion acquired HT Medical Systems, located in Gaithersburg, Maryland, renamed it Immersion Medical, and made it a subsidiary of Immersion Corporation. In the case of Haptic Technology and HT Medical Systems, the acquisitions brought into the company competitors' technologies in application

---

[2]The following informational sources informed the case study: interview at the company with Mr. Chris Ullrich, Director of Applied Research; company Web site: <*http://www.immersion.com*>; company 2004 Annual Report; company product brochures; Stanford University School of Engineering alumni profile of Dr. Rosenberg in its 1997–1998 Annual Report; Dun & Bradstreet Company Profile Report; and ownership information obtained from Charles Schwab Investment Service.

[3] Quote is taken from the Stanford University School of Engineering alumni profile of Dr. Rosenberg contained in its 1997–1998 Annual Report.

[4]Mr. Ullrich, who was interviewed for this case, joined the company in 2000, in conjunction with the acquisition of Virtual Technologies.

IMMERSION CORPORATION: COMPANY FACTS AT A GLANCE

- **Address:** 801 Fox Lane, San Jose, CA 95131
- **Telephone:** 408-467-1900
- **Year Started:** 1993
- **Ownership:** publicly traded on NASDAQ: IMMR
- **Revenue:** Approx. $23.8 million in 2004
    - Revenue share from SBIR/STTR grants & contracts: approx. 4 percent
    - Revenue share from sales, licensing, & retained earnings: 96 percent
- **Number of Employees:** 141
- **Patent Portfolio:** Over 550 issued or pending patents, U.S. and foreign
- **SIC:** Primary SIC: 3577, Computer Peripheral Equipment
            35779907, Manufacture Input/output Equipment, Computer
      Secondary SIC: 7374, Data Processing and Preparation
            73740000, Data Processing and Preparation, Computer
- **Technology Focus:** Touch-feedback technologies
- **Application Areas:** Computer peripherals, medical training systems, video and arcade games, touch-screens, automotive controls, 3-D modeling, and other
- **Funding Sources:** Licensing fees, product sales, contracts, stock issue, commercial loans, federal government grants, and reinvestment of retained earnings
- **Number of SBIR grants:**
    - From NSF: 10 (4 Phase I, 3 Phase II, and 3 Phase IIB)
    - From other agencies: 33 (20 Phase I and 13 Phase II)

areas new for Immersion. In the case of Virtual Technologies, the acquisition brought in a complementary technology.

THE TECHNOLOGY AND ITS USE

Of our five senses, the sense of touch differs from the others in that "it requires action to trigger perception." Development of a technology to sense touch draws on the disciplines of mechanical and electrical engineering, computer science, modeling of anatomy and physiology, and haptic content design. The technology uses extensive computer power to bring the sense of touch to many kinds of computer-based applications, making them more compelling or more informative processes.

As a company publication puts it, "At last, the world inside your computer can take on the physical characteristics of the world around you. . . . Tactile feedback makes software programs more intuitive."

The technology was brought to life for the interviewer by a series of demonstrations. The first demonstration was of a medical training simulator that teaches and reinforces the skills doctors need to perform a colonoscopy. Low grunts from "the patient" informed the performer that a small correction in technique was needed for patient comfort. "Stop, you are really hurting me!" informed the performer in no uncertain terms that her technique was in need of substantial improvement.

Immersion has developed five main AccuTouch® platforms for helping to teach medical professionals. The five platforms teach skills needed for endoscopy, endovascular, hysteroscopy, laparoscopy, and vascular access—all minimally invasive procedures.

The next demonstration was of a gaming application. The weight of a ball on the end of a string was "felt" to swing in different directions in response to manipulating a joystick. The technology is used also to enhance the computer feedback experience when using a mouse or other peripheral computer controllers for PC gaming systems, arcade games, and theme park attractions, as well as for other PC uses.

A third demonstration was of a "haptic interface control knob" to provide human-machine touch interface on an automobile dash to help manage the growing number of feedbacks from navigational, safety, convenience, and other systems. The purpose is to lessen the risk of overloading the driver.

A fourth demonstration was of Immersion's "Vibe-Tonz" system for mobile phones. The system expands the touch sensations for wireless communications by providing vibrotactile accompaniment to ringtones, silent caller ID, mobile gaming haptics and many other tactile features.

THE ROLE OF SBIR IN COMPANY FUNDING

Though the initial funding of Immersion Corporation was through private equity, the company applied for and received its first SBIR grant in its second year, 1994. In addition, the acquired companies, HT Medical and Virtual Technologies, had received SBIR grants prior to their acquisition by Immersion, and HT Medical had also received a grant from the Advanced Technology Program (ATP) that was nearing completion at the time Immersion acquired the company. All totaled, Immersion and its acquired companies have received 24 Phase I SBIR grants and 19 Phase II (including 3 Phase IIB) grants, summing to approximately $10.6 million. SBIR funding agencies include NIH, DoE, DoD, Navy, Army, and NSF. Table App-D-2 summarizes the company's SBIR and STTR grants in number and amount.

According to Mr. Ullrich, SBIR grants gave the company the ability to fur-

TABLE App-D-2 Immersion Corporation: SBIR/STTR Grants from NSF and Other Agencies

NSF Grants	Number	Amount ($)	Other Agency Grants	Number	Amount ($)	Total Number	Total Amount ($)
SBIR Phase I	4	400,000	SBIR Phase I	20	~1,400,000	24	~1,800,000
SBIR Phase II[a]	3	1,300,000	SBIR Phase II[a]	13	6,600,000	16	7,900,000
SBIR Phase IIB/ Enhancements	3	850,000	SBIR Phase IIB/ Enhancements	0	0	3	850,000
STTR Phase I	0	0	STTR Phase I	3	~200,000	3	~200,000
STTR Phase II	0	0	STTR Phase II	2	1,250,000	2	1,250,000
STTR Phase IIB/ Enhancements	0	0	STTR Phase IIB/ Enhancements	0	0	0	
Totals	10	2,550,000		38	9,450,000	48	12,000,000

[a]Exludes Phase IIB/Enhancement awards which are listed separately.

SOURCE: Immersion Corporation.

ther develop its intellectual property and to help to grow its intellectual property portfolio, which is the very core of the company's commercial success. The company has leveraged its government funding by investment funding from private sources in the amount of $12.7 million. The company attributes approximately $33 million in revenue to products directly derived from Phase II SBIR research projects, including licensing, direct sales of product, and product sales due to licensees. However, due to the company's licensing model, third-party revenues and tertiary economic activity, which are very significant, are not tracked directly by Immersion.

The company now receives only a small fraction of its annual revenue from SBIR/STTR funding, with the percentage ranging variously between 4 percent and 9 percent from 2001 to 2004. Its objectives for rapid commercialization growth are expected to reduce this percentage to an even lower level in the near future.

BUSINESS STRATEGY, COMMERCIALIZATION, AND BENEFITS

From its beginning, Immersion's prime business strategy has been to develop intellectual property in the field of touch sense and to license it. In addition, the company performs limited manufacturing operations in its 47,000 sq. foot facility in San Jose and in Gaithersburg, and arranges for some contract manufacturing. But far and away, the company's wealth generation depends on its ever-growing portfolio of patents which it licenses to others. At the time of this interview, the company had more than 270 patents issued in the United States and another 280 pending in the United States and abroad.

Important to identifying and developing relationships with new licensing partners is the company's participation in trade shows and conferences, and its ongoing interactions with industry associations and teaching universities. The company employs a business development specialist in each of its core business areas to cultivate these contacts.

Because direct sales for Immersion's technologies are derived from the much larger markets into which its licensees typically sell, estimating ultimate market size is considered "complicated" for Immersion, and it takes a more narrow view. For example, Immersion markets its cell phone vibration technology to a limited number of cell phone OEMs, and those OEMs in turn market to millions of customers. Estimating the larger consumer markets is not Immersion's focus.

Potential benefits of the technology include boosting the productivity of software use; enhanced online shopping experiences; enhanced entertainment from computer-based games; improved skills of medical professionals resulting, in turn, in improved outcomes for patients; increased automotive safety due to reduced visual distractions to drivers; and savings to industry through the ability to experience prototypes "first hand," but virtually, before building costly physical prototypes, and the ability to capture 3-D measurements from physical objects. In addition, visually impaired computer users may benefit from the tactile feedback of the mouse, keyboard, or touch-screen.

VIEWS ON THE SBIR PROGRAM AND PROCESSES

Mr. Ullrich made several observations about the SBIR program and its processes that may serve to improve the program. These are summarized as follows:

Difference in Agency Program Intent Helpful to Companies

Mr. Ullrich thought it was clear that there is "a difference in intent" among the various SBIR programs. In particular, DoD is focused on solutions to well-specified problems, while NSF and NIH are more interested in basic technology development that has commercial potential. This distinction is helpful to companies who may wish to develop technologies under both sets of condition. Given the need to respond to fast developing commercial markets, Mr. Ullrich finds the openness and flexibility of a program to accommodate where a company needs to go to find market acceptance to be a big advantage.

SBIR Application Process

According to Mr. Ullrich, there are only minor differences among the agencies in their proposal application processes, and these differences do not pose a major concern in terms of proposal logistics. At the same time, he noted that the last time the company proposed to NIH, there was no electronic submission process, and he expressed the hope that this lack has been remedied.

SBIR Proposal Review Process

Mr. Ullrich has found the review process in support of the various agencies' SBIR grant selection to be "tough but fair." He has found the NSF review to be "much more academic" than the others. Overall, he sees no need for change in the review process.

Turning more exclusively to the NSF's SBIR program, Mr. Ullrich offered the following comments:

Timing Issue—Funding Cycle Too Long for Software Providers

According to Mr. Ullrich, the biggest drawback in NSF's SBIR program is the two deadlines per year, with six months between application and grant and 18 months to Phase II grants. This can be too slow for a software developer.

Timing Issue—Funding Gap

Mr. Ullrich pointed to an associated gap in funding that arises in the NSF program, which he thought would be a real hardship for start-up companies that had not yet developed any sales to sustain them in the interval. He pointed to the Fast Track program at NIH and DoD as being very good ideas. At the some time, he noted that having to develop both Phase I and Phase II proposals at once entails a huge investment of time for an all or nothing outcome. He suggested that providing a supplement—as he recalled some parts of DoD do—to close the funding gap would likely be a preferable approach from the company's perspective.

Phase IIB Matching Funds Requirement

For Immersion, NSF's Phase IIB matching requirement of "cash in the bank" was an easy test to meet—once the company had partners. At the same time, he found the associated review awkward in one respect: the company was required to take its business partner (the investor) to a panel review at NSF. The problem was that the company was required to discuss certain financial issues in front of its investor that it would have preferred to have discussed with NSF in private. Furthermore, it found the need to insist that the investor attend the meeting to be cumbersome and, in its opinion, unnecessary.

Commercialization Assistance

The company participated earlier in the Dawnbreaker Commercialization Assistance Program, and found that "it made sense." However, given the company's current level of business experience, Mr. Ullrich does not think the company would wish to participate again, and is glad participation is optional. Currently, the company is participating in the Foresight Commercialization Assistance Program for the first time and is "seeing if it will help."

SUMMARY

This case study describes how SBIR grants helped a young company develop a large intellectual property portfolio centered on adding the sense of touch to diverse computer applications, and how the company grew the business over its first decade to approximately 141 employees and $24 million in annual revenue. It illustrates how government funding can be used by a university spin-off to leverage funding from private sources to achieve faster growth, eventually essentially eliminating the need for government R&D support. The case also illustrates how a basic idea—adding the sense of touch to computer applications—can be used to enhance entertainment experiences, increase productivity of computer use, train doctors, and more. Immersion's technology was inspired by a NASA system, but its growth centers on its embodiment in consumer products. The case provides a number of suggestions for improving the SBIR program.

ISCA Technology, Inc[5]

THE COMPANY

ISCA was founded in 1996 by Dr. Agenor Mafra-Neto, an entomologist performing basic research at the University of California-Riverside on pheromones, chemical substances produced by, in this case, insects that stimulate behavioral responses in other insects of the same species. From his background in basic research, Dr. Mafra-Neto took on the challenge of applying this knowledge to real-world applications. He contacted growers back in his native Brazil who were very supportive of putting his ideas for pest control into practice. His contacts wired up-front financing to cover a contract for pest control traps, and ISCA was born.

For the next two years, the company's principal business was export of pest control traps to Brazil. Then in January 1999, the company was caught up in a financial crisis that was a result both of the devaluation of the *Real*, the Brazilian currency, and the default by a customer on a large order of ISCA product. During the months that followed, the company was in severe financial distress, and Dr. Mafra-Neto was unsure of his company's ability to survive. It was with the help of an SBIR grant that he was able to restructure and reshape the company to provide more advanced product lines targeted at new domestic and foreign markets.

From its new start, the company has grown to 12 employees and annual revenue of $2 million. The company's offices and facilities are located in an industrial park in Riverside, CA, and occupy a combined area of approximately 8,500 square feet. The staff is a multidisciplinary team of specialized researchers, including synthetic organic chemists, engineers, entomologists, and information technologists.

THE TECHNOLOGIES AND THEIR USES

ISCA synthesizes and analyses sex pheromones for a variety of insects. These pheromones are species specific, occur in nature, are environmentally friendly, and do not result in the development of insecticide resistance by the pest. These properties make them an ideal alternative to insecticides for pest management.

The insect's response to pheromones and other attractants is often quantified through the use of electroantennograms (EAG), which measure the neural

---

[5]The following informational sources informed the case study: interviews conducted at the company's headquarters in Riverside, CA, with Dr. Agenor Mafra-Neto, President, Dr. Reginald Coler, Vice President, and Mr. Annlok Yap, Business and Finance Director; the company Web site: <*http://www.iscatech.com*>; company product brochures; and a recent Dun & Bradstreet Company Profile Report.

ISCA TECHNOLOGY, INC.: COMPANY FACTS AT A GLANCE

- **Address:** 2060 Chicago Ave., Suite C2, Riverside, CA 92507-2347
- **Telephone:** 951-686-5008
- **Year Started:** 1996; restructured in 1999
- **Ownership:** privately held
- **Revenue:** Approx. $2.4 million in 2004
  - — Revenue share from SBIR/STTR and other government grants: approx. 40 percent
  - — Revenue share from sale of product: approx. 60 percent
- **Number of Employees:** 12
- **SIC:** Primary SIC: 0721 Crop Planting, Cultivating, and Protecting
            2879 Insecticides and Agricultural Chemicals, NEC
  Secondary SIC: N/A
- **Technology Focus:** Pest management tools and solutions
- **Application Areas:** Insect semiochemicals (pheromones and kairomones, i.e., naturally occurring compounds that affect behavior of an organism); attractants and monitoring traps; pheromone synthesis and analysis; pheromone delivery and dispensing systems; pest management information systems; automated insect identification and field actuation devices; contract entomological R&D; insect rearing and bio-assays
- **Funding Sources:** Product sales domestic and foreign, contracts, federal government grants, and a small amount of licensing revenue
- **Number of SBIR Grants:**
  - — From NSF: 1 Phase I, 1 Phase II, and 1 Phase IIB
  - — From other agencies: 6 Phase I and 3 Phase II

activity originating from the insect's antenna. In addition to EAGs, biological assays are also used to determine a variety of performance metrics, such as the optimal pheromone release method and the optimal pheromone trap design and placement. ISCA then develops pheromone delivery and dispensing systems and monitoring traps, and integrates the traps with data collection systems, including automated sensors to give pest counts and GPS/GIS analytical tools that are internet accessible to give pest locations.[6]

The results of monitoring provide timely information about the type, num-

---

[6]GPS refers to Global Positioning System, which relies on a system of satellites orbiting the earth to provide precise location and navigation information. GIS is a technology that is used to view and analyze data from a geographic perspective. Looking at the distribution of features on a map can help reveal emerging patterns. The combination of GPS/GIS is an important component of the company's overall information system framework.

ber, and location of pests captured in time to predict pest population densities, identify alarm situations, and deliver limited targeted treatments of pheromones to disrupt mating patterns. The advent of ISCA's pest information management system, equipped with smart traps and wireless communication puts an end to hand counting and the all-too-familiar-to-counters tangled balls of deteriorating insects. The resulting information enables a timely response that avoids insect proliferation throughout a field or larger area and reduces the need for blanket applications of insecticides. Although these technologies individually are not new to the world, their application in the area of integrated pest management has broken new ground.

The company has lines of pheromones, attractants, and repellents to address agricultural pests, including the boll weevil, carob moth, European corn borer, corn earworm, Mediterranean fruit fly, tomato fruit borer, olive fruit fly, peachtree borer, pecan nut casebearer, potato tuber moth, tobacco budworm, and many others. Additionally, ISCA has a product line designed for urban pests, such as the cockroach, housefly, yellow jacket wasp, and mosquito. One of ISCA's most recent lure technologies is the development of "SPLAT™" (Specialized Pheromone & Lure Application Technology), a sprayable matrix that dispenses attractants over an extended time interval substantially greater than that provided by traditional dispensing technologies.

Information technology comprises a critical component of ISCA's approach to pest management. The information technology features modular scalability and GPS/GIS capability. At its core is Moritor, an integrated and automated Internet accessible monitoring system.

In support of its R&D, ISCA operates insect rearing chambers, testing rooms, wind tunnels, and olfactometers. It uses an artificial blood membrane system to maintain its mosquito colonies. The company tests its tools and solutions through rigorous field tests as well as by feedback from user groups.

THE ROLE OF SBIR IN COMPANY FUNDING

According to the company founder and president, Dr. Mafra-Neto, the SBIR program was essential to survival of the company after it hit a major financial setback in its third year of operation. He learned about the SBIR program by reviewing U.S. Department of Agriculture's SBIR proposals during his days in the university. He reasoned that if the company were to get an SBIR grant, it could use the research funding to improve its approach to pest control in terms of the chemicals produced, the lure and trap design and placement, and, eventually, data collection and analysis.

NSF put out a call for sensors. The company responded with a proposal to develop an Internet accessible pest monitoring system with automated traps that would count insects. It was subsequently granted SBIR Phase I and Phase II grants, including a Phase IIB supplement. At NSF, there was interest in bringing

TABLE App-D-3 ISCA Technology, Inc.: SBIR/STTR Grants from NSF and Other Agencies

NSF Grants	Number	Amount ($)	Other Agency Grants	Number	Amount ($)	Total Number	Total Amount ($)
SBIR Phase I	1	100,000	SBIR Phase I	6	560,000	7	660,000
SBIR Phase II[a]	1	499,700	SBIR Phase II[a]	3	1,756,000	4	2,255,700
SBIR Phase IIB/ Enhancements	1	250,000	SBIR Phase IIB/ Enhancements	0	0	1	250,000
STTR Phase I	0	0	STTR Phase I	0	0	0	0
STTR Phase II	0	0	STTR Phase II	0	0	0	0
STTR Phase IIB/ Enhancements	0	0	STTR Phase IIB/ Enhancements	0	0	0	0
Totals	3	849,700		9	2,316,000	12	3,165,700

[a]Exludes Phase IIB/Enhancement awards which are listed separately.

SOURCE: ISCA Technology, Inc.

innovation to a field not known for its use of technology. "The NSF SBIR gave us lots of prestige; it gave us credibility," said Dr. Mafra-Neto.

ISCA has received a total of 7 Phase I SBIR grants, 4 Phase II grants, and 1 Phase IIB supplemental grant. It has received SBIR grants from NSF, USDA, DoD, and NIH. The amount the company has received in SBIR grants since 1999 totals a little more than $3 million. Table App-D-3 summarizes the company's SBIR/STTR grants in number and amount.

According to Dr. Mafra-Neto, the receipt of additional SBIR grants in the future is expected to be important to the company as a means of continuing the innovations necessary to maintain its technical base.

In addition to its SBIR grants, the company received a grant from the Advanced Technology Program, for the period 2002 to 2005. The grant supports the integration of sensor technologies and information technology in a highly automated pest management system.

BUSINESS STRATEGY, COMMERCIALIZATION, AND BENEFITS

The company's competitive advantage lies in its innovations to make smarter traps, which are then linked wirelessly to a centralized database located on the Internet. Automated data collection and subsequent analysis and reporting enables targeted pest control strategies. In addition, the company derives strength from its internally developed "SPLAT" technology that extends the time interval needed for effective seasonal control of pests. Sales of SPLAT products are expected to increase dramatically in the near future.

Between 60–70 percent of the company's sales now are in the domestic U.S.

market. Remaining sales comprise exports to Brazil, Argentina, Chile, India, and other countries.

The company's approach to pest monitoring and control offers environmental benefits in terms of reductions in the need for and use of insecticides. These benefits result from early alerts of pest activity, targeted treatments, and use of strategies that do not involve the use of insecticides to disrupt mating patterns. Humans may benefit from reduced insecticides on products they consume, as well as from higher quality products due to less damage to fruit and vegetables from pest outbreaks.

This pest management approach benefits growers who can avoid multiple blanket spraying of fields with insecticides that may cost as much as 10 times more for pest control than ISCA's method. Avoidance of pest outbreaks may also increase growers' yield and quality of produce.

The recent development of smart traps that automatically count mosquitoes, together with the company's pest management information system, may also offer important health benefits by providing an early alert of threats of possible outbreaks of mosquito-borne disease. The widespread use of the system could enable a near instantaneous warning of threatening trends and activities.

VIEWS ON THE SBIR PROGRAM AND PROCESSES

"The SBIR program has allowed us to get where we are today," said Dr. Mafra-Neto, emphasizing the importance he places on the program. He went on to make the following several observations about the program and its processes, some of which focused on the NSF program.

Reporting Requirements

While noting that he did not particularly like reporting requirements, Dr. Mafra-Neto acknowledged that they forced the company to stay on track. No specific need for change was noted.

Financing Gap

Dr. Mafra-Neto spoke of the difficulties posed by gaps in financing between Phase I and Phase II funding, noting that other companies have died during the gap. "The gap creates uncertainty and breaks the research cycle," he noted. A mechanism is needed to bridge this gap in those agency programs, which have not already found a solution. He pointed to an approach used by the Army's SBIR program as an example of a workable bridge.

Resubmittal of Phase II Proposals and Appeal of Funding Decisions

ISCA interviewees were of the opinion that NIH allows submittal of Phase II proposals up to three times, while NSF allows only a single submittal. Similarly, the interviewees believed that NSF does not allow appeals of its SBIR funding decisions. "Often there is a small issue that we could quickly and easily fix if only we were given the chance," said Dr. Mafra-Neto, noting that "Reviewer critiques can vary substantially."

Value of Keeping Phase I Grants as Prerequisite to Phase II

Dr. Mafra-Neto stated emphatically that Phase II grants are critically important and should be continued largely as they exist today. The Phase I grants allow companies to test ideas; they may reveal multiple solutions; they may give companies early, though typically limited, insight into markets. "Phase I grants represent a good investment of public funding," he said.

Possibly a Premature Emphasis on Venture Capital Funding

Dr. Mafra-Neto expressed a concern that NSF may be pushing companies to attempt to obtain venture capital funding too early in the innovation cycle. In the case of ISCA's approach to show matching funds needed to obtain an NSF Phase IIB SBIR grant, he said the company used ATP funding, and, alternatively, could have used sales revenue. However, had the company been without existing sales and without an ATP grant, he thought that it would have been too early for his company to have attempted to obtain venture capital funding, making it very difficult to meet the Phase IIB requirement. Thus, the comment reflected an impression and a concern for the possible plight of other companies rather than the actual experience of ISCA.

Value of Commercialization Assistance

"It is very useful to train scientists to have business points of view," said Dr. Mafra-Neto, in commenting on his company's participation in both the Dawnbreaker Commercialization Assistance Program and the Foresight Program. At the same time, he commented that he would like to see participation continue to be optional, particularly for companies that have established a degree of business acumen.

Observation About NSF's SBIR Program Manager System

In the opinion of Dr. Mafra-Neto, NSF's program manager system is good. "The program manager becomes involved with the grantee. He or she can put you in touch with other sources to help meet your special needs. For example, we were put in touch with Iguana Robotics."

SUMMARY

This case study illustrates how SBIR grants helped a young company survive following the collapse of export sales several years after start-up. It further shows how SBIR-funded research brought needed innovation to the important but largely static field of pest monitoring and control. The development of better lures and smarter traps integrated with advanced communication tools is effectively cutting the grower's use of insecticides, and thus reducing the unwanted effects on insects (i.e., increasing their resistance to insecticides and impacting nontargeted organisms), lowering pollution, improving the quality of fruits and vegetables, and providing potential health benefits. The case also provides valuable company observations and opinions about the SBIR program and how to improve it.

Language Weaver[7]

THE COMPANY

Like a newly born gazelle, Language Weaver found its legs early. In two years it has developed a fully functional commercial software product from a novel, statistics-based translation technology brought to a research prototype by the company founders, professors and researchers in the University of Southern California's Information Sciences Institute. Of course, it should not be overlooked that approximately 20 person-years of university research, heavily funded by government agencies, were critical to establishing the scientific and technical underpinnings of Language Weaver's technology. Language Weaver gained exclusive licenses to past and future patents filed by the university in the field.

LANGUAGEWEAVER: COMPANY FACTS AT A GLANCE

- **Address:** 4640 Admiralty Way, Suite 1210, Marina del Rey, CA 90292
- **Telephone:** 310-437-7300
- **Year Started:** 2002
- **Ownership:** privately held
- **Revenue:**
    — Revenue share from government grants: approx. 60 percent
    — Revenue share from licensing fees: approx. 40 percent
- **Number of Employees:** 35
- **Technology Focus:** Statistically based automated machine language translation
- **Application Areas:** Language translation of documents, newscasts, and other source materials for defense and commercial purposes. Application languages include Arabic, Farsi, Somali, Hindi, Chinese, French, and Spanish
- **Funding Sources:** Federal government grants, venture capital, and licensing revenue
- **Number of SBIR grants:**
    — From NSF: 3
    — From other agencies: 1

---

[7]The following informational sources informed the case study: an interview conducted at the company's headquarters in Marina del Rey, CA, with Mr. William Wong, Director of Technology Transfer; the company Web site: <*http://www.languageweaver.com*>; company brochures; two articles: "Automated Translation Using Statistical Methods—A Technology that supports communication in Hindi and other Asian languages," *Mutilingual Computing & Technology*, Vol. 16, Issue 2, and "Breaking the Language Barrier," *Red Herring*, 02/28/05; and a recent Company Commercialization Report to the Department of Defense SBIR Program.

With a conception date of November 27, 2001, the company's annual revenue reached several million dollars in 2004, in a market whose potential is estimated in the billions. Having a technology that rather unexpectedly turned out to be much needed at just the right time is paying off.

The company founders are still professors at the university. The company now has about 35 employees, many of them attracted from the university's Information Sciences Institute. The company is headquartered in an office building with a grand view overlooking a marina just west of downtown Los Angeles.

The company's first funding came from NSF grants. "When we were trying to start the company," related Mr. Wong, "it was a little before 9/11, and no one cared about languages. There were no Senate hearings about languages. Then 9/11 happened, and at the time we had already submitted a proposal to NSF. But we didn't hear back until November, and by then the NSF was able to bootstrap us to get us working quickly, moving code from the university to the company. . . . It was after the SBIR grant that everything happened. We started getting government interest as it became apparent that we had something interesting. But we would not have been positioned to move quickly to respond to the need if it hadn't been for that first small amount of NSF funding and the confirmation of the technology." Subsequently the company was able to obtain venture capital funding.

"What we are trying to do," explained Mr. Wong, "is to create the best machine translation in the world, with the highest quality and readability. Our best selling system right now is Arabic to English —to the government. We have customers and we have partners."

THE TECHNOLOGY AND ITS USE

The state of the practice in commercial machine translation is rule-based, e.g., noun before verb. But 30 years of working with rules has reportedly shown that the approach does not do well in handling special cases. In contrast, Language Weaver's statistical learning approach to machine translation is designed to learn the appropriate linguistic context for distinguishing and handling words with multiple meanings. This is the kind of problem that confounds rule-based systems because it is impossible to capture all the necessary cases in rules. For example, the English word, "bank" may need to be translated differently in each of the following: "put the money in the bank;" "you can bank on it;" and "paddle the canoe near the bank."

The machine-based software uses computational algorithms and probability statistics to learn from existing translated parallel texts, analyze words and word groupings, and build translation parameters that will provide the highest statistical probability of providing a correct translation. The development of translation software for a given language entails performance of two analyses: First, a bilingual text analysis is performed using a corpus of text and statistical analysis to learn associations. For example, a bilingual corpus for Spanish/English may be found at Microsoft's Web site which gives the same material in both Spanish and in English. From such existing one-to-one translations, the system learns. A translation model is built from the resulting analysis. Second, a great deal of monolingual text is fed into it to increase translation fluency. The process is akin to a computer chess game, whereby the computer is playing chess with itself trying to find the best move, or, in this case, the best translation. The approach requires a lot of computing, but fortunately substantial computing power is available at a reasonable cost.

"Basically, we consume this text corpus. We learn from it and develop the parameters which will be used by the runtime module we call the 'decoder.' What we license to the customer is the parameters and the decoder," related Mr. Wong, describing Language Weaver's approach.

What about languages for which there is not much of an existing corpus of digital translations available? Language Weaver's first two contracts involved Somali and Hindi—both of them "electronically low density languages." In this case, the company had to employ human translators to generate a body of digital text that they could put into the learning system to create the parameters. "We don't get as high a quality taking this approach," said Mr. Wong, "but it's still readable."

"We have also developed a Chinese translator—for which a lot of existing data exists—but which is difficult because it uses characters instead of letters. And in the case of Arabic, there is a similar difficulty because we are dealing with script. Chinese and Arabic each presented special challenges that we met," noted Mr. Wong.

"In summary," explained Mr. Wong, "instead of rule based, our approach is code breaking. We just need a few weeks to create a new language set. We are getting to the point that we can deal with any language, translating it into another language by pressing buttons."

Mr. Wong brought the technology to life for the interviewer with some graphic depictions of how it can be used for defense applications: "Everyday many hours of potentially important information to U.S. efforts against terrorism are broadcast in Arabic. Our technology is used to provide near simultaneous translation. Similarly, hours of taped interviews with people from different cultures, speaking in a variety of languages, may contain insights and information important for the military effort. Our technology can perform the translations and allow specific questions to be searched."

Language Weaver's translation system, for example, has been incorporated into media monitoring systems by BBN, Virage, and Z-Micro Systems. These systems can capture an Arabic satellite news broadcast. The audio track containing speech is extracted in 5-second chunks. A speech recognition system does the transcription of the chunks into Arabic text. From there, the media monitoring system sends the Arabic text to the Language Weaver translation system, which translates it into English. "An exciting part of the media monitoring systems is the way they allow an analyst or viewer to track each translated text segment in synch with the broadcast video and audio. As the speaker speaks, the English text chunk is highlighted that corresponds to what the announcer is saying."

Continuing, Mr. Wong said, "Or imagine that you are searching a house where terrorists have been and you uncover a trove of dirty or degraded documents written in Arabic, or Farsi, or some other language. These documents may contain valuable information. They can be cleaned, scanned, and transcribed using optical character recognition software, and our system can be used quickly to generate translations. These translations can be stored and searches can be performed on key words as needed."

THE ROLE OF SBIR IN COMPANY FUNDING

The company founders, Dr. Kevin Knight and Dr. Daniel Marcu, were professors at the University of Southern California's Information Sciences Institute, when they saw business potential in a new approach to machine language translation they had brought to a research prototype stage.[8] This awareness came right at the time the Internet bubble was about to burst, but at a time everyone was still excited about technology, allowing the professors to get their foot in the door with venture capitalists. But they couldn't get funding—perhaps because the venture capitalists were just then realizing that conditions were about to take an unfavorable turn.

Next the professors went to see the Tech Coast Angels, a group of investors who provide seed and early stage capital in Southern California. They were able to present to the group, but they didn't get money. However, they did get a mentor who worked with the would-be company for about nine months.

Still, during this time, the professors were getting no traction from the commercial sector in terms of investment funding. It was at that point that Daniel Marcu, following something of a whim, decided to see if he could get STTR funding and start the company on that. At the university, he had worked on DARPA and NSF-funded research projects, and he knew from this experience

---

[8]With backgrounds in computer science, mathematics, and linguistics, Drs. Marcu and Knight, were, and are, two of the top researchers in the field of statistical machine translation. The interviewee, William Wong, was a student of Dr. Marcu at the University of Southern California (USC), and joined the company at its inception. Prior to coming to USC, Mr. Wong had worked at Intel. At USC his intention, he said, was to go into computer animation. He took a class from Daniel Marcu on natural language processing, "fell in love with the idea," and changed his plans.

TABLE App-D-4 Language Weaver: SBIR/STTR Grants from NSF and Other Agencies

NSF Grants	Number	Amount ($)	Other Agency Grants	Number	Amount ($)	Total Number	Total Amount ($)
SBIR Phase I	1	100,000	SBIR Phase I	1	50,000	2	150,000
SBIR Phase II[a]	1	500,000	SBIR Phase II[a]				
SBIR Phase IIB/ Enhancements	1	500,000	SBIR Phase IIB/ Enhancements				
STTR Phase I	[b]		STTR Phase I				
STTR Phase II			STTR Phase II				
STTR Phase IIB/ Enhancements			STTR Phase IIB/ Enhancements				
Totals							

[a]Exludes Phase IIB/Enhancement awards which are listed separately.
[b]An STTR Phase I grant was converted to an SBIR Phase I grant.

SOURCE: Language Weaver.

about the NSF STTR grant. He submitted a proposal to NSF and got the STTR grant.

In the words of Mr. Wong, "Getting the STTR grant was a real boon to us, because we were on the verge of saying, 'This technology is not ready. No one is interested in talking to us. Let's just shelve it for a while.' But then came the grant from the National Science Foundation and with it the confirmation and redemption of the technology—an indication that it was useful, or interesting at least."

"What did that do for us?" Mr. Wong continued. "It wasn't enough to support more than one person, but it forced us to actually incorporate—that's one of the rules. The other thing was that it forced us to find a CEO—someone who could drive the formation of a whole company as you see today. And it forced us to spend some time working out the details, including those of us who were volunteering our time. It reinforced that we had something interesting. So basically November 27, 2002 was our conception date, because that is when we got our first STTR. Then we were given a chance to convert the STTR into an SBIR, and we did because the SBIR offered more advantages.[9] So the STTR/SBIR from NSF created Language Weaver and what we are today. Without that we would have shelved the technology."

Language Weaver has received a total of $150,000 in Phase I SBIR grants and $1,500,000 in Phase II grants. It has received SBIR grants from NSF and

[9]At the time, research funds were extremely limited. The company found it advantageous to provide the university an ownership share of the company rather than give up the limited STTR research dollars to it.

the U.S. Army. The amount the company has received in SBIR grants since its founding in 2002 totals $1,150,000. Table App-D-4 summarizes the company's SBIR/STTR grants in number and amount. In addition to its SBIR grants, the company received a multiyear grant from the Advanced Technology Program, for a large scale syntax-based system, expected to bear fruit several years out.

BUSINESS STRATEGY, COMMERCIALIZATION, AND BENEFITS

The company is "a core technology house based on licensing its software." It licenses its product, statistical machine translation software (SMTS) directly to customers, and through partners, such as solution vendors who add multilingual capability to their applications with Language Weaver. Underpinning its ability to license are more than 50 patents pending worldwide on SMTS. These partners are important marketing vehicles. "We have a symbiotic relationship with our partners," Mr. Wong explained. "We promote our products and our partners' product lines that contain our technology; our partners do the same thing."

In its first two years, the company focused on heavy government funding. Grants and development contracts comprising 80–90 percent of company revenue were the company's "life's blood." In 2004, the company received more revenue from licensing, allowing it to get the government share down to 60–70 percent of revenues. According to Mr. Wong, the goal in 2005 is to cut the share of government grants to less than half of company revenue, with the majority coming from licensing.

The company is "not hanging onto research," explained Mr. Wong. "Yes, we will continue to have researchers internally, but the growth vehicle is the market now. We are sticking with the company's core technology and then building supporting technologies around it to make the core more useful." An additional avenue for research support, reminded Mr. Wong, is the fact that the company will continue to benefit from research advances made at the University of Southern California through Language Weaver's rights to future discoveries.

Mr. Wong elaborated on the ongoing relationship between Language Weaver and the University. "Whatever they are working on that improves the quality of automated language translation, Language Weaver will gain the rights to. This arrangement used to be unusual, especially unusual for research university institutions because they are not focused on commercial prospects. We were probably the poster child—actually the second one—to break this path where we not only receive current rights but also future rights (for five years) from the university. Because the university shares in the company's equity, and we are a healthy company, the university benefits from our success. In addition, Language Weaver hires a steady stream of graduates coming out of the university program."

Mr. Wong also provided perspective about the company's view of and use of venture capital and external business management, explaining, "Early on we made the decision that we wouldn't be control freaks, and that we couldn't handle all of the management aspects. We found our CEO—Bryce Benjamin—through the Tech Coast Angels. From then on we just wanted reasonable ownership of the company for the amount of money they were giving us. We decided not to worry about the fact that we are giving up shares. Rather we would worry about how much value is being added. We did not have to give up majority ownership in order to attract funding. We stayed away from people who didn't see the future of the technology, and just seemed out to get majority ownership. To develop a beneficial relationship with venture capitalists, you need to be in that phase when you are moving towards having customers, but need to grow more; in this phase venture capital can help and investors can see potential."

Language Weaver's technology offers societal benefits in several ways: First, it reportedly provides a significantly higher rate of accuracy in translation than counterpart rule-based machine translations, providing customers more value. Second, it is able to provide translation systems for languages for which there is a shortage of available translators and for which there is considerable demand for translations, particularly for defense purposes. Third, it can be a more cost-effective solution for translation of large volumes of information than human translators. Fourth, the technology may offer a faster means to obtain needed translations by its ability to process large volumes of data quickly. For example, it reportedly can process in one minute what a human translator would take several days to produce.

VIEWS ABOUT THE SBIR PROGRAM AND ITS PROCESSES

Because NSF SBIR grants make up most of Language Weaver's funding received through the SBIR program, Mr. Wong's views on the SBIR program, which follow, mainly reflect NSF's program.

Commercialization Assistance Program

Language Weaver chose not to participate because when the opportunity arose, the company had just brought in business management and a salesperson with an extensive directory of contacts. The company found the optional nature of participation to its liking.

NSF's Emphasis on Commercialization

"The NSF provides a lot of encouragement to companies to look more at the commercialization side," stated Mr. Wong. "You've got to find your marketing or sales guy; you've got to find your customers." They push you to ask "What do your customers really want."

Assessing this focus, Mr. Wong stated, "I think it was about the right emphasis in our case, being in the software industry. I can see how it might be too fast for some technology areas such as materials, manufacturing, and chemicals, or pharmaceuticals. The time required to commercialize a technology is definitely industry dependent. In the software area, minimizing the time to market is important. We have to worry that someone in some other place will copy our technology—even though at this point there are only a handful of people who understand statistical machine translation, and they are a very tight group, making it less easy to copy. In any case, speed is of the essence."

NSF's Phase IIB Matching Funds Requirement

Mr. Wong noted, "NSF's Phase IIB was very good; it worked well for Language Weaver." However, he expressed concern that other start-up companies not able to move as fast into commercialization would find it very difficult to meet the matching funds condition of Phase IIB. He explained that getting a Phase I, Phase II, and Phase IIB is not enough to enable a start-up company to bring in a marketing person and is not enough to allow a CEO like Mr. Benjamin to build a business. A start-up company must have already found additional funding in order to position itself to do what is being asked to do at the Phase IIB stage. Thus, the implied sequence of the SBIR phases feeding directly into one another as a tool to launch a business is most cases would be quite problematic. A company will need to go out and find more funding sources and partners very early on. In the words of Mr. Wong, "We were lucky that we were able to do that."

In response to a question about whether the company was able to use its government defense-related contracts as match for its Phase IIB grant, Mr. Wong responded as follows: "We needed commercial contracts. We could not use our DoD contracts as match. We used venture capital funding as our match. We had to show that we had the money in the bank. It meant actually showing bank receipts.

NSF's Flexibility and Empowerment of Program Managers

"NSF's flexibility was extremely helpful to us," noted Mr. Wong, explaining that because the company was just getting started, it needed to make some changes in its research plan in Phase II after getting underway. He observed that NSF empowers its program managers and provides them enough leeway to make decisions that allow changes from the original research plan if it seems warranted—provided the company reports on it and explains why it was beneficial and the results.

Another point Mr. Wong made about the advantages of NSF's flexibility was that the NSF allowed Language Weaver to go at an accelerated pace and finish early—without having to turn back money. The ability to accelerate research was reportedly very important to meeting the rapidly developing market demand for a better machine translation system.

Financing Gap

"A lot of us were willing to work for free during that early time and that helped relieve the financial stress. But I can see how surviving during the start-up phase would be a hardship for some companies," said Mr. Wong.

Opinion about Phase I Grants

In the words of Mr. Wong, "I think requiring Phase I is a good idea. Always a good idea! You don't know if an idea is even interesting to someone and if you can get it together in the beginning. So I totally see why we need to have the Phase I—especially when the follow-on phases are so much bigger. Of course it would be nice to get more money up front, but the Phase I is a way for the government to take a manageable risk."

NSF Proposal Review Process

In Mr. Wong's opinion, the process seemed fair and the quality of reviews seemed good. "It seemed more academic in nature," he said, "but that was good, because it reinforced that we had something new and interesting. And as a new company we needed that third party affirmation that our ideas were worthwhile. Because we received the NSF SBIR, and the affirmation it gave us, we were able to do follow-on work."

He noted that there seemed more comments on the business side at the Phase II review. And, the grading seemed pass/no pass. "Is the project sound; is there an inkling of business sense?"

"At the Phase IIB review stage, the emphasis in the review was definitely financial; nuts and bolts. The CEO, Mr. Benjamin and others gave a presentation before a panel for Phase IIB. The presentation was very business oriented, addressing why we have potential as a company."

Idea for "Phase IIC" Grant

"I think commercialization is very hard for people," said Mr. Wong. "If we hadn't been able to recruit our CEO from Tech Coast Angels, we would have had the same problems as everyone else. Making available a Phase IIC grant would be very helpful to extend the time to get to market for longer lead time technologies. If everything is looking good through the Phase IIB period, a little more money might make a big difference."

SUMMARY

This case study illustrates how an NSF SBIR grant was critical to bootstrapping a technology with national security and economic potential out of a university into use on a fast-track basis. The credibility afforded the technology by the SBIR grant enabled the company to get the management, additional funding, and strategic partners it needed to make a business. Without the NSF SBIR grant, a technology that turned out to be extremely timely would not have been developed in the same time frame. The technology is statistical machine translation that Language Weaver has applied to translating Arabic, Farsi, Chinese, and other languages. At this time, it is being used mainly for military-related purposes such as to create translations of Arabic broadcasts. From its founding in 2002, Language Weaver has moved from being almost entirely dependent on government grants to receiving the majority of its revenue from licensing fees. The case illustrates the speed capabilities of a software company as well as the speed imperatives that often characterize this field. The interview provided many observations that may help improve the SBIR program.

MER Corporation[10]

THE COMPANY

MER (Materials and Electrochemical Research) Corporation was created 20 years ago by two former employees of Arco Chemical Company in an R&D spin-off from Arco. The founders continue to own and operate the firm. Incorporated in Arizona and located in Tucson, MER now has about 75 employees. It operates as an engineering services company.

The firm seeks a leadership position in developing advanced and exotic materials. It has received a number of prestigious grants in recognition of its technical capabilities, including multiple R&D 100 grants—for its work in SiC fibers, VLS SiC whiskers, metal powders, and titanium extraction. It also has received grants from the Arizona Innovation Network in recognition of its prowess as an innovator. Moreover, it has received from the Small Business Administration (SBA) several "Tibbetts Grants," named for Roland Tibbetts, a former National Science Foundation (NSF) employee who is acknowledged as "the father of the SBIR program." The stated focus of the SBA in selecting firms to receive the Tibbetts Grant is on the economic impact of the technological innovation, business achievement and effective collaborations, and demonstrated state and regional impact.

THE TECHNOLOGY AND ITS USE

According to its website, "Researching and developing new materials is the backbone of MER." Its specialty research areas include ceramic composites, diamond and other specialty coatings, solid free-form fabrication, and custom development and evaluation of batteries. Using SBIR grants, MER has developed technologies in advanced composites, powders, coatings, reinforcements, fullerenes and nanotubes, near net shape metals and alloys, and energy conversion systems.

One current focus of the company is on commercialization through military channels of prototypes of its rapid manufacturing near net shape processing technology. This was the technology featured by MER at the Navy Opportunity Forum, the occasion of the interview in support of this case study. By depositing successive layers of molten metal based on CAD-file designs, a variety of metal shapes—including very complex shapes—can be produced without tooling, without waste of materials, with desirable joining features, and at a cost advantage to machining techniques.

---

[10]This case study was informed by the sources: interview with Dr. Roger Storm, MER Program Manager, conducted at the Navy Opportunity Forum, May 2-4, 2005, Reston, VA; information from the company Web site: <*http://www.mercorp.com*>; and a recent Dun & Bradstreet Company Profile Report.

MER CORPORATION: COMPANY FACTS AT A GLANCE

- **Address:** 7960 South Kolb Road, Tucson, AZ 85706
- **Telephone:** 520-574-1980
- **Year Started:** 1985 (Date Incorporated: 1985)
- **Ownership:** 100 percent of capital stock owned by the officers
- **Revenue:** $7.9 million in a recent year
  — Roughly 60 percent from government sources
  — Roughly 40 percent from engineering services and product sales
- **Number of Employees:** 75
- **SIC:** Primary SIC 8711 Engineering Services
- **Technology Focus:** Electrochemical systems, porous materials, coatings, fullerenes and nanotubes, composites, rapid manufacturing of near net shape metals/alloys, and energy conversion systems
- **Application Areas:** Fuel cell based power generators, biomedical applications, products fabricated from advanced ceramic composites, applications of fullerenes and nanotubes, and rapid manufacturing prototyping for military and civilian uses
- **Facilities:** Leases 45,000 sq. ft. of state-of-the-art laboratories
- **Funding Sources:** Engineering consulting fees, commercial sales, government sales, licensing fees, Federal government grants & contract R&D, and private investment of owners
- **Patent Position:** More than a dozen patents granted
- **Number of SBIR grants:** 31
  — From NSF: 4
  — From other agencies: 27

THE ROLE OF SBIR AND OTHER GOVERNMENT R&D FUNDING

Beginning early in its history and continuing to the present, the MER Corporation has relied heavily on SBIR grants and government engineering contracts for funding. The firm has also received funding from DARPA and, as part of a joint venture, from the ATP. Today, roughly 60 percent of the company's funding comes from government sources—mainly from SBIR grants. The remaining approximately 40 percent comes from engineering services and product sales.

The SBIR program has been important to the owners as a means for not losing control of the company. It has allowed the company steadily to improve and advance its R&D capabilities.

BUSINESS STRATEGY, COMMERCIALIZATION, AND BENEFITS

In parallel with its R&D activities, MER seeks new opportunities to commercialize these technologies through strategic alliances, joint ventures, licensing, and by production and sale of product. The firm has developed extensive cooperative arrangements with major universities and international corporations as part of its R&D efforts. Among its affiliate companies are Fullerene International Corporation, LiTech LLC, Tailored Materials Corporation Inc. (TMC), and Frontier Materials Corporation (FMC). Among its research partners are Mitsubishi Corporation and RC Technologies.

Focusing on the firm's rapid manufacturing prototyping technology, Dr. Storm, MER's Program Manager interviewed for this case, described the company's competitive advantage as consisting of the following elements: a) a minimum of waste, b) no need for tooling, c) very specific capability in joining and producing composite structures, and d) a cost advantage primarily in terms of low capital and operating costs. Currently at the prototype stage, the firm faces the challenges of scaling-up and moving into commercial applications. In the process, it looks for value-added opportunities in military and commercial markets.

VIEWS ON THE SBIR PROGRAM AND PROCESSES

Dr. Storm characterized the SBIR program overall as "a good program." He called out several agency program features as particularly noteworthy. He spoke positively about the Navy's approach to fostering SBIR Phase III activities. He noted the benefit that certain parts of DoD provide by eliminating the problem of a funding gap between Phase I and Phase II through an option to keep projects going if satisfactory progress is being made.

Phase I Grants as Prerequisite to Phase II

Dr. Storm commented that the current progression from a Phase I feasibility demonstration to a Phase II grant as quite reasonable, and sees no reason to change this arrangement.

Turning to the firm's experience with NSF's SBIR program, Dr. Storm noted that an on-going NSF project is "running good…with sensible management from NSF." He then identified the problems with NSF's SBIR program:

Impression that NSF's Proposal Review Process Relies Excessively on University Reviewers who Lack Business Savvy

Based on the firm's experience in applying to an NSF solicitation for which prominent goals were international competitiveness and cost effectiveness, the firm concluded that a reviewer, quite possibly from a university, failed to grasp the economic significance of forecasted reductions in material cost to be achieved through the proposed R&D project. Two of the reviewers agreed with the premise that the proposed program would dramatically decrease the cost of Ti components and provide a major advantage to U.S. industry that is currently limited by the cost of the Ti components. The third reviewer demonstrated a complete lack of understanding of the central issues. In another cited example, a proposal for low-cost manufacturing of Ti powder was rejected as not qualifying as a manufacturing proposal. Based on its years of experience with manufacturing, the firm felt that it was capable of identifying a manufacturing proposal, and concluded that the reviewer did not understand the manufacturing world. This was taken to imply an academic background. Referencing his many years of experience in writing proposals, Dr. Storm acknowledged that not every proposal should be expected to be funded, but, in the cases cited, he felt strongly that the review was faulty.

No Appeal of Proposal Review Decision

In light of the above cited experiences, Dr. Storm felt that it would be helpful if there were an appeal procedure when the reviewers so clearly split in their opinions.

Rumored NSF "Black List" for Firms Who Have Received "Too Many" Grants

According to Dr. Storm, there is a rumor in the SBIR community that a firm will be "black listed" by NSF if it has received "too many" grants. A problem is that no one knows what number triggers this action, or even if it is true. If it is true, it would be very helpful to companies if this policy could be made explicit and public, so that companies would better know how to interface with the NSF SBIR program. The company would accept an NSF decision to limit its grants per company, but it needs to know what the policy is. The current uncertainty leads to frustration among companies and possibly a waste of time and resources that could be avoided by greater program transparency. (Dr. Storm noted that other agencies do not appear to have a limit on grants, and that the rumor applies only to the NSF program.)

Restriction on the Number of Proposals that can be Submitted in a Year

NSF's restriction on the number of proposals a given company can submit during a year to four proposals favors the smallest companies that have less competition among staff members and technology areas. The implication is that a business like MER's with more employees and areas of technology than many other very small businesses is limited in the number of significant ideas for proposals it can identify. He commented that if the goal of NSF is improving our nation's competitiveness, this restriction is difficult to understand.

SUMMARY

This case study illustrates the continuing role played by SBIR grants in the research of a grant winning company started 20 years ago. The SBIR program was said to be particularly important to the owners as a means for not losing control of the company. It has allowed the company steadily to improve and advance its R&D capabilities in advanced composites, powders, coatings, reinforcements, nanotubes, manufacturing processes to produce near net shape metals and alloys, and energy conversion systems. In parallel with its R&D activities, MER is commercializing its technologies, primarily through military channels. One current focus of the company is on commercializing its rapid manufacturing near net shape processing technology. The process technology allows a variety of very complex shapes to be produced without tooling, without waste of materials, with desirable joining features, and at a cost advantage to machining techniques. Today, roughly 60 percent of the company's funding comes from government sources, and the remaining from engineering services and product sales. The case draws on the experience of the company to identify potential problems with the SBIR program

MicroStrain, Inc.[11]

THE COMPANY

While pursuing a graduate degree in mechanical engineering at the University of Vermont, Steve Arms witnessed an incident that led him to his future business. During a horse vaulting gymnastics competition, a friend flipping off the back of a horse injured the anterior cruciate ligaments in both knees when she landed. That set Steve, an avid sportsman himself, wondering about the amount of strain a human knee can take and how to measure that strain. Soon he was making tiny devices called "sensors" in his dorm room to measure biomechanical strain, and soon afterwards he was making money for graduate school by selling sensors around the world—the first a tiny sensor designed for arthroscopic implantation on human knee ligaments.

In 1985, Steve Arms left graduate school to start his company, MicroStrain, Inc. Its business: sensors. "In many ways," he said, "an excellent time to start a business is when you first leave school and it is easier to take the risk, the opportunity cost is small, and one is used to living on a budget." He operated the business out of his home at first.

The company is not a university spin-off, but the company has a number of academic collaborators. Among them are the University of Vermont, Carnegie Mellon, the University of Arizona, Penn State, and Dartmouth University.

He located the company in Vermont to be close to family and friends, and to continue to enjoy the excellent quality of life offered by that location. In the longer run, the location has proven positive for high employee retention.

From its initial focus on micro sensors with biomechanical applications, MicroStrain moved into producing micro sensors for a variety of applications. Its sensor networks are in defense applications, security systems, assembly line testing, condition-based maintenance, and in applications that increase the smartness of machines, structures, and materials.

The company has grown to approximately 22 employees, including mechanical and electrical engineers. It occupies 4,200 square feet of industrial space near Burlington, Vermont. Its annual sales revenue was recently reported as $3.0 mil-

---

[11]The following informational sources informed the case study: interview with Mr. Steve Arms, President of MicroStrain, conducted at the Navy Opportunity Forum, May 2–4, 2005, Reston, VA; the company Web site: <*http://www.microstrain.com*>; company product brochures; a paper, "Power Management for Energy Harvesting Wireless Sensors," by S. W. Arms, C. P. Townsend, D. L. Churchill, J. H. Galbreath, S. W. Mundell, presented at SPHASE IE International Symposium on Smart Structures and Smart Materials, March 9, 2005, San Diego, CA; a book chapter, "Wireless Sensor Networks: Principles and Applications," *Sensor Technology Handbook*, ed.: Jon S. Wilson, pub.: Elsevier Newnes, ISBN: 0-7506-7729-5, Chapter 22, pp. 575–589, 2005; a University of Vermont Alumnus Profile of Mr. Arms; a presentation by Mr. Arms at a DHHS SBIR conference; and a recent Dun & Bradstreet Company Profile Report.

MICROSTRAIN, INC.: COMPANY FACTS AT A GLANCE

- **Address:** 310 Hurricane Lane, Suite 4, Williston, VT 05495-3211
- **Telephone:** 802-862-6629
- **Year Started:** 1985
- **Ownership:** privately held
- **Revenue:** Approx. $3.0 million in 2004
  - — Revenue share from SBIR/STTR and other government grants: approx. 25 percent
  - — Revenue share from sale of product and contract research: approx. 75 percent
- **Number of Employees:** 22
  SIC: Primary SIC: 3823 Industrial Instruments for Measurement, Display, and Control of Process Variables, and Related Products

  Secondary SICs:

3625	Relays and Industrial Controls
3679	Electronic Components, not elsewhere classified
3812	Search, Detection, Navigation, Guidance, Aeronautical, and Nautical Systems and Instruments
3823 8711	Engineering Services

- **Technology Focus:** Wireless sensors and sensor networks for monitoring strain, loads, temperature, and orientation
- **Application Areas:** Condition-based maintenance; smart machines, smart structures, and smart materials; vibration and acoustic noise testing; sports performance and sports medicine analysis; security systems; assembly line testing
- **Funding Sources:** Product sales, contract research, and federal government grants
- **Number of SBIR Grants:**
  - — From NSF: 3 Phase I, 3 Phase II, and 3 Phase IIB
  - — From other agencies: 6 Phase I, 2 Phase II, 1 Phase III

lion in 2004, with revenues growing at about 30 percent per year. Revenues are expected to reach $4.0 million in 2005.

THE TECHNOLOGY AND ITS USES

A "sensor" is a device that detects a change in a physical stimulus, such as sound, electric charge, magnetic flux, optical wave velocity, thermal flux, or mechanical force, and turns it into a signal that can be measured and recorded.

Often, a given stimulus may be measured by using different physical phenomena, and, hence, detected by different kinds of sensors. The best sensor depends on the application and consideration of a host of other variables.

MicroStrain focuses on producing smarter and smaller sensors, capable of operating in scaleable networks. Its technology goal is to provide networks of smart wireless sensing nodes that can be used to perform testing and evaluation automatically and autonomously in the field and to report resulting data to decision makers in a timely and convenient manner. The data can be used to monitor structural health and maintenance requirements of such things as bridges, roads, trains, dams, buildings, ground vehicles, aircraft, and watercraft. The resulting reports can alert those responsible to problems before they become serious or even turn into disasters. They can eliminate unnecessary maintenance and improve the safety and reliability of transportation and military system infrastructure, while reducing overall costs.

Among the features that determine how useful sensors will be for the type of system monitoring function described above are the degree to which the sensors are integrated into the structures, machinery, and environments they are to monitor; the degree to which the systems are autonomous, i.e., operate on their own with little need for frequent servicing; and the degree to which they provide efficient and effective delivery of sensed information back to users. MicroStrain's research has focused on improving its technology with respect to each of these performance features.

Another way to look at it is that MicroStrain has addressed barriers that were impeding the wider use of networks of sensors. For example, MicroStrain was one of the first sensor companies to add wireless capability. Wireless technology overcomes the barrier imposed by the long wire bundles that are costly to install, tend to break, have connector failures, and are costly to maintain. A recently passed international standard for wireless sensors (IEEE 802.15-4) is expected to facilitate wider acceptance of wireless networks.

A barrier to the use of wireless sensor networks is the time and cost of changing batteries. MicroStrain is an innovator in making its networks autonomous, without need of battery changes, by pursuing two strategies: First, it has adopted various passive energy harvesting systems to supply power, such as by using piezoelectric materials to convert strain energy from a structure into electrical energy for powering a wireless sensing node, or by harvesting energy from vibrating machinery and rotating structures, or by using solar cells. Second, the company has reduced the need for power consumption by such strategies as using sleep modes for the networks in between data samples.

A recent newsworthy application of MicroStrain's sensors was to assist the National Park Service move the Liberty Bell into a new museum. The bell has a hairline fracture that extends from its famous larger crack, making the bell quite frail. MicroStrain applied its wireless sensors developed as part of an NSF SBIR grant to detect motion in the crack and fracture as small as 1/100th the width of a human hair. During a lifting operation at the end of the move, the sensors detected shearing motions of about 15 microns (roughly half the width of a human hair) at the visible crack with simultaneous strain activity at the hairline crack's tip. MicroStrain's engineers stopped the riggers during this activity, and the sensor readings returned to baseline. Further lifting proceeded very slowly, and no further readings of concern were observed. The bell was protected by this early warning detection system, which saved it by literally splitting hairs.

Another newsworthy application by the company of a sensor network was to the Ben Franklin Bridge which links Philadelphia and Camden, NJ, across the Delaware River. The bridge carries automobile, train, and pedestrian traffic. At issue was the possible need for major and costly structural upgrades to accommodate strains on the bridge from high-speed commuter trains crossing the bridge. MicroStrain placed a wireless network of strain sensors on the tracks of the commuter train to generate the data needed to assess the added strain to the bridge. "For a cost of only about $20,000 for installing the wireless sensor network, millions were saved in unnecessary retrofit costs," explained Mr. Arms.

In the future, military systems will benefit from the cost-saving information from MicroStrain's sensor networks. Current development projects include power-harvesting wireless sensors for use aboard Navy ships, and damage-tracking wireless sensors for use on Navy aircraft. Mr. Arms explained that the data collected in this application is expected to result in recognition that the lives of the aircraft can be safely extended, avoiding billions of dollars of replacement costs.

THE ROLE OF SBIR IN COMPANY FUNDING

Early on, SBIR funding played an important role in supporting company research. While in graduate school at the University of Vermont, Mr. Arms was involved in proposal writing. He also had learned of the SBIR program. "Were it not for this," he said, "the application process may have seemed intimidating." He tapped Vermont's EPSCoR[12] Phase O grants to leverage his ability to gain Federal SBIR grants. EPSCoR Phase O grants provide about $10,000 per grant. According to Mr. Arms, these Phase O grants helped the company get preliminary data for convincing results and helped it write competitive proposals. The company has leveraged a total of $40,500 in EPSCoR grants to obtain $3.6 million in SBIR funds.

According to Mr. Arms, he found the NSF SBIR program with its "more open topics" particularly helpful in the early stages when the company was building capacity. "The open topics allowed the company to pursue the technical development that best fit its know-how," he explained. "Now the company is

---

[12]EPSCoR (Experimental Program to Stimulate Competitive Research Program) is aimed currently at 25 states, Puerto Rico and the U.S. Virgin Islands—jurisdictions that have historically received lesser amounts of federal R&D funding.

TABLE App-D-5 MicroStrain, Inc.: SBIR/STTR Grants from NSF and Other Agencies

NSF Grants	Number	Amount ($)	Other Agency Grants	Number	Amount ($)	Total Number	Total Amount ($)
SBIR Phase I	3	224,800	SBIR Phase I	6	445,800	9	670,600
SBIR Phase II[a]	3	1,198,800	SBIR Phase II[a]	2	1,349,100	5	2,547,900
SBIR Phase IIB/ Enhancements	3	346,900	SBIR Phase III	1	63,000	4	409,900
STTR Phase I	0	0	STTR Phase I	0		0	0
STTR Phase II	0	0	STTR Phase II	0		0	0
STTR Phase IIB/ Enhancements	0	0	STTR Phase IIB/ Enhancements	0		0	0
Totals	9	1,770,500		9	1,857,900	18	3,628,400

[a]Exludes Phase IIB/Enhancement awards which are listed separately.

SOURCE: MicroStrain, Inc.

better able to respond to the solicitations of the Navy and the other agencies that issue very specific topics."

The company regards receipt of an SBIR grant as "a strong positive factor that is helpful in seeking other funding," said Mr. Arms. "It is used not only to fund the development of new products, but as a marketing tool," he continued, pointing out that the company issues a press release whenever it receives an SBIR grant.

MicroStrain has received a total of 9 Phase I SBIR grants, 5 Phase II grants, 3 Phase IIB supplemental grants, and 1 Phase III grant. It has received SBIR grants from National Science Foundation (NSF), Navy, Army, and the Department of Health and Human Services. The amount the company has received in SBIR grants since its founding in 1985 totals about $3.6 million. Table App-D-5 summarizes the company's SBIR/STTR grants in number and amount.

According to Mr. Arms, the receipt of additional SBIR grants in the future is hoped for as a means to enable it to continue to innovate and stay at the forefront of its field. The company is targeting about 25 percent of its total funding to come from SBIR grants in coming years.

BUSINESS STRATEGY, COMMERCIALIZATION, AND BENEFITS

The company operates at an applied R&D level, and, unlike most R&D-based companies, has had sales from its beginning. Mr. Arms, the company founder and president, emphasized his belief in the need to produce product "to make it real as soon as possible." Continuing, Mr. Arms said, "Having products lets people know you know how to commercialize and that you intend to do it."

Mr. Arms sees the company's main competitive advantage as its role as an integrator of networked sensors. "Our goal is to produce the ideal wireless sensor networks," he explained, "smart, tiny in size, networked and scaleable in number, able to run on very little power, software programmable from a remote site, capable of fast, accurate data delivery over the long run, capable of automated data analysis and reporting, low in cost to purchase and install, and with essentially no maintenance costs." These features are important because they help to overcome the multiple barriers that were impeding the wider acceptance of sensors.

While the company sells its sensors mainly in domestic markets, it has from the beginning shipped sensors to customers around the world. Now the company sees market potential particularly in Japan and China. Patenting is reportedly very important to the company's commercialization strategy.

MicroStrain has received a number of grants in recognition of outstanding new product development in the sensors industry. It has received seven new product grants in the "Best of Sensors Expo" competition. Products that have been recognized by grants include the company's V-Link/G-Link/SG-Link microdatalogging transceivers for high speed sensor datalogging and bidirectional wireless communications; its WWSN wireless Web sensor networks for remote, Internet enabled, ad hoc sensor node monitoring; its FAS-G gyro enhanced MEMS based inclinometer; its MG-DVRT microgauging linear displacement sensor; its 3DM-G gyro enhanced MEMS based orientation sensor; its EMBEDSENSE embeddable sensing RFID tag; AGILE-Link frequency agile wireless sensor networks; and INERTIA-LINK wireless inertial sensor.

Society stands to benefit in a variety of ways from improved sensors and networks of sensors. Structures, such as buildings, bridges, and dams, as well as transportation and industrial equipment should have fewer catastrophic failures, because managers will be alerted to emerging problems in time to take preventative action. Homeland security should be enhanced by smarter networks of sensor-based warning systems. Manufacturing productivity may be increased by better planning of required maintenance and avoidance of costly, unplanned downtime. In general, integration of smart sensor networks into civilian and military structures and infrastructure, transportation equipment, machinery, and even the human body can conserve resources, improve performance, and increase safety.

VIEWS ON THE SBIR PROGRAM AND ITS PROCESSES

Mr. Arms made the following several observations about the SBIR program and its processes, some of which focused on the NSF program, some on the Navy program.

Topic Specification

Mr. Arms contrasted the "open topics" of NSF with the "very specific topics" of the Navy and other agencies, noting the former is particularly important to a company when it is "building capacity," while the latter is important when the company is positioned to generate a variety of new products.

Financing Gap

Mr. Arms noted that "early on in the life of the company the funding gap was very difficult, but now the company is able to bridge the gap using its sales revenue."

Value of Keeping Phase I Grants as Prerequisite to Phase II

"Phase I grants are important for getting a reaction to an area; to understanding better a technology's potential," said Mr. Arms. "I would not want to see this phase eliminated or bypassed."

Size of Grants

"It is great that the agencies are beginning to increase the size of their grants," commented Mr. Arms. "I especially like the NSF's Phase IIB match grant; it fits well with my company's commercial emphasis."

Application Process

Mr. Arms finds the Navy's SBIR application process particularly agreeable, calling it "the best!"

Value of Commercialization Assistance

The company has not participated in an NSF-sponsored commercialization assistance program, but it has participated in Navy-sponsored opportunity forums and in NSF conferences. It has found the networking provided by these forums and conferences to be very valuable. In fact, it was at an NSF-sponsored conference that MicroStrain made contact with Caterpillar Company, leading it to become a participant in a joint venture led by Caterpillar and sponsored by the Advanced Technology Program.

Observation about NSF's and Navy's SBIR Program Manager Systems

"The way NSF conferences facilitate face to face meetings between program managers, who have extensive business experience, with budding entrepreneur-scientists is excellent," Mr. Arms said. He expressed special enthusiasm for the Navy program managers, calling them "extremely knowledgeable and focused."

NSF's Student and Teacher Programs (outside SBIR)

Like several of the other companies interviewed, MicroStrain has used the NSF students program, "but, regretfully, not the teacher program." Like the other companies that have used these programs, Mr. Arms said MicroStrain had found the NSF students program valuable. "I think it would be a great thing to expand this idea to the other agencies," he suggested.

SUMMARY

This case shows a still-small company that has emphasized product sales since its inception in 1985. It has leveraged $40,500 of Vermont's EPSCoR "Phase O" grants to obtain $3.6 million in Federal SBIR grants. With SBIR support it developed an innovative line of microminiature, digital wireless sensors, which it is manufacturing. These sensors can autonomously and automatically collect and report data in a variety of applications. Unlike most research companies, MicroStrain, started by a graduate student, has emphasized product sales since its inception in 1985. Its sensors have been used to protect the Liberty Bell during a move and to determine the need for major retrofit of a bridge linking Philadelphia and Camden. Current development projects include power-harvesting wireless sensors for use aboard Navy ships, and damage-tracking wireless sensors for use on Navy aircraft. Although annual revenues are relatively small ($3 million in 2004), the company can document many millions of dollars of savings achieved by users of its wireless sensor networks. A little more than a quarter of the company's revenue come from government sources.

National Recovery Technologies, Inc.[13]

THE COMPANY

At the time National Recovery Technologies (NRT) was founded, the growth potential for municipal solid waste recycling looked promising. To develop municipal recycling technology, Dr. Ed Sommer applied for an SBIR grant from the U.S. Department of Energy (DoE) soon after starting NRT in 1981. NRT applied first to DoE's SBIR program for funding, because of the energy implications of municipal waste recycling. After being granted a Phase II DoE SBIR grant, Dr. Sommer said he was advised by a DoE program manager that further research to develop the plastics sorting technology would better fit the mission of EPA because of the environmental benefits of reducing PVC plastic waste in incineration. According to Dr. Sommer, having a close fit with EPA's mission made it more likely to receive SBIR grants.

Dr. Sommer described his company's location in Tennessee as very positive from a business standpoint. However, he noted that a drawback is the lack of technology infrastructure in the state. In developing proposals to the SBIR, "you are on your own," he said. There are not the incubators and other institutional assistance provided by some of the states that have a stronger technology infrastructure. He noted that NRT is the first- or second-largest recipient of SBIR grants in Tennessee.

NRT developed and commercialized several innovative processes for high-speed, accurate analyzing and sorting of municipal solid waste streams. The initial customer base had its origins in state recycling laws. Demand for the company's initial product was politically driven, not economically driven. In 1991, with venture capital funding, the company installed process lines in large sorting plants located in states with recycling requirements.

An EPA-granted SBIR project was to remove chlorine-bearing PVC from municipal waste streams prior to incineration for the purpose of emissions control. The company was successful in bridging to the new application, and quickly became a world leader in the recycling equipment industry, providing equipment and automated systems for analyzing and sorting plastics and curbside collected materials. It continues to have worldwide equipment sales, mainly in North America, Europe, Japan, and Australia. As a result of this technology development, NRT received EPA's National Small Business of the Year grant in 1992.

In 1994, the U.S. Supreme Court ruled on a case brought by waste haulers that found that a city violated their rights by requiring them to take collected waste to recycling plants. Because it was cheaper to take it to landfills, many haulers stopped taking it to the sorting plants—taking it to landfills instead—

---

[13]The following informational sources informed the case study: Interview at the company with company founder, president and CEO, Dr. Ed Sommer, Jr.; company Web site (*<http://www.nrt-inc. com>);* company brochures; and a recent Dun & Bradstreet Company Profile Report.

NATIONAL RECOVERY TECHNOLOGIES, INC.: COMPANY FACTS AT A GLANCE

- **Address:** 566 Mainstream Drive, Suite 300, Nashville, TN 37228
  Telephone: 615-734-6400
- **Year Started:** 1983
- **Ownership:** private
- **Annual Sales:** ~$4 million
- **Number of Employees:** 14 total; 5 in R&D
- **3-year Sales Growth Rate:** 67 percent
- **SIC:** Primary SIC 3589, Manufacture Service Industry Machinery
  358890300, Manufacture Sewage and Water Treatment Equipment
  Secondary SIC: 8731, Commercial Physical Research
  87310202, Commercial Research Laboratory
- **Technology Focus:** Initial Focus—Mixed municipal solid waste recycling system. Later focus—Automated process for sorting plastics by type with high throughput and accuracy for cost-effective recycling; electronics-driven metals recycling system; inspection technology for security checking in airports and other security check points; and continuation of the plastics sorting business.
- **Funding Sources:** Internal revenue mainly from sales of plastics sorting equipment, venture capital, SBIR funding, and ATP (as subcontractor pass-through).
- **Number of SBIR Grants:** From NSF, 3 Phase I, 3 Phase II, and 2 Phase IIB, plus additional grants from DoE and EPA.

causing large numbers of sorting plants to shut down. Plants that employed their own haulers were more likely to survive. At the same time, there was a move for presorting of curbside waste pick-up, requiring less sorting by secondary processors, further reducing demand for the company's equipment.

As a result of these developments, the company was in trouble. It needed to move into others areas or go out of business. As it looked for new ways to leverage its existing intellectual capital and technology capabilities, it identified new areas in which to apply its technical capabilities: metals recycling and airport security. It continued to pursue automated process technology for high throughput sorting of plastics.

Today the company maintains sales of plastics analysis and sorting equipment with annual sales running about $4 million. At the same time, it is developing new lines of business that were not yet generating sales at the time of the interview. The company employs a staff of 14, 5 of whom are in R&D.

THE TECHNOLOGIES AND THEIR USES

For plastics recycling, NRT used such technologies as IR spectroscopy for polymer identification, machine vision for color sorting, concurrent parallel processing for rapid identification, quick real-time sorting, and precision air jet selection of materials.

An idea that emerged from discussions with potential customers was metals sorting, smelting, and refining. NRT undertook a research effort now in its seventh year to develop metals processing technologies. It is collaborating with another company, wTe Corporation of Bedford, MA, which has an automobile shredder division, to develop a novel optoelectronic process for sorting metals at ultra-high speeds into pure metals and alloys. It also joined with wTe to form a new company, Spectramet LLC, to serve as the operator of metals reprocessing plants.

An idea for a new technology/business platform that emerged just in the past several years is in the security area. The stream of objects moving along a conveyer belt at an airport security system resembles in many ways a mixed waste stream in the recycling business. NRT's approach combines fast-throughput materials detection technology with data compilation, retrieval, analysis, storage, and reporting to provide an improvement over the current non-automated, manual inspection system. The Transportation Security Administration (TSA) is evaluating NRT's system, a necessary step in qualifying it for use in airport security. According to Dr. Sommer, NRT anticipates product sales in 2005.

THE ROLE OF SBIR IN COMPANY FUNDING

According to Dr. Sommer, "Without the SBIR program, NRT wouldn't have a business. We couldn't have done the necessary technical development and achieved the internal intellectual growth." The SBIR program was critical, he explained, both in developing NRT's initial technology, and in responding to market forces to develop new technologies after a Supreme Court decision caused many municipal solid waste sorting plants to close and the growth potential of the initial waste recycling technology to decline. "SBIR saved our bacon," said Dr. Sommer. As a result of the intellectual capabilities built within the company through SBIR-funded research, "we continue to be able to contribute."

In 1985, the company had received a Phase II grant from DoE to pursue development of an automated process for sorting municipal solid waste. Subsequently the company received a series of SBIR grants from EPA, including eight completed Phase II grants between 1989 and 1996, aimed at developing and refining plastics recycling technologies. In 1996, NRT received SBIR grants again from DoE for plastics recycling and mixed radioactive waste recycling. Since 1996, the company has received SBIR grants from both EPA and NSF. Its first Phase II grant from NSF was received in 1999, when the company was in its second decade of operation. The funded project was aimed at developing new technologies in scrap metals processing. Later NSF SBIR funding was aimed at developing new technologies in the security area.

From the NSF, the company has received a total of three Phase I grants, totaling $0.3 million, three Phase II grants, totaling $1.4 million, and, reportedly two Phase IIB grants. Altogether, the company has received more than $5 million in SBIR funding combined from DoE, EPA, and NSF over the past 20 years.

As time passed and the growth potential of the plastics recycling business flattened, Dr. Sommer said that he turned to NSF's SBIR program to develop new lines of technology. He said that NSF's SBIR program was particularly appealing because its solicitation is the broadest among the agency programs. NSF's solicitations allowed the company more leeway to propose new technology development projects that it believed would lead to business opportunities with higher growth potential. At the same time, he characterized NSF's SBIR program as "very competitive" and its grants "the hardest to get." With NSF funding, NRT was able to develop metals recycling technology and, more recently, the detection system for airport security.

The "openness" of NSF, Dr. Sommer said, was critical to his business model, which entails first finding a need for a new product, performing market analysis, and then looking for funding to perform research needed to bring a product to the prototype stage. Contrasting NSF's openness with the "narrow" solicitations of the Department of Defense (DoD), Dr. Sommer explained that he did not apply to the DoD SBIR program because "for us it is not conducive to developing products aimed at a general market."

Dr. Sommer related how his company (in a subcontractor position to the wTe Corporation, Bedford, MA) had subsequently looked to the ATP for larger amounts of funding to help further develop the metals reprocessing technology. In describing the sequence, he said there was a "spring off from SBIR to ATP."

BUSINESS STRATEGY, COMMERCIALIZATION, AND BENEFITS

While it develops the metals reprocessing and security product lines, NRT has maintained a steady revenue stream on the order of $2–4 million/year, primarily from sales of plastics analysis and sorting equipment. Dr. Sommer sees a larger market potential in metals reprocessing—which includes partnerships to operate as well as provide equipment—with projected annual sales revenue in the range of $10–30 million. He sees a larger potential in the security market, with projected sales revenue in the range of $100 million/year. Dr. Sommer expressed his intention to take the company into a faster growth mode with the development of these new lines of business.

Annual sales revenue is currently running approximately $4 million. Revenues reportedly generated as a result of SBIR grants, referred to as "Phase III revenues," totaled approximately $44 million from the start of the company up to November 2003.

At least eight products are on the market derived from DoE and EPA SBIR-funded research, including, for example, the following:

- NRT VinylCycle® system—a grant winning sorting system for separating PVC from a mixed stream of plastic bottles introduced in 1991
- NRT MultiSort®IR System—an advanced plastic bottle sorting system for separating specific polymers from a mixed stream of materials
- NRT Preburn™ Mixed Waste Recycling System—a facility with integrated technologies to provide a system for achieving maximum material recovery from waste streams otherwise slated for landfill

Products funded by the NSF SBIR program are still in the development stage. The metal alloy sorting technology under development with NRT's commercialization partner wTe Corporation is planned to be used by the Spectramet LLC spin-off, jointly owned by NRT and wTe Corp, for processing of metals as opposed to the technology being made available as a commercial equipment product. The advanced third generation of this sorting technology is installed and in initial commercial operation processing selected loads of scrap metals from various suppliers.

Two patents resulting from the NSF funded research have been issued. Four additional patents are pending.

Dr. Sommer identified four types of social benefits that have resulted, or may result, from the SBIR-funded technologies. (1) Knowledge creation and dissemination result from patents the company filed on the intellectual capital coming out of its NSF research, and from presentations. Patents signal the creation of new knowledge and provide a path of dissemination. Company researchers have also presented at conferences in the fields of metallurgy and plastics recycling. (2) Safety effects arise from the automation of sorting machines in recycling plants which appear to have reduced injuries as compared with conveyer belts using labor-intensive hand-picking techniques that bring the worker in close interface with potentially unhealthy waste streams and possibly injurious equipment. (3) Environmental effects result because NRT's sensing and sorting technologies are based in electronics, not chemicals. Using "dry processes" rather than "wet processes" avoids the runoff of chemicals into waste streams and the associated pollution. Additionally, environmental benefits result directly from recycling plastics and metals into reuse instead of dumping them into landfills. The availability of automated systems that increase the efficiency of the process helps to enable cost-effective recycling of diverse materials around the world. (4) National security benefits may result if NRT is successful in leveraging its automated materials sensing technology into the security arena, improving the efficiency—and more important—the effectiveness of security at airports and other security check points. NRT's technology is currently under evaluation for airport security applications, and not yet in use.

VIEWS ON THE SBIR PROGRAM AND PROCESSES

Submitting Proposals Through NSF's FastLane

Discussing the SBIR application process and how the application process compares among agencies, Dr. Sommer noted that the answer is very time dependent given that the agencies have recently developed more computerized applications processes. He noted that NSF's FastLane system is "very slick." It is also very complex, he said, with many modules, which make navigating around the system hard on a newcomer. However, once one becomes familiar with it, it becomes more useful, he concluded.

NSF's Review Process

Dr. Sommer's view was that NSF does a "fabulous job" with its review of SBIR proposals. He noted that earlier there was an issue—"too heavy a reliance on university reviewers"—but believes that now there is more use of reviewers who come from the commercial sector who are better able to assess proposals for technology development with commercial potential. He noted that he was so impressed that he wanted to give back to the system, and volunteered to serve on a review panel for NSF. The experience, he said, gave him confidence in the process as being fair. He also saw the experience as a good way to learn the ins and outs of preparing higher quality proposals.

NSF's Feedback on Reviews

Dr. Sommer also found useful NSF's feedback system to give applicants information from review results. He said his company had resubmitted a rejected proposal, taking into account feedback received, and was successful with the resubmitted proposal. Asked if he felt his company's proprietary ideas had ever been threatened during the proposal and review process, he responded with a definite no with respect to DoE, EPA, and NSF.

NSF Program Managers

In speaking of NSF program managers, Dr. Sommer praised those with whom he had direct experience as "extremely dedicated."

Grant Size

When asked if he thought the size of SBIR grants should be increased, possibly in trade-off to a decrease in the number of grants, Dr. Sommer responded that he thought the size of Phase I grants is about right, and noted that "it is good to spread around the funding," rather than concentrate it in fewer grants. He said that he would, however, like to see somewhat larger Phase II grants. In this regard, he characterized NSF's Phase IIB grant as "a very good tool, providing a boost to finding other dollars." He noted that the Phase IIB requirements fit well with his business model. He also reiterated that it is good that the ATP is available to provide larger research grants.

Funding Gap

Dr. Sommer noted that there often is a lag—a funding gap—between Phase I and Phase II grants that can "put the brakes on research." He explained that he is fortunate in having an ongoing business with a revenue stream that can help him bridge the gap with internal funding, rather than shut down the research as he would otherwise have to do.

Phase I as a Prerequisite to Phase II

When asked if he would like the opportunity to bypass the Phase I grant and go directly to Phase II, Dr. Sommer responded that often research funded in-house positions him to have the ability to apply directly for a Phase II grant. "Phase I," he said, "makes you do your feasibility analysis more thoroughly than you might otherwise do, but this can slow you down." He saw both pros and cons to keeping Phase I as a prerequisite; his response was inconclusive.

Commercialization Assistance Program

A company founder and CEO, Dr. Sommer, holds a doctorate degree in physics from Vanderbilt University. He went into business soon after receiving the degree. According to Dr. Sommer, he earlier resisted participating in the commercialization assistance programs sponsored by SBIR programs, and dropped out of a program that he had begun. He thought that the time requirements were excessive, and he resisted a diversion from his focus on technical issues. In 2000, however, he enrolled for a second time in the commercialization assistance program provided by Dawnbreaker Company. This time he completed the program, which he described as highly beneficial, providing him with insights, vision, know-how, and tools to more aggressively pursue business opportunities. He said he needed the training.

NSF's Solicitation Topics

Dr. Sommer emphasized that he would like to see preserved the "openness" of NSF's solicitation, which he called "critical" to maintain.

NSF's Emphasis on Commercialization

Another important feature of the NSF program that Dr. Sommer thought should be kept is the emphasis of the program on commercialization.

NSF's Review Process

Dr. Sommer expressed the view that it is very important that NSF achieve a balance in the use of university reviewers and those knowledgeable and experienced in business.

SUMMARY

This case study shows how SBIR is used not only by a start-up company to help it establish a technology platform from which it can launch a business, but also by a mature company that needs to rejuvenate its technology platform in order to meet changing markets. The case study company, NRT, used SBIR grants initially to support R&D underlying its first line of business. In its second decade, it used SBIR grants to leverage its existing technological base in a directional change that offers potential for future robust growth.

The NSF SBIR grant solicitation with its broad topic areas and emphasis on commercialization fits particularly well the company's business strategy of first identifying a potential market opportunity, developing a research plan to bring a product/process to the prototype stage, and then looking for early stage research funding to make it happen.

This case also illustrates a business trajectory that, although up to this time, is not dramatic in terms of its growth, is significant when considered as one of many such companies enabled by the SBIR program. The company has survived for over 20 years, remaining small but steadily employing approximately 10 to 15 people, generating on the order of $4 million per year, meeting niche market needs that have energy and environmental implications, and poised to make further, potentially substantial technological contributions and to achieve growth. It may be argued that this company represents a component of the R&D landscape, which in its aggregate is an extremely important part of the nation's capacity to innovate.

NVE Corporation[14]

THE COMPANY

This small company traces its origins to a very large company, Honeywell, Inc. The company founder, Dr. James Daughton, co-invented "Magnetoresistive" Random Assess Memory (MRAM) while at Honeywell. On retiring from Honeywell, he licensed the technology for pursuit of civilian development and applications, and, in 1989, founded Nonvolatile Electronics, Inc. (NVE).[15] Other former Honeywell employees joined NVE following a downsizing at Honeywell and continue with NVE today.

Initially the company operated out of the founder's home in a Minneapolis suburb, but in the early 1990s after receiving research funding from the SBIR program and from the Advanced Technology Program (ATP), it found space in a nearby Eden Prairie, MN, industrial park. Today it leases a facility of approximately 21,000 square feet, which includes offices and a clean-room for research and fabrication of semiconductor devices. It employs a staff of 70, including 24 in R&D—12 at the Ph.D. level, and the rest in product development, manufacturing, and administration.

The company has licensing arrangements with other companies, among them Honeywell, Motorola, Cypress Semiconductor, and Agilent Technologies. It also has affiliations with a number of universities, including the University of Minnesota, Iowa State University, and the University of Alabama, and it has sponsored a university student through an NSF program.

THE TECHNOLOGY AND ITS USE

Since its founding, the company has continued development of MRAM, a revolutionary technology that fabricates memory with nanotechnology and uses electron spin to store data. MRAM computer chips could prevent accidental losses of information when the power is interrupted, extend battery life, and replace essentially all RAM technology in use today. A website devoted to MRAM news (*<http://mram-info.com>*) calls the technology the "holy-grail" of memory, and states that it "promises to provide non-volatile, low-power, high-speed and low-

---

[14]The following informational sources informed the case study: interview at the company with Robert Schneider, Director of Marketing, Richard George, Chief Financial Officer, and John Myers, Vice President of Development; information from the company Web site: *<http://www.nve.com>*; the company's Annual Report for 2004; company press releases and selected press clippings; an article about the company's technology in Sensors, Vol. 21, No. 3, March 2004; Dun & Bradstreet Company Profile Report; earlier interview results compiled by Ritchie Coryell, NSF (retired); and an earlier "status report" developed by ATP.

[15]In 2000, through a reverse merger, the company officially took the corporate name of NVE Corp. It trades on the Nasdaq SmallCap Market under the symbol "NVEC."

NVE CORPORATION: COMPANY FACTS AT A GLANCE

- **Address:** 11409 Valley View Road, Eden Prairie, MN 55344
- **Telephone:** 952-829-9217
- **Year Started:** 1989 (Incorporated 1989)
- **Ownership:** Traded on the NASDAQ Small Cap Market as "NVEC"
- **Revenue:** $12 million annually
    — Share from SBIR/STTR grants: 35 percent
    — Share from product sales, R&D contracts, and licensing: 65 percent
- **Number of Employees:** 70
- **SIC:** Primary SIC 3674
- **Technology Focus:** Spintronics-based semiconductors, a nanotechnology which utilizes electron spin rather than electron charge to acquire, store and transmit information
- **Application Areas:** Electronic memory, sensors, and isolated data couplers
- **Facilities:** Leases 21,362 sq. ft.
- **Funding Sources:** Commercial sales, government sales, licensing fees, federal government grants & contracts, stock issue, private investment, venture capital funding, and reinvestment of retained earnings.
- **Issued Patent Portfolio:** 34 issued U.S.; more than 100 patents worldwide issued, pending, or licensed from others
- **Number of SBIR grants:**
    — From NSF: 31
    — From other agencies: 90

cost memory." MRAM is based on effects named "Giant MagnetoResistance" (GMR) and "spin-dependent tunneling," whereby a sandwich of metals shows a substantially greater change in resistance than a single metal of the same size when exposed to a magnetic field. These effects enabled researchers to increase signal strength while increasing density and decreasing size.

The technology has proven a challenging pursuit indeed. A decade later MRAM remains largely a promise still falling short of the realization of its huge commercial potential. However, NVE has developed significant intellectual property in MRAM that it has licensed to both Motorola and USTC for initial license fees and future royalty payments when it is put into commercial use. Other companies, including IBM Corporation, Hewlett-Packard Company, Infineon Technologies AG, NEC Corporation, Fujitsu Limited, Sony Corporation, and Samsung Electronics are also seeking to develop MRAM chips.

It should be noted that there are currently available nonvolatile memories, such as "flash" memories and ferroelectric random access memories (FRAMs), and there are also emerging technologies that are expected to compete with MRAM, such as polymeric ferroelectric random access memory (PFRAM), ovonic unified memory (OUM), and carbon nanotubes. However, according to its developers, MRAM offers advantages over the existing nonvolatile memories in terms of speed, lower power use, longer life expectancy, and freedom from other limitations, and also advantages over the emerging technologies, including being closer to market.

As NVE pursued MRAM development, it saw other potential commercial applications in the GMR effect. By the mid-1990s, NVE was making and selling GMR-based sensing products for such diverse applications as automotive braking systems, medical devices, and portable traffic monitoring instruments.

NVE now describes its technical focus as "spintronics," a nanotechnology based on MRAM research, which, like MRAM, takes advantage of the property of electron spin. The technology combines quantum mechanical tunneling with magnetic scattering from the spin of electrons, resulting in a new phenomenon called "spin-dependent tunneling (SDT)" or "magnetic tunnel junctions (MTJ)."(*SENSORS*, March 2004). The targeted application area is magnetic field sensing for which very small, inexpensive sensors with high sensitivity to small changes in the magnetic field are required. Standard silicon microprocessing methods can be used to fabricate SDT devices. The company expects to enter the market within several years with SDT sensors in complex magnetometer systems, in small simple event detectors, in arrays for perimeter security, vehicle detection, and other security systems, and also for detection of deep cracks, corrosion, and other deeply buried flaws. Further in the future, applications are anticipated for physiological monitoring, advanced magnetic imaging, and other areas not yet identified.

THE ROLE OF SBIR AND OTHER GOVERNMENT R&D FUNDING

Federal grants for R&D—both from SBIR and ATP—have played an essential role in the company's start-up, survival, and growth. During its early days, the company's founder credited government R&D funding with preventing the company from failing and improving its ability to attract capital from other sources.

More recently, NVE's vice president called the SBIR program "the mother of invention." The company currently derives approximately half of its funding from government funding, including SBIRs and BAAs (Broad Area Announcements that federal agencies may use to solicit contract work), and the remaining half from commercial sales, up-front license fees and royalties, stockholder investment, and retained earnings. It views SBIR and other government R&D funding programs as essential to being able to perform the advanced R&D that

TABLE App-D-6 NVE Corporation: SBIR/STTR Grants from NSF and Other Agencies

NSF Grants	Number	Amount ($)	Other Agency Grants	Number	Amount ($)	Total Number	Total Amount ($)
SBIR Phase I	19	1,487,000	SBIR Phase I	51	4,512,000	70	5,999,000
SBIR Phase II[a]	9	3,540,000	SBIR Phase II[a]	34	22,428,000	43	25,968,000
SBIR Phase IIB/ Enhancements	2	582,000	SBIR Phase IIB/ Enhancements	2	488,000	4	1,070,000
STTR Phase I	1	100,000	STTR Phase I	1	70,000	2	170,000
STTR Phase II	0	0	STTR Phase II	2	1,120,000	2	1,120,000
STTR Phase IIB/ Enhancements	0	0	STTR Phase IIB/ Enhancements	0	0	0	0
Totals	31	5,709,000		90	28,618,000	121	34,327,000

[a]Exludes Phase IIB/Enhancement awards which are listed separately.

SOURCE: NVE Corporation.

has allowed the company subsequently to produce products for sale and to license intellectual property.

Table App-D-6 summarizes the company's SBIR/STTR grants from the National Science Foundation (NSF) and other agencies. The 121 SBIR and STTR grants received have totaled $34.3 million in R&D funding. Approximately a fourth of the 121 grants have come from NSF.

BUSINESS STRATEGY, COMMERCIALIZATION, AND BENEFITS

NVE President and CEO, Daniel Baker, has been recently quoted as saying, "We believe that NVE is well positioned with critical intellectual property covering a broad range of near-term and long-term MRAM designs. Our MRAM strategy, therefore, will be to focus on an intellectual property business model, providing technology to enable revolutionary memory designs rather than both providing technology and selling devices." (NVE Press Release, April 19, 2005) Regarding NVE's strategy regarding sensors and signal couplers, it appears that NVE will continue to build its intellectual property base in SDT and GMR sensors, as well as continue to design, fabricate, and sell a variety of sensor and signal coupler devices for both commercial and defense applications.

Product revenue from sales of spintronic sensors and couplers has steadily increased, reaching $5.4 million in FY 2004, and the company projects its commercial product revenues to continue to grow. R&D revenue for FY 2004 rose to $6.6 million, as government contract revenue increased. Total revenue for FY 2004 totaled $12 million. As of the end of FY 2004, NVE had been profitable for two years, and the firm was projecting continued profitability into FY 2005.

The spintronic sensors and couplers sold by NVE offer value-added benefits to users in terms of accuracy and data rates. The firm's unique components provide up to three-to-four times the accuracy and twice the data rate of conventional electronics, allowing users to make better products at lower costs.

NVE has stated recently that the commercial viability of MRAM technology is now more assured. It expects FY 2005 to be pivotal for MRAM commercialization. In additional to nonvolatility, MRAM's potential benefits to users are high speed, small size, increased life expectancy, lower power use, and scalability. Nevertheless, as indicated earlier, the competition among firms and among technologies is intense.

VIEWS ON THE SBIR PROGRAM

The interviewees strongly supported the SBIR program in general. They noted that it has fostered the development of a large number of small R&D companies like itself that "collectively comprise the modern-day equivalents of the Bell Labs of the past." They emphasized the advantages of performing R&D in a small-company environment where "there is much more freedom to innovate" and "R&D is not viewed as an unwelcome tax on what is considered the productive part of the corporation." They expressed the hope that program officials and policy makers will not underrate the importance of this collective group of small firms by assuming that only if they individually grow into large companies are they worthwhile from the standpoint of the economy and its innovative capacity.

At the interviewer's request, the three NVE officials shifted their focus to NSF's SBIR program, beginning with a positive comment and then turning to areas for potential improvement:

Praise for NSF Portfolio Mangers

NVE officials emphasized the high quality of NSF's program managers. They identified several of the program managers by name, calling them as a group "a class act."

Concern about an Unofficial Limit on Number of Grants to a Firm

The NVE team voiced concern that NSF (NIST was also mentioned) appears to be imposing an unofficial criterion on top of the official proposal eligibility criteria, in the form of a limit on the number of grants a given company can receive. According to the company representatives, the company is sometimes informed that it must choose a subset of the total number of grants for which it has been deemed eligible, as evaluated against the published criteria.

Moreover, since the company is aware of firms that it believes have received many more SBIR grants than NVE, it has the impression that the agency may be applying an unofficial limit on grants unevenly among applicant firms. At the same time, other agencies do not appear to have such a criterion—either officially or unofficially.

NVE's position is that if the company has submitted multiple proposals that meet an agency's published SBIR eligibility requirements, it should be able to receive the grants for which it is eligible without limit, given there is no overlap among them and adequate funding. If there is a limit on the number of grants that a company can receive, NVE believes this limit should be made official, explicit, and be evenly applied so that companies can know up front exactly what rules apply before they incur the considerable costs of proposing.

Concern about Limit on Number of Proposals Allowed

NVE also noted a limit on the number of proposals a company can submit to NSF in a given year. If topical solicitations are spread out over the year, as they are for NSF, this limit means that a company must make decisions about how it will spend its proposal "quota" among the topic areas before it may be ready to do so strategically.

Need to Overcome a Timing Problem

NVE noted that NSF's timing of its solicitations mean that if you miss a key date, you miss a whole grant cycle. Extending the proposal submittal window or opening it several times a year would help to overcome this timing problem.

Comments on Commercialization Assistance

According NVE's Director of Marketing, participation in the Foresight activity was of greater value to the company than was the Dawnbreaker activity. A reason given was that the Foresight staff appeared to have more industry experience.

Lack of NSF Travel Funds

NVE officials expressed concern that NSF program managers are not able to conduct company site visits. "On-site visits would be good." At the same time, they noted that the annual SBIR/STTR Phase II meeting provides them the opportunity to discuss their projects with the NSF program managers.

SUMMARY

This case study shows how a company, which traces its origins to a large company, used SBIR and other federal grants to help launch the company, to keep it from failing, and to improve its ability to attract capital from other sources. Since its founding, the company has pursued development of MRAM technology that uses electron spin to store data and promises nonvolatile, low-power, high-speed, small-size, extended-life, and low-cost computer memory. As NVE pursued MRAM development, it saw related potential applications such as magnetic field sensors. NVE has developed substantial intellectual property in MRAM technology. The company has licensing arrangements with a number of other companies. Approximately half of the company's funding comes from government funding, and the remainder from commercial sales of magnetic field sensors, up-front license fees, and royalties. The company is now traded on the NASDAQ Small Cap Market.

T/J Technologies, Inc.[16]

THE COMPANY

Maria Thompson, with an M.B.A. degree and industry experience in product development and marketing, and her husband, Levi Thompson, with a Ph.D. in chemical engineering, combined their expertise to start T/J Technologies in 1991. They started the company by taking over a Department of Defense SBIR Phase I grant from a company that was divesting contractual obligations. The Thompsons rented a lab bench in a friend's small laboratory space, hired an employee, and put their savings into the enterprise. The SBIR grant was to develop ultrahard coatings for armor. Because the funding was quite limited, Ms. Thompson, "kept her day job" as an IBM marketing representative, and Dr. Thompson, who is an officer in the company, continued at the University of Michigan where he is now a tenured professor in the College of Engineering, and Associate Dean for Undergraduate Education. They ran T/J Technologies as a part-time business for the next three years, working evenings and weekends. Then, in 1994, Ms. Thompson left IBM and focused on being president of T/J Technologies.

T/J TECHNOLOGIES, INC.: COMPANY FACTS AT A GLANCE

- **Address:** 3850 Research Park Drive, Ann Arbor, MI 48108
- **Telephone:** 734-213-1637
- **Year Started:** 1991
- **Ownership:** privately held; minority-owned and operated
    - **Revenue:** Confidential
- **Number of Employees:** 24
- **Primary SIC:** 8731 Commercial Physical Research
- **Technology Focus:** Nanomaterials for alternative energy devices such as batteries and fuel cells.
- **Application Areas:** Rechargeable batteries, ultra-capacitors, and fuel cells
- **Funding Sources:** Government grants and grants, contract research, joint development programs, angel investment, and sales of material samples
- **Number of SBIR Grants:**
    - From NSF: 10 Phase I, 7 Phase II, and 2 Phase IIB
    - From other agencies: 23 Phase I, 8 Phase II, 1 Phase III

---

[16]The following informational sources informed the case study: interview with Maria Thompson, President and CEO of T/J Technologies, conducted at the company's offices in Ann Arbor, MI; the company Web site: <*http://www.tjtechnologies.com*>; company product brochures and press releases; two papers, "Improving the Power Density of PEM Fuel Cells" and "Ultra-High-Rate Batteries Based on Nanostructured Electrode Materials," both presented at the 41st Power Sources Conference, June, 14–17, 2004, and available at the company's Web site; a profile of Ms. Thompson appearing in the *Detroit Free Press*, July 28, 2003; and a recent Dun & Bradstreet Company Profile Report.

From its initial work in ultrahard coatings, the company moved into developing ultracapacitors, and that led them into a specialty in advanced materials and devices for electrochemical energy storage and conversion. Particularly notable are its advances in rechargeable batteries, ultracapacitors, and fuel cells. The company holds seven patents and has a number of others pending in the area of nanomaterials for alternative energy technologies.

The technical and business achievements of T/J Technologies have been recognized in grants at the national and state level. In 2005, the company was designated one of the "50 Companies to Watch in Michigan." In 2000, the company received the DoD Nunn Perry Grant for its ultra-capacitor development with Lockheed Martin, and it also received the "Product of the year Grant" from the Michigan Small Business Association. In 2000, Ms. Thompson was named one of Metro Detroit's Innovators by Crain's Detroit Business. Among other grants, T/J Technologies and Ms. Thompson received Black Enterprise Magazine's "Innovator of the Year" grant in 1998. The company has been featured in Fortune Small Business, Black Enterprise, the Detroit News, and other business journals and newspapers.

The company has grown to approximately 25 employees, most of whom are scientific researchers and many of whom have doctorate degrees. Employee disciplines include chemical engineering, polymer science, chemistry, and electrochemistry. The company occupies 12,500 square feet of industrial space in Ann Arbor, Michigan, the home state of Ms. Thompson and where both Dr. and Ms. Thompson attended the University of Michigan.

THE TECHNOLOGY AND ITS USES

T/J Technologies designs advanced materials for energy storage and energy conversion devices. It has a nanocomposite cathode composite material for lithium-ion (Li-ion) batteries. It has developed a series of electrocatalysts for fuel cells that have implications for portable power applications.

The characteristics of the materials developed by T/J Technologies are expected to enable a new class of proprietary, high power Li-ion batteries that are suited for a wide variety of high rate commercial and military applications.

The company also provides expertise in materials synthesis and processing and materials testing services. The company offers consulting services in nanomaterials, energy storage materials, and fuel cell battery research, development, and commercialization.

THE ROLE OF SBIR IN COMPANY FUNDING

Despite the fact that Ms. Thompson did not know of the SBIR program in advance of starting the company, it was a Phase I SBIR grant that got the company started. When her husband identified an opportunity to start their company by picking up research from another company, she regarded it as an opportunity. During the course of arranging the transfer, she learned about the SBIR. Her teacher was MERRA, a Michigan-based organization aimed at boosting the State's technology businesses. "MERRA explained the concept of SBIR and how the various phases worked." Afterwards, T/J Technologies went on to propose and win new SBIR grants.

"The SBIR grants served as building blocks for us," explained Ms. Thompson. With the technical capacity developed in a series of SBIR, the company applied for an award from the Advanced Technology Program (ATP). "When we went to apply for an ATP award, even though MERRA was disbanded, the people were still around and they helped us with our ATP proposal by giving us feedback. When we reached the finalist stage, they volunteered to do a dry-run with us in preparation for the oral review. As a result, we were very well prepared. So now we try to give back to the community the kind of support we received when we needed it. At this time, most of this kind of assistance in Michigan is on an Ad Hoc basis."

Continuing with the building block model, Ms. Thompson explained that the ATP and SBIR grants in turn served as building blocks for pursuing research contracts with the Army under competitive-issued Broad Agency Announcements (BAA) for contract research, the company's current major source of funding. And the BAA work in turn is serving as a building block allowing the company to develop partnerships with global firms who are testing T/J Technologies' materials and building batteries collaboratively for demonstration purposes.[17] The company is executing its commercialization strategy with its strategic partners.

"We've been granted these larger contracts because of some of the technologies developed with SBIR. All of these things are very synergistic. Without the SBIR, we couldn't have won the ATP. And, without the ATP and SBIR, we may not have had the technology with which to earn the larger contracts and the joint development agreements. So they are all linked."

Ms. Thompson noted that SBIR grants are still part of the company's funding mix, "helping us to round out our capabilities in certain areas, such as hydrogen storage and high temperature membranes." About 15 to 20 percent of the company's funding reportedly is coming from SBIR grants in 2005, and most of the remainder is from contract research. A small amount is from sales of material

---

[17]A recent report commissioned by the Advanced Technology Program investigates the question, "Why are there no volume Li-ion battery manufacturers in the United States?" (See Ralph J. Brodd, *Factors Affecting U.S. Production Decisions: Why are there No Volume Lithium-Ion Battery Manufacturers in the United States?* ATP Working Paper Series, Working Paper 05-01, June 2005.

TABLE App-D-7 T/J Technologies, Inc.: SBIR/STTR Grants from NSF and Other Agencies

NSF Awards	Number	Amount ($)	Other Agency Awards	Number	Amount ($)	Total Number	Total Amount ($)
SBIR Phase I	10	949,999	SBIR Phase I	23	1,737,113	33	2,687,112
SBIR Phase II[a]	7	3,525,361	SBIR Phase III	8	5,247,169	15	8,772,530
SBIR Phase IIB	2	575,000	SBIR Phase IIB/ Enhancements	1	223,864	3	798,864
REU	2	12,000				2	12,000
STTR Phase I	0		STTR Phase I	6	598,499	6	598,499
STTR Phase II	0		STTR Phase II	1	500,000	1	500,000
STTR Phase IIB/ Enhancements	0		STTR Phase IIB/ Enhancements	0			
Totals	21	5,062,380		39	8,306,645	60	13,369,005

[a]Exludes Phase IIB/Enhancement awards which are listed separately.

SOURCE: T/J Technologies, Inc.

samples in conjunction with the joint work with global companies that is expected to lead to commercialization of the materials.

Emphasizing that the building block model may need to be repeated as they look at different materials, Ms. Thompson said, "We need a family of materials for multiple applications. Multiple materials are needed for batteries, and also for the various components. You have to demonstrate a whole system, and that takes time and money," she said.

"STTRs also can help a company like ours," noted Ms. Thompson, "because you may need a piece of technology from a university to help put together a broader system." Also, along the way, the company has received funding from angel investors.

Thus far, T/J Technologies has received a total of 33 Phase I SBIR grants, 15 Phase II grants, 2 Phase IIB supplemental grants, and 1 Phase III grant. It has received SBIR grants from multiple agencies, including the Army, Air Force, NASA, DoE, and NSF. The amount the company has received in SBIR grants since its founding in 1991 totals about $12.3 million, and the total in SBIR and STTR grants amounts to approximately $13.4 million. Table App-D-7 summarizes the company's SBIR/STTR grants in number and amount.

BUSINESS STRATEGY, COMMERCIALIZATION, AND BENEFITS

As a materials development company, T/J Technologies typically faces a 5 to 7 year cycle to develop a material. With materials development, it is unrealistic to expect to advance from the discovery of a nanomaterial to an end product with one cycle of SBIR Phase I, II, and IIB funding. "It just doesn't happen that way," said Ms. Thompson. She noted that the company is now focusing on "moving further up the value chain. We have demonstrated the value of our proprietary technology to the end user in specific applications. When the customer understands that our materials will enable them to enter new markets, the conversations with suppliers in that value chain are much easier."

"T/J Technologies has a pipeline of cutting edge alternative energy technologies which has made the company interesting to larger corporations seeking to get into this market. While forming partnerships can be beneficial, they need to be negotiated with great care. There must be real need for both parties," she said. "The increasing need and interest in alternative energy technologies, and the cutting edge intellectual property that we have developed through the SBIR and ATP programs, have attracted multiple players to us. Small companies have a stronger negotiating position when more than one company competes for their technology."

"It is essential that these small companies—particularly those run by scientists—not be forced into a position where they will be taken advantage of," she said, commenting on public policies that might lead to this outcome. . . . Yes, the requirement for a match for the Phase IIB grant is useful in ruling out dumb technologies which will never be commercialized. But, too often the companies that you go to for matching funds want too much in return—if you go too early to the table. . . . A better approach is to work with them, develop a relationship. Find out what the end customer wants; what the relevant problems are. Let them test your materials. They will spend resources to do so, and they will not do so unless there is real interest. It is usually possible to get them to put a dollar value on those resources."

The significance, Ms. Thompson explained, of obtaining such valuation of resource expenditures comes into play if it is acceptable for meeting SBIR matching funds requirements of the Phase II grant. This is a good indicator of interest in commercialization, she explained, "Because companies are busy; they are not going to give you in-kind support unless they are interested." However, she went on to say that letters of commitment and estimates of resource committed for testing are also very difficult to get. "You might get an agreement from the scientist or business person, only to be shot down by the company's legal department. They may agree to do the testing, but refuse to put it in writing."

T/J Technologies' novel materials offer potential benefits in terms of meeting the need for alternative energy sources that can be used to power automobiles and other vehicles, as well as to meet stationary power needs. The materials offer high-rate performance and reductions in the cost, size, and weight of batteries. The materials also offer environmental benefits in that, unlike most other lithium ion cathodes, T/J Technologies materials do not contain cobalt, an undesirable component from a safety and environmental perspective. In addition, if the company's technology furthers the adoption of alternative energy sources, it stands to yield broad environmental and national security benefits associated with reduced dependency on conventional energy sources.

VIEWS ON THE SBIR PROGRAM AND PROCESSES

Ms. Thompson expressed her views about the SBIR program and its processes. She also made relevant comments about broader public policy concerning government support of innovative companies in the United States. These comments are also summarized.

Different Roles of the SBIR Programs of NSF, DoE, and DoD

"NSF is unique in that it is willing to fund basic materials research," explained Ms. Thompson. "Even though the military is an early adopter customer, a lot of DoD-funded SBIR grants tend to be focused at the systems end versus the basic-materials-research end."

Difficulty in Getting SBIR Grants

"I think it has gotten harder to get SBIR grants," said Ms. Thompson, "because other sources of money have dried up or gotten harder to get. This seems to have resulted in more competition for SBIR grants.

Funding Gap

Early on, there was a problem with the gap between the different SBIR funding phases, noted Ms. Thompson. But the state [Michigan] had a program—which was very short lived but very helpful while it was there—that provided some funding to help keep you going until Phase II came through. She further noted that now the company is big enough and has enough of a variety of funding sources, so sustaining the gap is no longer a problem.

Value of Keeping Phase I Grants as Prerequisite to Phase II

Ms. Thompson sees several reasons to keep the Phase I grant as a prerequisite to Phase II. One concern she expressed was that allowing companies to bypass Phase I may give larger (small) companies, which tend to have more internal funds and more experienced managers, an advantage over smaller (small) companies. The larger companies could self-fund entry level research and be positioned to win more Phase II grants. "Keeping the Phase I requirement may help keep the playing field more level," she said, "because the only companies you are competing with at the Phase II stage are all the others who, like you, won a Phase I."

Size of Grants

"Writing proposals for small amounts of funding is a distraction from the research. So I think I'd rather see fewer but larger Phase II SBIR grants," she noted. "Even with the cost share, a benefit of the ATP is that the money was sufficient to allow us to focus on getting the research done to a stage where we could get commercial interest."

Time Between Solicitations

"Having no more than six months, instead of a year, between solicitations would help," she stated. "It would allow the SBIR program to generate better and more ideas."

Proposal Review

"I think the agencies need more people who have headed small innovative companies as reviewers," stated Ms. Thompson. "I think this because sometimes you get the reviews back and you get two 'excellent' ratings that clearly explain why, and one 'poor' rating with comments that show that the reviewer just doesn't get it. This happens to everybody. It is frustrating for a company that has put a lot of resources into its proposal. It's just your tough luck. There currently is no remedy for this problem."

Reviewer Feedback

Feedback from the reviewer process was described as "very useful." Most agencies were said to give feedback only if the company loses. "It would be helpful to give it to you if you win also as it makes sense to continually improve your work."

Application Process

Ms. Thompson noted that differences in formatting styles among the various agencies can cause extra effort, and there does not seem to be any apparent value in having the differences. However, she also noted that having variations in styles is only a minor problem.

Value of Commercialization Assistance

Ms. Thompson (who herself has an MBA and years of industry experience) found some things about the Dawnbreaker Program useful, but she also saw opportunities for improvement. She found the quality of the assistance provided to be highly variable depending on the particular staff assigned. At the same time, she said, "I think for a scientist who doesn't have anybody who is business oriented, it would be a very good training ground. It would let them know that it is not just the technology that they are betting on. They need to hear this and also learn how to put together a business plan." Assuming the appropriate companies attend the matchmaking event, the program can be very helpful.

Ms. Thompson found that the networking events sponsored by the various agencies could be very beneficial. These allowed her to go and meet companies who are interesting in possible partnering arrangements. T/J's relationship with Lockheed Martin started at one of these events that was held in Michigan.

She also commented that training sessions at the NSF conferences were very useful, particularly the session on patents. She suggested that adding more educational events aimed at business topics could be very helpful to companies. Commenting on NSF's online Matchmaker service, Ms. Thompson said she thought it was more focused on fostering matches with venture capitalists.

Company Site Visits by Agency Program Managers

"Having site visits would help the SBIR program staff get closer to the companies and understand better what they are doing," she said. "SBIR program managers have a tough job dealing with a variety of technologies. Getting out to the companies would help them develop more depth." She gave as an example of an excellent model the company's experience working with ATP's program manager for battery research. "[This ATP program manager] knows everybody who is working in the field. He knows the business. He performs matchmaking very naturally and very skillfully. It is very valuable."

Assumption of Inappropriate Business Models

Ms. Thompson urged that some public policy makers reconsider the business models many appear to hold. "Some policy makers may think a small company like ours can get a Phase I SBIR grant, then a Phase II, then get venture capital and build a plant, and then produce product. There are other business models that can be equally as successful.

Assistance to Innovators

Continuing, Ms. Thompson noted that other countries—providing fierce competition to the United States—have set up their own technology funding programs like the ATP . . . "at a time that our own ATP program is under attack. They are positioning to eat our lunch! Not only are our manufacturing jobs at stake, but now our research jobs are also at stake. We are in a time-critical race. And it is tough to get American companies to invest in a market that is still down the road. So when you look at all these small companies toiling away with SBIR help, and making it in spite of the situation, that is very exciting. And whatever from a policy standpoint we can do to help these companies that are developing new technologies and markets—activities that ensure our economic future and competitiveness—then we should do it."

SUMMARY

This case study features an innovative materials research company facing a relatively long time to commercialization and a need to form partnerships with global companies to reach civilian markets. It shows a company moving up the value chain in order to increase the value it can receive for its technology in an environment where funding is scarce and negotiations difficult. The case illustrates a "building block" strategy, where SBIR grants enabled the company to start and build capacity; that capacity was leveraged into an ATP grant; the ATP grant leveraged the company's ability to go after government research contracts; and research contracts leveraged its ability to form commercial partnerships with much larger companies for testing and demonstrating its materials. The next step is commercialization, which will be achieved through a partner. At the helm of the company is a grant-winning minority woman, during a time that woman-owned and minority-owned businesses have received a relatively low share of SBIR grants.

WaveBand Corporation[18]

THE COMPANY

WaveBand Corporation emerged in 1996 as a spin-off of Physical Optics Corporation (POC). POC is a small, employee-owned company that specializes in optoelectronic solutions and products. WaveBand is one of several companies that POC has spun off. Six POC employees moved to WaveBand to assist with the spin-off, and POC retained major ownership of the new company. According to Ms. Quintana, WaveBand's Director of Business Development and a former POC employee, "a goal of WaveBand was to become truly independent of POC."

Based in Irvine, CA, WaveBand Corporation became known for innovation in the field of millimeter wave (MMW) technologies, including beam-steering antenna and imaging radar systems. In early 2005, the company had 24 employees, half of whom had Ph.Ds. Its revenue in 2004 was $5.5 million.

Today, WaveBand is a wholly owned subsidiary of Sierra Nevada Corporation (SNC). SNC acquired WaveBand Corporation in May 2005. SNC is a rapidly growing innovative systems integrator, located in Sparks, NV, specializing in the design, development, production, installation and servicing of defense electronics engineering systems. Founded in 1963, SNC is the parent company of a group of more than six companies with the following four areas of focus: air traffic control; unmanned aerial vehicle systems; instrumentation, test, and training systems; and intelligence, surveillance, and reconnaissance systems. SNC employs more than 750 employees, and, hence, is not eligible for SBIR grants.

THE TECHNOLOGY AND ITS USES

WaveBand's antennas provide rapid beam steering and beam forming, and they do this without the use of bulky mechanical steering. Rather, the WaveBand antenna technology in one of its implementations relies on a grating formed on a spinning drum to steer the radar beam. In this way, it avoids the use of mechanically moved reflectors, which are slow, and electronically steered phase shifters, which are fast but expensive. In short, WaveBand's antennas were developed to overcome problems of slowness or expense that characterize traditional antennas. Their more advanced antenna technology implementation is designed to meet the need for electronically steerable antennas that are comparable to phased array

---

[18]The following informational sources informed the case study: interview conducted with Ms. Toni Quintana, WaveBand's Director of Business Development, at the U.S. Navy Opportunity Forum, May 2–4, 2005, in Reston, VA; WaveBand's Web site: <*http://www.waveband.com*>; Sierra Nevada Corporation's Web site: <*http:www.sncorp*>; Sierra Nevada Corporation's press release announcing the acquisition of WaveBand; and WaveBand's Department of Defense SBIR Commercialization Report.

WAVEBAND CORPORATION: COMPANY FACTS AT A GLANCE

- **WaveBand Address:** 17152 Armstrong Ave., Irvine, CA 92614
- **Telephone:** 949-253-4019
- **Year Started:** Spun off from Physical Optics Corporation (POC) in 1996; acquired by Sierra Nevada Corporation in 2005
- **Ownership:** Sierra Nevada Corporation, 444 Salomon Circle, Sparks, NV, itself a privately held corporation
- **Revenue:** Approx. $5.5 million in 2004
    — Revenue share from SBIR/STTR grants and contracts: approx. 50 percent
    — Revenue share from sale of products, including DoD sales: approx. 50 percent
- **Number of Employees:** 24 prior to the acquisition
- **SIC:** Primary SIC: 3812 Search, Detection, Navigation, Guidance, Aeronautical, and Nautical Systems and Instruments
  Secondary SIC: N/A
- **Technology Focus:** Millimeter wave (MMW) technologies
- **Application Areas:** Beam-steering antenna and imaging radar systems for aviation, transportation, and security use
- **Funding Sources:** Product sales in military and commercial markets, and federal government grants and contracts
- **Number of SBIR grants:**
    — From NSF: 3 Phase I grants, of which all three went to Phase II and one to Phase IIB
    — From other agencies: 45 Phase I's and 19 Phase II's

antennas in performance, but are highly compact, light weight, robust, with low power needs, without requirements for continuous calibration, and comparatively inexpensive. The antennas are smart; they can be set to provide multiple beams, each steerable. They offer a price advantage 100 times more favorable to buyers than traditional systems.

The antennas can help meet guidance needs for aircraft landing, missile seekers, and surveillance sensors. In civilian markets, they may also be useful for adaptive cruise control on cars, allowing cars to regulate their own speeds in response to traffic congestion. The advantage is that WaveBand's antenna scans, enabling it to cover more area than the fixed-beam radar antennas in current adaptive cruise control systems, and, therefore, enabling it provide the car's computer with more data on which to base its calculations. The millimeter wave radar beam can penetrate fog, rain, or snow, making it ideal for a variety of autonomous guidance and landing systems. WaveBand is working with both the automobile industry and avionics suppliers to develop prototype systems. It is in the process of validating application of the antenna to the Navy fleet.

THE ROLE OF SBIR IN COMPANY FUNDING

When WaveBand spun-out of POC, it continued to work—in a subcontractor role—on some of the SBIR programs its personnel had been involved with prior to the spin-out. Later on, funding for the company came directly from SBIR, according to Ms.Quintana.

At the time of its take-over by SNC, WaveBand was receiving approximately half of its funding from SBIR grants. The remainder came primarily from studies and sales, particularly sales to defense agencies.

Earlier, WaveBand had pursued SBIRs that were mainly aimed at highly focused military objectives. But around 2000, the company made a decision to put increased attention on attracting commercially driven R&D to broaden the technology's applications into nondefense markets, and to deemphasize the attention given to winning highly focused SBIR grants and contracts. It also went through a cycle of Phase 1, Phase 2, and Phase 2B NSF SBIR grants. More recently, it switched back to the highly focused defense SBIRs. According to Ms. Quintana, a result of this strategy was the development not only of technical prowess but also commercial strength for WaveBand.

According to WaveBand's SBIR Commercialization Report to the Department of Defense, "the technologies that WaveBand has developed under Phase II SBIR research are all vital to our commercialization success. Each research project contributed to our commercial and Non-SBIR revenue."

WaveBand received a total of 48 Phase 1 SBIR grants and 22 Phase 2 SBIR grants. It has received SBIR grants mainly from Navy, Army, Air Force, NASA, and DoE, and to a lesser extent from NSF. It received 2 Phase 1 STTR grants and 2 Phase 2 STTR grants. The amount the company has received in SBIR/STTR grants since 1996 totals $19.8 Million. Table App-D-8 summarizes the company's SBIR/STTR grants in number and amount. Eligibility for SBIR grants has ended with WaveBand's acquisition by SNC, a company with more than 500 employees.

BUSINESS STRATEGY, COMMERCIALIZATION, AND BENEFITS

WaveBand's beam-steering antennas are expected to find about equal market applications in civilian and military applications. Civilian applications are expected to include landing systems for commercial jets and helicopters, as well as intelligent cruise control for automobiles. Military applications include landing systems for military aircraft, guidance for missile seekers, and surveillance systems for Navy ships.

WaveBand has sold test units of its MMW steering antenna to major auto-

TABLE App-D-8 Waveband Corporation: SBIR/STTR Grants from NSF and Other Agencies

NSF Awards	Number	Amount ($)	Other Agency Awards	Number	Amount ($)	Total Number	Total Amount ($)
SBIR Phase I	3	274,946	SBIR Phase I	45	4,479,906	48	4,754,852
SBIR Phase II[a]	2	892,291	SBIR Phase II[a]	18	12,167,924	20	13,060,215
SBIR Phase IIB/ Enhancements	1	249,961	SBIR Phase IIB/ Enhancements	1	249,488	2	499,449
STTR Phase I	—	—	STTR Phase I	2	199,984	2	199,984
STTR Phase II	—	—	STTR Phase II	2	1,249,924	2	1,249,924
STTR Phase IIB/ Enhancements	—	—	STTR Phase IIB/ Enhancements	—	—	—	—
Totals	6	1,417,198		68	18,347,226	74	19,764,424

[a]Exludes Phase IIB/Enhancement awards which are listed separately.

SOURCE: WaveBand Corporation.

motive companies and to a major manufacturer of avionics instrumentation. MMW steering antenna units have been installed on unmanned ground vehicles; a unit has been demonstrated for a railroad grade-crossing monitoring system; it has been demonstrated for bird detection along airport runways; and it has been demonstrated to "see through" various atmospheric conditions to facilitate aircraft landing.

WaveBand's acquisition by SNC has reportedly not changed WaveBand's technology direction. However, the acquisition has changed its funding situation. Without the ability to receive SBIR grants and contracts to support its research, WaveBand is expected to develop more of a product orientation, even as it attempts to stay on its R&D path. The parent company is expected to provide some direct R&D funding, it is including WaveBand as a subcontractor on military contracts, and it is expected to encourage the group to attract R&D funding from other sources. Accelerated growth of sales revenue is expected to provide another source of support for continued R&D.

WaveBand's antennas offer performance benefits and cost advantages. They stand to increase automotive and aviation safety in civilian and defense applications and improve the effectiveness of a variety of military-defense systems.

VIEWS ABOUT THE SBIR PROGRAM AND ITS PROCESSES

Ms. Quintana made the following several observations about the program and its processes, some of which focused on the NSF program:

Reporting Requirements

Each agency has its own reporting requirements, noted Ms. Quintana. "Uniformity in reporting would help."

Financing Gap

The existence of a financing gap varies by agency SBIR program, according to Ms. Quintana. The Army bridges over the gap, making the transition from Phase I to Phase II relatively seamless. The Navy has a funding gap between phases of its program, making it very hard for small companies without alternative financing.

Requirements for Phase IIB Matching Funds

Air Force and Navy require that matching funds for Phase IIB grants be contingent on approval of the contract, as specified in a letter by the organization to provide the match. This causes trouble: One problem is that the timing is tricky. It requires that a contingency pledge be made not too soon and not too late—timing which may not suit the potential provider of the matching funds. Another problem is that many companies who would provide matching funds do not think in terms of contingencies—they decide either to provide the funding or not to provide it. In contrast, NSF does not require that the matching funds be expressed as a contingency. Therefore, companies are able to use a purchase order or sales revenue in the bank. NSF's approach, according to Ms. Quintana, makes it easier for companies to comply with the Phase IIB matching funds requirement.

Solicitation Cycles

Unlike some of the other interviewees, Ms. Quintana does not believe that more solicitations each year would be advantageous. In fact, she sees more solicitations and more changes in topics as a potential burden, as company staff must be constantly monitoring the situation and trying to respond to the changes. Once yearly posting of topics allows companies more time to plan their research programs around the announced topics.

SUMMARY

This case study illustrates the role played by SBIR grants in the creation of a company as a spin-off of another small company. It also shows how the company used SBIR and other research funding sources to develop a portfolio of technologies attractive to a larger company that recently acquired it. The case illustrates the dual special roles played by highly targeted SBIR grants from defense agencies and by less targeted grants from NSF. It describes SBIR-funded innovations important from both a military standpoint and important in civilian markets. The company has used its SBIR funding to develop antennas that rely on an electron-hole plasma grating to provide rapid beam steering and beam forming without the use of bulky mechanically moved reflectors, which are slow, and without electronically steered phase shifters, which are fast but expensive. WaveBand's antennas reportedly offer a price advantage 100 times more favorable to buyers than traditional systems. Approximately half of WaveBand's revenue in 2004 came from SBIR grants and contracts. The company provided helpful comments for improving the SBIR program.

Appendix E

Structured Interview Guide Used in Coryell Study of Phase IIB Grants

NSF SBIR COMMERCIAL RESULTS INTERVIEW

1. Identifying information
Company Name: <text>
Current Address: <text>
Year Company Founded: <year>
Year of First NSF SBIR/STTR grant: <year>
President/CEO: <text>
Contact Name: <text>
Title: <text>
Telephone: <text>
Fax: <text>
E-mail: <text>
Company Web address: <text>
Date (or dates) of interview: <month xx, yyyy>
Designated lead for marketing in co., e.g., President, Marketing VP, other: <text>

2. Benefits
Describe benefits to the company, derived from NSF SBIR grant(s) by checking
  any of the following items: <Y/N>
Enabled research that otherwise would not have been undertaken
Enabled significantly faster development of product or proof of concept
Opened new market opportunities with new applications/uses for products
Enabled access to critical private equity capital
Enabled access to critical private debt capital
Enabled licensing agreements for intellectual property

Led to key strategic alliance(s) with other firm(s)
Led to new business segments of the firm and/or spin-offs
Enabled hiring or retention of key professionals/technical staff
Other benefits: _____ <text> _____(explain)
Comments: <text> _____

3. Commercial Products

Have you had commercial sales, products, processes, or services (excluding any government R&D grants) that you would attribute, directly or indirectly, to one or more NSF SBIR grants? <Y/N> Yes___ No___

In your opinion, if there had there been no SBIR program, would the product, process, or service been developed? If so, within the same time frame? (Use table below.)

Please list in the table below each product, service, or process developed that has generated revenues of at least $100,000, and describe each briefly under "Comments." Please show estimates of sales for each product, service, or process, accounting accuracy not needed. (Do not include R&D or IP revenues here.)

Product / Product Line	Year SBIR Phase II Start	Year First Sales	Develop Anyway (Y/N)	Within Same Time Frame (Y/N)	Estimate Sales in Current FY	Estimate Cumulative Sales
<text>	<year>	<dollars>	<Y/N>	<Y/N>	<dollars>	<dollars>
<text>	<year>	<dollars>	<Y/N>	<Y/N>	<dollars>	<dollars>

Comments (including product descriptions): <text>

_____

_____

4. Company Sales

Please provide some information on your company sales and those sales that relate to SBIR. Please identify specific sources of revenue, and if confidential, only identify by category and not name the entity.

List major/significant customers for NSF/SBIR-related products, services, or processes under "Comments."

	Total Company Sales (in dollars)	Attributed to Internally Funded R&D or Acquired Technology	Attributed to NSF/ SBIR	Attributed to Other Agency SBIR
Total Actual Sales for Last Fiscal Year	<dollars>	$ amount: <dollars> % sales:	$ amount: <dollars> % sales:	$ amount: <dollars> % sales:
Total of Cumulative Sales since (insert date)	<dollars>	$ amount: <dollars> % sales: <percentage>	$ amount: <dollars> % sales: <percentage>	$ amount: <dollars> % sales: <percentage>

Comments (including revenue sources): <text>

_____

_____

Are there any products, services, or processes in your pipeline scheduled to generate new sales within the next 12 months? If so, projected lifetime sales? <text>

_____

_____

EXPANDED SECTION 4. Sales/Revenues (Highlighted in italic for emphasis) (Developed in response to Advisory Committee feedback and administered to first 15 companies, as well as additional companies interviewed. Shown highlighted in italics for emphasis.)

a. Company Sales
Please provide estimates of sales for your company in all of its parts, including all divisions and majority-owned subsidiaries. Do not include joint ventures, spin-offs, licensees, and strategic partners. Use table below. If not confidential, please enter under "Comments" the names of major/significant customers for NSF/SBIR-related products, services, or processes as well as IP recipients.

	Est. product, service, and process sales by company	Est. royalties to company and sales of IP	Est. company R&D sales to government and industry	Total sales of company
Actual sales for most recent FY	<dollars>	<dollars>	<dollars>	<dollars>
Cumulative sales since ____ (insert year)	<dollars>	<dollars>	<dollars>	<dollars>

Comments (including customer names): <text>

_____

_____

b. New Sales in Economy

Please provide some information on sales of those companies that relate to your R&D and technologies. Consider only first-tier users of your technologies, and disregard wider uses and multiplier effects. Use table below. If not confidential, please enter under "Comments" the names of firms that are included in each of category outside your company.

	Total sales of company	Est. sales by joint ventures	Est. sales by spin-off firms	Est. sales by licensees	Est. sales by other partners	Total economy sales
Actual sales for most recent FY	<dollars>	<dollars>	<dollars>	<dollars>	<dollars>	<dollars>
Cumulative sales since ____ (insert year)	<dollars>	<dollars>	<dollars>	<dollars>	<dollars>	<dollars>

Comments (including firm names): <text>

_____

_____

c. Technology Sources of Sales in Economy

Please identify funding sources of technology that generated revenues in the economy identified above, and apportion the contribution to total economy sales by each funding source category. Please list under "Comments" the SBIR agencies outside NSF that have been principal R&D funding sources.

	Total economy sales	Attributed to company funded R&D or acquired technology	Attributed to NSF/ SBIR	Attributed to other agency SBIR	Unattributed to company-performed R&D
Actual sales for most recent FY	*<dollars>*	*$ amount:* *<dollars>* *% sales:*	*$ amount:* *<dollars>* *% sales:*	*$ amount:* *<dollars>* *% sales:*	*$ amount:* *<dollars>* *% sales:*
Cumulative sales since ____ *(insert year)*	*<dollars>*	*$ amount:* *<dollars>* *% sales:* *<percentage>*	*$ amount:* *<dollars>* *% sales:* *<percentage>*	*$ amount:* *<dollars>* *% sales:* *<percentage>*	*$ amount:* *<dollars>* *% sales:* *<percentage>*

Comments (including SBIR funding agencies): <text>

_____

_____

d. Next Year

Are there any products, services, or processes in your pipeline scheduled to generate new sales within the next 12 months? If so, what are their projected lifetime sales? <text>

_____

_____

5. Investments:

List sources and dollar amounts of investment capital to date that are attributable, directly or indirectly, to results gained from SBIR grants.

Identify specific sources of investment, and if confidential, only identify by category and not name the entity.

Source	Total $ for Company	NSF/SBIR $	Other SBIR $
Friends, Family, Founders	<dollars>	<dollars>	<dollars>
High Net Worth Individuals—"Angels"	<dollars>	<dollars>	<dollars>
Organic Growth— "Bootstrapping"	<dollars>	<dollars>	<dollars>
Customer Funding	<dollars>	<dollars>	<dollars>
Strategic Partners	<dollars>	<dollars>	<dollars>
Venture Capital	<dollars>	<dollars>	<dollars>
Governmental Entities (Federal, state, local)	<dollars>	<dollars>	<dollars>
Initial Public Offering (IPO)	<dollars>	<dollars>	<dollars>
Later Stage Stock Sales	<dollars>	<dollars>	<dollars>
Other Entities (specify)	<dollars>	<dollars>	<dollars>
Total:	<dollars>	<dollars>	<dollars>

Comments: (Sources, including City/State) <text>

_____

_____

6. Number of Patents:

	Total Company	NSF/SBIR Related	Other SBIR Related
U.S. Issued Patents	<number>	<number>	<number>
Issued Foreign Patents	<number>	<number>	<number>
Applied for but not issued —U.S.	<number>	<number>	<number>
Applied for but not issued —Foreign	<number>	<number>	<number>
Total Patents received and applied for	<number>	<number>	<number>

Comments (assess strength of company IP): <text>

_____

_____

7. Licenses/Other Strategic Alliance Agreements

Have you licensed SBIR-related technology to other entities? <Y/N>

    Yes__ No____

A. Number of Licensing Agreements

	Total Company	NSF/SBIR Related	Other Agency SBIR Related
# IP Protected Technologies Licensed	<number>	<number>	<number>
# of Licensing Agreements with U.S. Firms	<number>	<number>	<number>
# of Licensing Agreements with Foreign-Owned Firms	<number>	<number>	<number>
# of Marketing or Other Agreements	<number>	<number>	<number>
# IP sales	<number>	<number>	<number>

B. Dollar Value of Licensing Agreements

	Total $ for Company	NSF/SBIR Related $	Other Agency SBIR Related $
U.S. Licensing Revenues for Last Fiscal Year (Best estimate of actual, not projected)	<dollars>	<dollars>	<dollars>
Foreign Licensing Revenues for Last Fiscal Year (Best estimate of actual, not projected)	<dollars>	<dollars>	<dollars>
Cumulative Licensing & Royalty Revenue from U.S. Firms	<dollars>	<dollars>	<dollars>
Cumulative Licensing & Royalty Revenue from Foreign-Owned Firms	<dollars>	<dollars>	<dollars>

Comments (Licensees, City/State, Nationality): <text>

_____

_____

8. Joint Ventures/Spin-offs/Acquisitions

Has SBIR technology developed by your firm been responsible for or led to any joint ventures, spin-offs, or acquisitions involving your firm? <Y/N> Yes__ No___

If "Yes," provide information on related cumulative products, sales, investment, employment, and founding year date for each, if possible.

JV/ Spin-offs/ Acquisitions	Company Partner/ Spin-off	Product Industry or Sector	Year Venture Established or Acquired	Estimated Cumulative Sales	Estimated Total Investment	Current # of New Jobs Created to Date
<text>	<text>	<text>	<year>	<dollars>	<dollars>	<number>
<text>	<text>	<text>	<year>	<dollars>	<dollars>	<number>
<text>	<text>	<text>	<year>	<dollars>	<dollars>	<number>

Comments: <text>

_____

_____

9. Employment

Has your company experienced an increase in the number of employees as a direct result of work related to SBIR Grants? <Y/N> Yes __ No ___

Employment at time of first SBIR/STTR grant:	<number>
Current employment:	<number>

Comments: <text>

_____

_____

10. University and Research Laboratory Collaborations:

Have you or your company collaborated with any universities either in the U.S. or internationally on NSF SBIR research? <Y/N> Yes ____ No _____

Have you or your company collaborated with any National Laboratories or private research laboratories either in the U.S. or internationally on NSF SBIR research? <Y/N> Yes ____ No _____

(Please enter numbers in table below.)

SBIR Projects	Consultant/Subcontract/Other
Current projects with collaboration (estimate number)	\<number\>
Previous projects with collaboration (estimate number prior to current calendar year)	\<number\>
Names of collaborating academic institutions on NSF research	\<number\>
Names of collaborating national laboratories and private research labs on NSF research	\<number\>

Comments: (Collaborating institutions, i.e., universities) \<text\>

_____

_____

11. Grants Received

Number and dollar amounts of SBIR Grants Received. (Use table below.)

Number and dollar amounts of STTR Grants Received. (Use table below.)

NSF Grants	Number of Grants	Dollar Amount
SBIR Phase I	\<number\>	\<dollars\>
SBIR Phase II	\<number\>	\<dollars\>
Total Grants:	\<number\>	\<dollars\>
STTR Phase I	\<number\>	\<dollars\>
STTR Phase II	\<number\>	\<dollars\>
Total Grants:	\<number\>	\<dollars\>

12. Narrative questions: Open-ended opportunity to expand on questions and for unstructured feedback.

A. Was an SBIR grant instrumental in the founding or survival of the company at any point? \<text\>

B. Are individual employees within your company rewarded or formally recognized for completing proposals and receiving SBIR grants? \<text\>

C. Is there anything you would like to add that has not been covered in this survey? \<text\>

Appendix F

Required NSF Postgrant Annual Commercialization Report (Deemed Ineffective by the NSF)

(Reference: <*http://www.eng.nsf.gov/sbir/annual_commercialization.htm*>)

Prepare the Annual Commercialization Report ensuring that all information is provided. The Annual Commercialization Report is required to be submitted annually for 5 years following the completion of the Phase II grant.

Attach a completed Annual Commercialization Report (a Blank Annual Commercialization Report Format is provided) that should not exceed 10 pages and consists of the following five parts:

- **Part 1: Company Data** (basic data about the company and the SBIR project)
  SBIR Grant Number:
  Name of Company:
  Company Address:
  Telephone Number:
  Fax Number:
  Email:
  Name and Title of the person preparing the report:

- **Part 2: List any products and/or processes currently in the marketplace, or patents resulting from the SBIR project.**

- **Part 3: What investments/activities were pursued to accomplish commercialization of the SBIR project?**

- **Part 4: Please furnish the revenues received from the commercialization of this SBIR project: include Sales, Manufacturing, Product Licensing, Royalties, Consulting, Contracts, Other.**

- **Part 5: Company Employment and Revenue Data**

	Start of SBIR Grant	Current
Number of Employees		
Revenue (Total $s)		
Percent of Revenue from SBIRs (from all agencies)		

IMPORTANT NOTE: Failure to submit annual commercialization reports may deter selection for future grants.

The remaining two components (listed below) may be used if a company has information to report.

- Click on Publications and Products (if applicable). (Report scientific articles or papers appearing in scientific, technical, or professional journals. Include any publication that will be published in a proceedings of a scientific society, a conference, or the like. Report any significant product, other than a publication (or electronic publication), that you have developed under this project.)

- Click on Contributions (if applicable). (A primary function of NSF support for research and education—along with training of people—is to help build a base of knowledge, theory, and technique in the relevant fields. That base will be drawn on many times and far into the future, often in ways that cannot be specifically predicted to meet the needs of the nation and of people. Most NSF-supported research and education projects should be producing contributions to the base of knowledge and technique in the immediately relevant field(s).)

Appendix G

Bibliography

Acs, Z., and D. Audretsch. 1991. *Innovation and Small Firms*. Cambridge, MA: MIT Press.

Advanced Technology Program. 2001. *Performance of 50 Completed ATP Projects, Status Report 2*. National Institute of Standards and Technology Special Publication 950-2. Washington, DC: Advanced Technology Program/National Institute of Standards and Technology/U.S. Department of Commerce.

Alic, John A., Lewis Branscomb, Harvey Brooks, Ashton B. Carter, and Gerald L. Epstein. 1992. *Beyond Spinoff: Military and Commercial Technologies in a Changing World*. Boston, MA: Harvard Business School Press.

American Association for the Advancement of Science. "R&D Funding Update on NSF in the FY2007." Available online at <*http://www.aaas.org/spp/rd/nsf07hf1.pdf*>.

Archibald, R., and D. Finifter. 2000. "Evaluation of the Department of Defense Small Business Innovation Research Program and the Fast Track Initiative: A Balanced Approach." In National Research Council. *The Small Business Innovation Research Program: An Assessment of the Department of Defense Fast Track Initiative*. Charles W. Wessner, ed. Washington, DC: National Academy Press.

Arrow, Kenneth. 1962. "Economic welfare and the allocation of resources for Invention." Pp. 609–625 in *The Rate and Direction of Inventive Activity: Economic and Social Factors*. Princeton, NJ: Princeton University Press.

Arrow, Kenneth. 1973. "The theory of discrimination." Pp. 3–31 in *Discrimination in Labor Market*. Orley Ashenfelter and Albert Rees, eds. Princeton, NJ: Princeton University Press.

Audretsch, David B. 1995. *Innovation and Industry Evolution*. Cambridge, MA: MIT Press.

Audretsch, David B., and Maryann P. Feldman. 1996. "R&D spillovers and the geography of innovation and production." *American Economic Review* 86(3):630–640.

Audretsch, David B., and Paula E. Stephan. 1996. "Company-scientist locational links: The case of biotechnology." *American Economic Review* 86(3):641–642.

Audretsch, D., and R. Thurik. 1999. *Innovation, Industry Evolution, and Employment*. Cambridge, MA: MIT Press.

Baker, Alan. No date. "Commercialization Support at NSF." Draft.

Barfield, C., and W. Schambra, eds. 1986. *The Politics of Industrial Policy*. Washington, DC: American Enterprise Institute for Public Policy Research.

Baron, Jonathan. 1998. "DoD SBIR/STTR Program Manager." Comments at the Methodology Workshop on the Assessment of Current SBIR Program Initiatives, Washington, DC, October.

Barry, C. B. 1994. "New directions in research on venture capital finance." *Financial Management* 23 (Autumn):3–15.

Bator, Francis. 1958. "The anatomy of market failure." *Quarterly Journal of Economics* 72:351–379.

Bingham, R. 1998. *Industrial Policy American Style: From Hamilton to HDTV.* New York: M.E. Sharpe.

Birch, D. 1981. "Who Creates Jobs." *The Public Interest* 65 (Fall):3–14.

Branscomb, Lewis M. 2000. *Managing Technical Risk: Understanding Private Sector Decision Making on Early Stage Technology Based Projects.* Washington, DC: Department of Commerce/National Institute of Standards and Technology.

Branscomb, Lewis M., and Philip E. Auerswald. 2001. *Taking Technical Risks: How Innovators, Managers, and Investors Manage Risk in High-Tech Innovations,* Cambridge, MA: MIT Press.

Branscomb, Lewis M., and J. Keller. 1998. *Investing in Innovation: Creating a Research and Innovation Policy.* Cambridge, MA: MIT Press.

Brav, A., and P. A. Gompers. 1997. "Myth or reality?: Long-run underperformance of initial public offerings; Evidence from venture capital and nonventure capital-backed IPOs." *Journal of Finance* 52:1791–1821.

Brodd, R. J. 2005. *Factors Affecting U.S. Production Decisions: Why Are There No Volume Lithium-Ion Battery Manufacturers in the United States?* ATP Working Paper No. 05-01, June 2005.

Brown, G., and Turner J. 1999. "Reworking the Federal Role in Small Business Research." *Issues in Science and Technology* XV, no. 4 (Summer).

Bush, Vannevar. 1946. *Science—the Endless Frontier.* Republished in 1960 by U.S. National Science Foundation, Washington, DC.

Cassell, G. 2004. "Setting Realistic Expectations for Success." In National Research Council. *SBIR: Program Diversity and Assessment Challenges.* Charles W. Wessner, ed. Washington, DC: The National Academies Press.

Caves, Richard E. 1998. "Industrial organization and new findings on the turnover and mobility of firms." *Journal of Economic Literature* 36(4):1947–1982.

Clinton, William Jefferson. 1994. *Economic Report of the President.* Washington, DC: U.S. Government Printing Office.

Clinton, William Jefferson. 1994. *The State of Small Business.* Washington, DC: U.S. Government Printing Office.

Coburn, C., and D. Bergland. 1995. *Partnerships: A Compendium of State and Federal Cooperative Technology Programs.* Columbus, OH: Battelle.

Cohen, L. R., and R. G. Noll. 1991. *The Technology Pork Barrel.* Washington, DC: The Brookings Institution.

Congressional Commission on the Advancement of Women and Minorities in Science, Engineering, and Technology Development. 2000. *Land of Plenty: Diversity as America's Competitive Edge in Science, Engineering and Technology.* Washington, DC: National Science Foundation/U.S. Government Printing Office.

Cooper, R.G. 2001. *Winning at New Products: Accelerating the process from idea to launch.* In Dawnbreaker, Inc. 2005. "The Phase III Challenge: Commercialization Assistance Programs 1990–2005." White paper. July 15.

Council of Economic Advisers. 1995. *Supporting Research and Development to Promote Economic Growth: The Federal Government's Role.* Washington, DC.

Cramer, Reid. 2000. "Patterns of Firm Participation in the Small Business Innovation Research Program in Southwestern and Mountain States." In National Research Council. 2000. *The Small Business Innovation Research Program: An Assessment of the Department of Defense Fast Track Initiative.* Charles W. Wessner, ed. Washington, DC: National Academy Press.

Davis, S. J., J. Haltiwanger, and S. Schuh. 1994. "Small Business and Job Creation: Dissecting the Myth and Reassessing the Facts," *Business Economics* 29(3):113–122.

Dawnbreaker, Inc. 2005. "The Phase III Challenge: Commercialization Assistance Programs 1990–2005." White paper. July 15.

Dertouzos. 1989. *Made in America: The MIT Commission on Industrial Productivity.* Cambridge, MA: MIT Press.

DoE Opportunity Forum. 2005. "Partnering and Investment Opportunities for the Future." Tysons Corner, VA. October 24–25.

Eckstein, 1984. *DRI Report on U.S. Manufacturing Industries.* New York: McGraw Hill.

Eisinger, P. K. 1988. *The Rise of the Entrepreneurial State: State and Local Economic Development Policy in the United State.* Madison, WI: University of Wisconsin Press.

Feldman, Maryann P. 1994a. "Knowledge complementarity and innovation." *Small Business Economics* 6(5):363–372.

Feldman, Maryann P. 1994b. *The Geography of Knowledge.* Boston, MA: Kluwer Academic.

Feldman, M. P., and M. R. Kelley. 2001. *Winning an Award from the Advanced Technology Program: Pursuing R&D Strategies in the Public Interest and Benefiting from a Halo Effect.* NISTIR 6577. Washington, DC: Advanced Technology Program/National Institute of Standards and Technology/U.S. Department of Commerce.

Fenn, G. W., N. Liang, and S. Prowse. 1995. *The Economics of the Private Equity Market.* Washington, DC: Board of Governors of the Federal Reserve System.

Flamm, K. 1988. *Creating the Computer.* Washington, DC: The Brookings Institution.

Freear, J., and W. E. Wetzel Jr. 1990. "Who bankrolls high-tech entrepreneurs?" *Journal of Business Venturing* 5:77–89.

Freeman, Chris, and Luc Soete. 1997. *The Economics of Industrial Innovation.* Cambridge, MA: MIT Press.

Galbraith, J. K. 1957. *The New Industrial State.* Boston: Houghton Mifflin.

Geroski, Paul A. 1995. "What do we know about entry?" *International Journal of Industrial Organization* 13(4):421–440.

Gompers, P. A. 1995. "Optimal investment, monitoring, and the staging of venture capital." *Journal of Finance* 50:1461–1489.

Gompers, P. A., and J. Lerner. 1996. "The use of covenants: An empirical analysis of venture partnership agreements." *Journal of Law and Economics* 39:463–498.

Gompers, P. A., and J. Lerner. 1998. "Capital formation and investment in venture markets: A report to the NBER and the Advanced Technology Program." Unpublished working paper. Harvard University.

Gompers, P. A., and J. Lerner. 1998. "What drives venture capital fund-raising?" Unpublished working paper. Harvard University.

Gompers, P. A., and J. Lerner. 1999. "An analysis of compensation in the U.S. venture capital partnership." *Journal of Financial Economics* 51(1):3–7.

Gompers, P. A., and J. Lerner. 1999. *The Venture Cycle.* Cambridge, MA: MIT Press.

Good, M. L. 1995. Prepared testimony before the Senate Commerce, Science, and Transportation Committee, Subcommittee on Science, Technology, and Space (photocopy, U.S. Department of Commerce).

Goodnight, J. 2003. Presentation at National Research Council Symposium. "The Small Business Innovation Research Program: Identifying Best Practice." Washington, DC May 28.

Graham, O. L. 1992. *Losing Time: The Industrial Policy Debate.* Cambridge, MA: Harvard University Press.

Greenwald, B. C., J. E. Stiglitz, and A. Weiss. 1984. "Information imperfections in the capital market and macroeconomic fluctuations." *American Economic Review Papers and Proceedings* 74:194–199.

Griliches, Z. 1990. *The Search for R&D Spillovers.* Cambridge, MA: Harvard University Press.

Hall, Bronwyn H. 1992. "Investment and research and development: Does the source of financing matter?" Working Paper No. 92–194, Department of Economics/University of California at Berkeley.

Hall, Bronwyn H. 1993. "Industrial research during the 1980s: Did the rate of return fall?" Brookings Papers: *Microeconomics* 2:289–343.

Hamberg, Dan. 1963. "Invention in the industrial research laboratory." *Journal of Political Economy* (April):95–115.

Hao, K. Y., and A. B. Jaffe. 1993. "Effect of liquidity on firms' R&D spending." *Economics of Innovation and New Technology* 2:275–282.

Hebert, Robert F., and Albert N. Link. 1989. "In search of the meaning of entrepreneurship." *Small Business Economics* 1(1):39–49.

Himmelberg, C. P., and B. C. Petersen. 1994. "R&D and internal finance: A panel study of small firms in high-tech industries." *Review of Economics and Statistics* 76:38–51.

Hubbard, R. G. 1998. "Capital-market imperfections and investment." *Journal of Economic Literature* 36:193–225.

Huntsman, B., and J. P. Hoban Jr. 1980. "Investment in new enterprise: Some empirical observations on risk, return, and market structure." *Financial Management* 9 (Summer):44–51.

Jaffe, A. B. 1996. "Economic Analysis of Research Spillovers: Implications for the Advanced Technology Program." Washington, DC: Advanced Technology Program/National Institute of Standards and Technology/U.S. Department of Commerce).

Jaffe, A. B. 1998. "Economic Analysis of Research Spillovers: Implications for the Advanced Technology Program." Washington, DC: Advanced Technology Program/National Institute of Standards and Technology/U.S. Department of Commerce.

Jaffe, A. B. 1998. "The importance of 'spillovers' in the policy mission of the Advanced Technology Program." *Journal of Technology Transfer* (Summer).

Jewkes, J., D. Sawers, and R. Stillerman. 1958. *The Sources of Invention*. New York: St. Martin's Press.

Kauffman Foundation. About the Foundation. Available online at <*http://www.kauffman.org/foundation.cfm*>.

Kleinman, D. L. 1995. *Politics on the Endless Frontier: Postwar Research Policy in the United States*. Durham, NC: Duke University Press.

Kortum, Samuel, and Josh Lerner. 1998. "Does Venture Capital Spur Innovation?" NBER Working Paper No. 6846, National Bureau of Economic Research.

Krugman, P. 1990. *Rethinking International Trade*. Cambridge, MA: MIT Press.

Krugman, P. 1991. *Geography and Trade*. Cambridge, MA: MIT Press.

Langlois, Richard N., and Paul L. Robertson. 1996. "Stop Crying over Spilt Knowledge: A Critical Look at the Theory of Spillovers and Technical Change." Paper prepared for the MERIT Conference on Innovation, Evolution, and Technology. Maastricht, Netherlands, August 25–27.

Lebow, I. 1995. *Information Highways and Byways: From the Telegraph to the 21st Century*. New York: Institute of Electrical and Electronic Engineering.

Lerner, J. 1994. "The syndication of venture capital investments." *Financial Management* 23 (Autumn):16–27.

Lerner, J. 1995. "Venture capital and the oversight of private firms." *Journal of Finance* 50:301–318.

Lerner, J. 1996. "The government as venture capitalist: The long-run effects of the SBIR program." Working Paper No. 5753, National Bureau of Economic Research.

Lerner, J. 1998. "Angel financing and public policy: An overview." *Journal of Banking and Finance* 22(6–8):773–784.

Lerner, J. 1999. "The government as venture capitalist: The long-run effects of the SBIR program." *Journal of Business* 72(3):285–297.

Lerner, J. 1999. "Public venture capital: Rationales and evaluation." In *The SBIR Program: Challenges and Opportunities*. Washington, DC: National Academy Press.

Levy, D. M., and N. Terleckyk. 1983. "Effects of government R&D on private R&D investment and productivity: A macroeconomic analysis." *Bell Journal of Economics* 14:551–561.

Liles, P. 1977. *Sustaining the Venture Capital Firm*. Cambridge, MA: Management Analysis Center.

Link, Albert N. 1998. "Public/Private Partnerships as a Tool to Support Industrial R&D: Experiences in the United States." Paper prepared for the working group on Innovation and Technology Policy of the OECD Committee for Science and Technology Policy, Paris.

Link, Albert N., and John Rees. 1990. "Firm size, university based research and the returns to R&D." *Small Business Economics* 2(1):25–32.

Link, Albert N., and John T. Scott. 1998. "Assessing the infrastructural needs of a technology-based service sector: A new approach to technology policy planning." *STI Review* 22:171–207.

Link, Albert N., and John T. Scott. 1998. *Overcoming Market Failure: A Case Study of the ATP Focused Program on Technologies for the Integration of Manufacturing Applications (TIMA)*. Draft final report submitted to the Advanced Technology Program. Gaithersburg, MD: National Institute of Technology. October.

Link, Albert N., and John T. Scott. 1998. *Public Accountability: Evaluating Technology-Based Institutions*. Norwell, MA: Kluwer Academic.

Longini, P. 2003. "Hot buttons for NSF SBIR Research Funds," Pittsburgh Technology Council, *TechyVent*. November 27.

Malone, T. 1995. *The Microprocessor: A Biography*. Hamburg, Germany: Springer Verlag/Telos.

Mansfield, E. 1985. "How Fast Does New Industrial Technology Leak Out?" *Journal of Industrial Economics* 34(2).

Mansfield, E. 1996. *Estimating Social and Private Returns from Innovations Based on the Advanced Technology Program: Problems and Opportunities*. Unpublished report.

Mansfield, E., J. Rapoport, A. Romeo, S. Wagner, and G. Beardsley. 1977. "Social and private rates of return from industrial innovations." *Quarterly Journal of Economics* 91:221–240.

Martin, Justin. 2002. "David Birch." *Fortune Small Business* (December 1).

McCraw, T. 1986. "Mercantilism and the Market: Antecedents of American Industrial Policy." In C. Barfield and W. Schambra, eds. *The Politics of Industrial Policy*. Washington, DC: American Enterprise Institute for Public Policy Research.

Mervis, Jeffrey D. 1996. "A $1 Billion 'Tax' on R&D Funds." *Science* 272:942–944.

Moore, D. 2004. "Turning Failure into Success." In National Research Council. *The Small Business Innovation Research Program: Program Diversity and Assessment Challenges*. Charles W. Wessner, ed. Washington, DC: The National Academies Press.

Mowery, D. 1998. "Collaborative R&D: how effective is it?" *Issues in Science and Technology* (Fall):37–44.

Mowery, D., ed. 1999. *U.S. Industry in 2000: Studies in Competitive Performance*. Washington, DC: National Academy Press.

Mowery, D., and N. Rosenberg. 1989. *Technology and the Pursuit of Economic Growth*. New York: Cambridge University Press.

Mowery, D., and N. Rosenberg. 1998. *Paths of Innovation: Technological Change in 20th Century America*. New York: Cambridge University Press.

Myers, S., R. L. Stern, and M. L. Rorke. 1983. *A Study of the Small Business Innovation Research Program*. Lake Forest, IL: Mohawk Research Corporation.

Myers, S. C., and N. Majluf. 1984. "Corporate financing and investment decisions when firms have information that investors do not have." *Journal of Financial Economics* 13:187–221.

National Research Council. 1986. *The Positive Sum Strategy: Harnessing Technology for Economic Growth*. Washington, DC: National Academy Press.

National Research Council. 1987. *Semiconductor Industry and the National Laboratories: Part of a National Strategy*. Washington, DC: National Academy Press.

National Research Council. 1991. *Mathematical Sciences, Technology, and Economic Competitiveness*. James G. Glimm, ed. Washington, DC: National Academy Press.

National Research Council. 1992. *The Government Role in Civilian Technology: Building a New Alliance*. Washington, DC: National Academy Press.

National Research Council. 1995. *Allocating Federal Funds for R&D*. Washington, DC: National Academy Press.

National Research Council. 1996. *Conflict and Cooperation in National Competition for High-Technology Industry*. Washington, DC: National Academy Press.

National Research Council. 1997. *Review of the Research Program of the Partnership for a New Generation of Vehicles: Third Report*. Washington, DC: National Academy Press.

National Research Council. 1999. *The Advanced Technology Program: Challenges and Opportunities*. Charles W. Wessner, ed. Washington, DC: National Academy Press.

National Research Council. 1999. *Funding a Revolution: Government Support for Computing Research*. Washington, DC: National Academy Press.

National Research Council. 1999. *Industry-Laboratory Partnerships: A Review of the Sandia Science and Technology Park Initiative*. Charles W. Wessner, ed. Washington, DC: National Academy Press.

National Research Council. 1999. *New Vistas in Transatlantic Science and Technology Cooperation*. Charles W. Wessner, ed. Washington, DC: National Academy Press.

National Research Council. 1999. *The Small Business Innovation Research Program: Challenges and Opportunities*. Charles W. Wessner, ed. Washington, DC: National Academy Press.

National Research Council. 2000. *The Small Business Innovation Research Program: A Review of the Department of Defense Fast Track Initiative*. Charles W. Wessner, ed. Washington, DC: National Academy Press.

National Research Council. 2000. *U.S. Industry in 2000: Studies in Competitive Performance*. Washington, DC: National Academy Press.

National Research Council. 2001. *The Advanced Technology Program: Assessing Outcomes*. Charles W. Wessner, ed. Washington, DC: National Academy Press.

National Research Council. 2001. *Attracting Science and Mathematics Ph.Ds to Secondary School Education*. Washington, DC: National Academy Press.

National Research Council. 2001. *Building a Workforce for the Information Economy*. Washington, DC: National Academy Press.

National Research Council. 2001. *Capitalizing on New Needs and New Opportunities: Government-Industry Partnerships in Biotechnology and Information Technologies*. Charles W. Wessner, ed. Washington, DC: National Academy Press.

National Research Council. 2001. *A Review of the New Initiatives at the NASA Ames Research Center*. Charles W. Wessner, ed. Washington, DC: National Academy Press.

National Research Council. 2001. *Trends in Federal Support of Research and Graduate Education*. Washington, DC: National Academy Press.

National Research Council. 2002. *Government-Industry Partnerships for the Development of New Technologies: Summary Report*. Charles W. Wessner, ed. Washington, DC: The National Academies Press.

National Research Council. 2002. *Measuring and Sustaining the New Economy*. Dale W. Jorgenson and Charles W. Wessner, eds. Washington, DC: National Academy Press.

National Research Council. 2002. *Partnerships for Solid-State Lighting*. Charles W. Wessner, ed. Washington, DC: The National Academies Press.

National Research Council. 2004. *An Assessment of the Small Business Innovation Research Program: Project Methodology*. Washington, DC: The National Academies Press.

National Research Council. 2004. Capitalizing on Science, Technology, and Innovation: An Assessment of the Small Business Innovation Research Program/Program Manager Survey. Completed by Dr. Joseph Hennessey.

National Research Council. 2004. *Productivity and Cyclicality in Semiconductors: Trends, Implications, and Questions.*. Dale W. Jorgenson and Charles W. Wessner, eds. Washington, DC: The National Academies Press.

National Research Council. 2004. *The Small Business Innovation Research Program: Program Diversity and Assessment Challenges.* Charles W. Wessner, ed. Washington, DC: The National Academies Press.

National Research Council. 2006. *Beyond Bias and Barriers: Fulfilling the Potential of Women in Academic Science and Engineering.*

National Research Council. 2006. *Deconstructing the Computer.* Dale W. Jorgenson and Charles W. Wessner, eds. Washington, DC: The National Academies Press.

National Research Council. 2006. Capitalizing on Science, Technology, and Innovation: An Assessment of the Small Business Innovation Research Program/Firm Survey.

National Research Council. 2006. Capitalizing on Science, Technology, and Innovation: An Assessment of the Small Business Innovation Research Program/Phase I Survey.

National Research Council. 2006. Capitalizing on Science, Technology, and Innovation: An Assessment of the Small Business Innovation Research Program/Phase II Survey.

National Research Council. 2006. *Software, Growth, and the Future of the U.S. Economy.* Dale W. Jorgenson and Charles W. Wessner, eds. Washington, DC: The National Academies Press.

National Research Council. 2006. *The Telecommunications Challenge: Changing Technologies and Evolving Policies.* Dale W. Jorgenson and Charles W. Wessner, eds. Washington, DC: The National Academies Press.

National Research Council. 2007. *Enhancing Productivity Growth in the Information Age: Measuring and Sustaining the New Economy.* Dale W. Jorgenson and Charles W. Wessner, eds. Washington, DC: The National Academies Press.

National Research Council. 2007. *India's Changing Innovation System: Achievements, Challenges, and Opportunities for Cooperation.* Charles W. Wessner and Sujai J. Shivakumar, eds. Washington, DC: The National Academies Press.

National Research Council. 2007. *Innovation Policies for the 21st Century.* Charles W. Wessner, ed. Washington, DC: The National Academies Press.

National Research Council. 2007. *SBIR and the Phase III Challenge of Commercialization.* Charles W. Wessner, ed. Washington, DC: The National Academies Press.

National Science Foundation. Committee of Visitors Reports and Annual Updates. Available online at *<http://www.nsf.gov/eng/general/cov/>*.

National Science Foundation. Emerging Technologies. Available online at *<http://www.nsf.gov/eng/sbir/eo.jsp>*.

National Science Foundation. Guidance for Reviewers. Available online at *<http://www.eng.nsf.gov/sbir/peer_review.htm>*.

National Science Foundation. National Science Foundation at a Glance. Available online at *<http://www.nsf.gov/about>*.

National Science Foundation. National Science Foundation Manual 14, *NSF Conflicts of Interest and Standards of Ethical Conduct.* Available online at *<http://www.eng.nsf.gov/sbir/COI_Form.doc>*.

National Science Foundation. The Phase IIB Option. Available online at *<http://www.nsf.gov/eng/sbir/phase_IIB.jsp#ELIGIBILITY>*.

National Science Foundation. Proposal and Grant Manual. Available online at *<http://www.inside.nsf.gov/pubs/2002/pam/pamdec02.6html>*.

National Science Foundation. 2005. Synopsis of SBIR/STTR Program. Available online at *<http://www.nsf.gov/funding/pgm_summ.jsp?Phase Ims_id=13371&org=DMII>*.

National Science Foundation. 2006. "News items from the past year." Press Release. April 10.

National Science Foundation, Office of Industrial Innovation. 2006. "SBIR/STTR Phase II Grantee Conference, Book of Abstracts." Louisville, Kentucky. May 18–20, 2006.

National Science Foundation, Office of Industrial Innovation. Draft Strategic Plan, June 2, 2005.

National Science Foundation, Office of Legislative and Public Affairs. 2003. SBIR Success Story from News Tip. Web's "Best Meta-Search Engine," March 20.

National Science Foundation, Office of Legislative and Public Affairs. 2004. SBIR Success Story: GPRA Fiscal Year 2004 "Nugget." Retrospective Nugget–AuxiGro Crop Yield Enhancers.

Nelson, R. R. 1982. *Government and Technological Progress*. New York: Pergamon.

Nelson, R. R. 1986. "Institutions supporting technical advances in industry." *American Economic Review, Papers and Proceedings* 76(2):188.

Nelson, R. R., ed. 1993. *National Innovation System: A Comparative Study*. New York: Oxford University Press.

Office of Management and Budget. 1996. "Economic analysis of federal regulations under Executive Order 12866."

Office of the President. 1990. *U.S. Technology Policy*. Washington, DC: Executive Office of the President.

Organization for Economic Cooperation and Development. 1982. *Innovation in Small and Medium Firms*. Paris: Organization for Economic Cooperation and Development.

Organization for Economic Cooperation and Development. 1995. *Venture Capital in OECD Countries*. Paris: Organization for Economic Cooperation and Development.

Organization for Economic Cooperation and Development. 1997. *Small Business Job Creation and Growth: Facts, Obstacles, and Best Practices*. Paris: Organization for Economic Cooperation and Development.

Organization for Economic Cooperation and Development. 1998. *Technology, Productivity and Job Creation: Toward Best Policy Practice*. Paris: Organization for Economic Cooperation and Development.

Pacific Northwest National Laboratory. SBIR Alerting Service. Available online at <*http://www.pnl.gov/edo/sbir*>.

Perko, J. S., and F. Narin. 1997. "The Transfer of Public Science to Patented Technology: A Case Study in Agricultural Science." *Journal of Technology Transfer* 22(3):65–72.

Perret, G. 1989. *A Country Made by War: From the Revolution to Vietnam—The Story of America's Rise to Power*. New York: Random House.

Powell, J. W. 1999. *Business Planning and Progress of Small Firms Engaged in Technology Development through the Advanced Technology Program*. NISTIR 6375. National Institute of Standards and Technology/U.S. Department of Commerce.

Powell, Walter W., and Peter Brantley. 1992. "Competitive cooperation in biotechnology: Learning through networks?" In N. Nohria and R. G. Eccles, eds. *Networks and Organizations: Structure, Form and Action*. Boston, MA: Harvard Business School Press. Pp. 366–394.

Price Waterhouse. 1985. *Survey of small high-tech businesses shows Federal SBIR awards spurring job growth, commercial sales*. Washington, DC: Small Business High Technology Institute.

Roberts, Edward B. 1968. "Entrepreneurship and technology." *Research Management* (July):249–266.

Romer, P. 1990. "Endogenous technological change." *Journal of Political Economy* 98:71–102.

Rosenbloom, R., and Spencer, W. 1996. *Engines of Innovation: U.S. Industrial Research at the End of an Era*. Boston, MA: Harvard Business School Press.

Rubenstein, A. H. 1958. *Problems Financing New Research-Based Enterprises in New England*. Boston, MA: Federal Reserve Bank.

Ruegg, Rosalie, and Irwin Feller. 2003. *A Toolkit for Evaluating Public R&D Investment Models, Methods, and Findings from ATP's First Decade*. NIST GCR 03-857.

Ruegg, Rosalie, and Patrick Thomas. 2007. *Linkages from DoE's Vehicle Technologies R&D in Advanced Energy Storage to Hybrid Electric Vehicles, Plug-in Hybrid Electric Vehicles, and Electric Vehicles*. U.S. Department of Energy/Office of Energy Efficiency and Renewable Energy.

Sahlman, W. A. 1990. "The structure and governance of venture capital organizations." *Journal of Financial Economics* 27:473–521.

Saxenian, Annalee. 1994. *Regional Advantage: Culture and Competition in Silicon Valley and Route 128*. Cambridge, MA: Harvard University Press.

SBIR World. SBIR World: A World of Opportunities. Available online at <*http://www.sbirworld. com*>.

Scherer, F. M. 1970. *Industrial Market Structure and Economic Performance*. New York: Rand Mc-Nally College Publishing.

Schumpeter, J. 1950. *Capitalism, Socialism, and Democracy*. New York: Harper and Row.

Scott, John T. 1998. "Financing and leveraging public/private partnerships: The hurdle-lowering auction." *STI Review* 23:67–84.

Small Business Administration. 1992. *Results of Three-Year Commercialization Study of the SBIR Program*. Washington, DC: U.S. Government Printing Office.

Small Business Administration. 1994. *Small Business Innovation Development Act: Tenth-Year Results*. Washington, DC: U.S. Government Printing Office (and earlier years).

Sohl, Jeffrey. 1999. *Venture Capital* 1(2).

Sohl, Jeffery, John Freear, and W.E. Wetzel Jr. 2002. "Angles on Angels: Financing Technology-Based Ventures—An Historical Perspective." *Venture Capital: An International Journal of Entrepreneurial Finance* 4 (4).

Stiglitz, J. E., and A. Weiss. 1981. "Credit rationing in markets with incomplete information." *American Economic Review* 71:393–409.

Stowsky, J. 1996. "Politics and Policy: The Technology Reinvestment Program and the Dilemmas of Dual Use." Mimeo. University of California.

Tassey, Gregory. 1997. *The Economics of R&D Policy*. Westport, CT: Quorum Books.

Tirman, John. 1984. *The Militarization of High Technology*. Cambridge, MA: Ballinger.

Tyson, Laura, Tea Petrin, and Halsey Rogers. 1994. "Promoting entrepreneurship in Eastern Europe." *Small Business Economics* 6:165–184.

U.S. Congress, House Committee on Science, Space, and Technology. 1992. *SBIR and Commercialization: Hearing Before the Subcommittee on Technology and Competitiveness of the House Committee on Science, Space, and Technology, on the Small Business Innovation Research [SBIR] Program*. Testimony of James A. Block, President of Creare, Inc. Pp. 356–361.

U.S. Congress. House Committee on Small Business. Subcommittee on Workforce, Empowerment, and Government Programs. 2005. *The Small Business Innovation Research Program: Opening Doors to New Technology*. Testimony by Joseph Hennessey. 109th Cong., 1st sess., November 8.

U.S. Congress. Senate Committee on Small Business. 1981. Small Business Research Act of 1981. S.R. 194, 97th Congress.

U.S. Congressional Budget Office. 1985. *Federal financial support for high-technology industries*. Washington, DC: U.S. Congressional Budget Office.

U.S. General Accounting Office. 1987. *Federal research: Small Business Innovation Research participants give program high marks*. Washington, DC: U.S. General Accounting Office.

U.S. General Accounting Office. 1989. *Federal Research: Assessment of Small Business Innovation Research Program*. Washington, DC: U.S. General Accounting Office.

U.S. General Accounting Office. 1992. *Small Business Innovation Research Program Shows Success but Can Be Strengthened*. RCED–92–32. Washington, DC: U.S. General Accounting Office.

U.S. General Accounting Office. 1997. *Federal Research: DoD's Small Business Innovation Research Program*. RCED–97–122, Washington, DC: U.S. General Accounting Office.

U. S. General Accounting Office. 1998. *Federal Research: Observations on the Small Business Innovation Research Program*. RCED–98–132. Washington, DC: U.S. General Accounting Office.

U.S. General Accounting Office. 1999. *Federal Research: Evaluations of Small Business Innovation Research Can Be Strengthened*. RCED–99–198, Washington, DC: U.S. General Accounting Office.

U.S. Public Law 106-554, Appendix I–H.R. 5667, Section 108.

U.S. Senate Committee on Small Business. 1981. Senate Report 97–194. *Small Business Research Act of 1981*. September 25. Washington, DC: U.S. Government Printing Office.

U.S. Senate Committee on Small Business. 1999. Senate Report 106–330. *Small Business Innovation Research (SBIR) Program.* August 4. Washington, DC: U.S. Government Printing Office.

U.S. Small Business Administration. 1994. *Small Business Innovation Development Act: Tenth-Year Results.* Washington, DC: U.S. Government Printing Office.

U.S. Small Business Administration. 2003. "Small Business by the Numbers." SBA Office of Advocacy. May.

Venture Economics. 1988. *Exiting Venture Capital Investments.* Wellesley, MA: Venture Economics.

Venture Economics. 1996. "Special Report: Rose-colored asset class." *Venture Capital Journal* 36 (July):32–34 (and earlier years).

VentureOne. 1997. National Venture Capital Association 1996 annual report. San Francisco: VentureOne.

Wallsten, S. J. 1996. The Small Business Innovation Research Program: Encouraging Technological Innovation and Commercialization in Small Firms. Unpublished working paper. Stanford University.

Wessner, Charles W. 2004. *Partnering Against Terrorism.* Washington, DC: The National Academies Press.